DATE			

© THE BAKER & TAYLOR CO.

The Aging Body

Preface

The objective of this book is to provide information that will be useful to people in a variety of disciplines who wish to learn more about normal aging processes in the human body. Although gerontologists in the biological sciences are making great strides in research on human aging and documenting this work in monographs, texts, and review chapters, this information is generally not easily accessible nor is it comprehensible to nonprofessionals in these fields. This book is intended to provide a summary of this work, along with its implications for psychological functioning of the aging individual.

The majority of the book is devoted to describing the results of research on the physiological changes in the human body with aging and to seeking explanations for these age effects. This description has been approached in such a way as to make it readable for the nonspecialist, but also to focus on research issues that will be useful reading for those who are currently working in these particular areas. In addition, throughout the book, I have tried to develop some themes regarding physiological and psychological adaptation during adulthood. In each chapter I have included a section on the "psychological consequences" of the aging of a particular system being reviewed. These sections should provide impetus for research on specific ways in which physical and psychological processes are related in adulthood. It is my hope that, from the integrative approach I have tried to follow, psychologists and other social scientists will gain an appreciation of the importance of knowing about physical functioning and the aging process and, in addition, those in the fields of medicine and physiology will gain an appreciation of areas outside their own specialty and how they relate to each other.

I would like to thank my colleagues—Stephen Buell, Andy Coyne, Paul Coleman, Frank Katch, Mark Ordy, and my brother, Ronald Krauss—who have provided me with valuable references, access to their own labs, and critiques of particular chapters. I would also like to thank the graduate students at the University of Rochester who shared with me their reactions to the manuscript and provided valuable assistance, in particular, Lisa Elliot. The editorial and production staff at Springer-Verlag have my gratitude for their encouragement and help. My children, Stacey Whitbourne and Jennifer O'Brien, deserve recognition for their admirable patience and understanding of their mother's obsession with the word processor. Finally, I would like to acknowledge the support and advice of my husband, Richard D. O'Brien, whose talents as a scientist and author are incorporated throughout this book.

Susan Krauss Whitbourne

Contents

CHAPTER 1

Introduction

This book is intended to provide an overview of the aging of the human body in adulthood, from postadolescence through old age, integrated with an exploration of the possible consequences of these changes for the individual's psychological adaptation and sense of well-being. It is the central thesis of this book that changes in the body's functioning with age have the potential to influence the adult's adaptation to the physical and social environment, which in turn can affect the adult's sense of competence and self-esteem. The material addressed in this chapter forms the basis for development of this central thesis throughout the book in sections concerning "psychological consequences" of the various physiological changes described in each of the subsequent chapters. In addition, this chapter will include a discussion of research issues in the physiology of human aging. These are issues to which the reader needs to be sensitized before approaching the individual chapters covering specific physiological systems.

Adaptation

The main focus of this book is on adaptation, both physiological and psychological. Physiological adaptation involves the maintenance of an internal environment in which the body's systems can carry out their activities without threat of destruction of any or all of the organism's cellular elements. The body's organ systems are oriented toward maintaining life at all costs. If stress is placed upon the body, from a disease entity or from severe environmental conditions, the body's systems engage in counterregulatory tactics to preserve the integrity of the organism's life. In addition to maximizing the ability of the organism to exercise its physiological functions, physiological adapatation involves achieving the body's maximum of sensory acuity, strength, energy, coordination, dexterity, recuperative power, and immunity (Foote & Cottrell, 1955).

The relationship of aging to physiological adaptation would appear to be an inverse one. Aging, by definition, involves the movement of the organism toward death over the course of time. It is not known why the body ages or what causes the body to age, except that all living systems must die (which is not a very satisfactory answer). The empirical study of the physiology of aging is largely oriented toward answering these very basic questions, but within the context of specific organ systems. In the course of this research, what has emerged is a picture of the human body as making numerous physiological adaptations to the changes in its integrity over time. Rather than simply accelerating downward year by year until the end of life is reached, the body is actively integrating the deleterious changes in its cells into new levels of organization in an attempt to preserve life for as long as possible. While much of the physiology of aging invariably involves description of deleterious changes, there is also an accumulation of literature on compensation within and across bodily systems.

Psychological adaptation is a concept not unlike physiological adaptation, in that both processes are oriented toward maintaining the well-being of the individual. Both processes are also ones of which the individual in whom the adaptation is taking place can be aware, particularly when one of them fails to operate effectively. However, they differ in that the substrate for psychological adaptation is not limited to the outcome of the activities of the body's life-support systems. Instead, psychological adaptation is the outcome of the activities of the individual within the individual's environment. The specific outcomes concern whether or not the individual is able to operate effectively as an agent in that environment and achieve desired objectives of mastering the challenges that the individual perceives within that environment. This definition of psychological adaptation is based on the concept of competence, defined by White (1959, p. 297) as the "organism's capacity to interact effectively with its environment."

The position will be developed in this book that physiological adaptation provides an important set of inputs to psychological adaptation, since the body's ability to adapt to the stresses and changes associated with the aging process directly affect the individual's ability to master the environment. Moreover, it is assumed throughout this discussion that the product of the psychological adaptation process feeds directly into the individual's self-concept. This feeding in of information about the self occurs in the form of descriptive statements ("I am this type of person," "I have handled this situation in this manner") and evaluative statements ("I am an effective person", "I am an incompetent person"). It is the latter, evaluative, statements that are assumed to have the strongest impact on feelings of well-being, since they are what contribute to psychological adaptation. This position regarding the contribution of feelings of competence to psychological adaptation is basic to the literature on self-esteem (e.g., Franks & Marolla, 1976; Smith, 1974). Its applicability is particularly useful to the aging process, because it can be assumed that much of the relevance to psychological adaptation of the aging process is its effect on feelings of competence, a position that will be elaborated on throughout this book.

The Bodily Self-Concept

The particular way in which psychological adaptation affects self-esteem is thought to occur through the mechanism of the body image, or bodily self-concept. This component of the self-concept is one that has been addressed by many theorists who write about the self, but its exact nature is still very elusive, and even more so when considered in the context of how it is affected by the physiological changes associated with aging. Nevertheless, the definitions of body image, or bodily self-concept, together provide a sense of what is meant with respect to psychological adaptation and feelings of bodily competence.

Cooley (1902), who is best known for inventing the term "looking-glass self" seems to have literally meant bodily self, although this is not the usual interpretation of the term. According to Cooley (1902, p. 184), the looking-glass self may be regarded as being composed of the person's views of how others regard his or her physical appearance: one's "face, figure, and dress." Thus, how we think we look, and how we think others judge our appearance, are two of the components of the self-image as it is reflected off others, according to Cooley.

Theorists on the self-concept since the time of Cooley have iterated this theme that the "me" that is reflected from our own and others' views of our physical attributes forms an important part of our overall self-esteem (Wylie, 1974). Bodily identity also includes the sense that will be developed most explicitly in this book, the self-evaluation of the effectiveness with which one's body is functioning (Back & Gergen, 1968; Epstein, 1973; L'Ecuyer, 1981).

Systematic treatment of bodily identity is found in theories and research dealing specifically with the body image, a tradition of empirical investigation that dates back to Schilder's (1935/1950) work. The individual's image of his or her body is considered to be a reflection of self-concept and personality and also to be influenced by social factors, according to Schilder (1935/1950):

A body is always the expression of an ego and of a personality, and is in a world. Even a preliminary answer to the problem of the body cannot be given unless we attempt a preliminary answer about personality and world. (p. 304)

Conversely, Secord and Jourard (1953) regarded the individual's attitudes toward the body to have influence over personality, and to be "integrally related to self-concept although identifiable as a separate aspect thereof" (p. 343).

Following this line of reasoning are Fisher and Cleveland (1958):

If by the term self-concept one means the whole range of complicated attitudes and fantasies an individual has about his identity, his life role, and his appearance, then, the two constructs undoubtedly overlap considerably . . . . We definitely consider the body image to be a condensed formulation or summary in body terms of a great many experiences the individual has had in the course of defining his identity in the world . . . the body image is a sensitive indicator which registers many of the individual's basic social relationships, especially those early involved in his development of a sense of identity. (p. 111)

The bidirectional role of the body image as intervening between the environment and the self-concept, alluded to by Schilder (1935/1950) and also by Fisher and Cleveland (1958), was stressed by Werner (1965): "One may see in it [the body schema] a device which . . . mediates between the concrete tangible body and the abstract self-concept; between the visual tangible environment and the abstractly elusive self" (p. 6). The position put forth in the volume on body perception edited by Wapner and Werner (1965) was expressed most clearly by

Witkin (1965) who described the fully articulated body image as one that is differentiated from its environment: "Progress toward self-differentiation entails awareness of one's own needs, feelings and attributes as one's own and not others'. . . . We refer to the outcome of this development as a 'sense of separate identity'" (p. 39). Witkin (1965) related his well-known experiments on field dependence–independence as a cognitive style dimension to body image through this principle of self-environment differentiation.

Definitions of body self-concept or identity reveal that it is a multifaceted construct. A fairly straightforward meaning of body image is given by Abend (1974) as "more or less concrete anatomical self-representations" (p. 618). The body self was defined by Epstein (1973) as "the individual's biological self, his possessions, and those individuals, groups, and symbols he identifies with" (p. 412). Included in the class of "self-experience" described by Gergen (1977) are various bodily feelings such as sensations of pleasure and pain, and overt and covert physical movement. The body image, according to Schilder (1935/1950) is a schema, a mental representation of the body's position in space, the "picture of our body which we form in our mind" (p. 11). This "postural" body image is essentially content-free and emotionally neutral. In contrast, "body cathexis" is the form of the bodily self described by Secord and Jourard, defined as "the degree of feeling of satisfaction or dissatisfaction with various parts or processes of the body" (1953, p. 343), which intimately links its objective and subjective features.

The definition that will be adopted here of body image, or bodily self-concept, is that it is the individual's self-appraisal of the adequacy of the body to cope effectively with the environment in which the individual carries out his or her daily activities. As such, it includes the physical functioning of the body, and also the way that the body is judged by others, that is, its appearance as socially valued. Although bodily self-concept has been operationalized in more or less this manner and studied in relation to overall self-esteem (Goldberg & Folkins, 1974; Leonardson, 1977; O'Brien & Epstein, 1982; Zion, 1965), the extension of this process to the study of aging has been made in only a very limited fashion.

The effects of aging on bodily self-concept have been alluded to by some authors but not really adequately dealt with in any depth. Murphy (1947) speculated that physical changes can change the "bodily matrix of selfhood so that one does not recognize himself" (p. 520). However, this comment does not pertain to the question of self-esteem. Back and Gergen (1968) speculated that as a person ages, he or she may feel a diminishing influence on the environment because the body is less capable of action, but their research was inconclusive on this issue. Finally, Thompson (1972) summarized research on older adults using the Tennessee Self-Concept Scale, which includes a measure of physical self-concept, and reported no relationship between declining scores on physical self and overall self-satisfaction. However, there is reason to suspect that scores on this measure were unduly influenced by a social desirability bias, so the results are not really conclusive.

In this book, an attempt will be made to achieve a greater level of specificity than has been presented before on how the changes within each bodily system can potentially influence bodily self-concept and in this way pose a challenge for psychological adaptation in adulthood. Although it is necessary to go well beyond the empirical literature, in most cases, in order to achieve that level of specificity, the inferences that are made are at least theoretically consistent with the concepts of adaptation presented here. In addition, there is adequate support, from at least the research on the psychological effects of aerobic exercise training, to proceed with some confidence throughout the rest of the chapters.

Methodology in the Study of Human Aging

In this section, some of the specific problems involved in studying the aging process will be discussed. These problems include the use of age as a variable in empirical research and the design concerns that arise in trying to control for competing explanations of age effects. The resolution of these issues that is arrived at forms an important framework for interpreting the material in subsequent chapters on age effects on physiological functions in adulthood.

Age Effects in Bodily Functioning in Adulthood

In research on adults, age is often used to classify people into groups for the purposes of studying the effects of the aging process. Typically, people are sorted according to age, and their performance on various measures of functioning analyzed in relation to age. The measures chosen are thought to be sensitive to some aspect of the aging process in the system being studied. Classification according to age is done in one of two ways: comparisons within a cross-section of adults at different ages studied at one point in time (a "cross-sectional" study), or in following the same adult across several time points, observing changes from one age to another (a "longitudinal" study). These two methods are illustrated in Figure 1.1.

The empirical study of the aging process almost by necessity would appear to entail the use of age as the indicator of an individual's status with regard to bodily

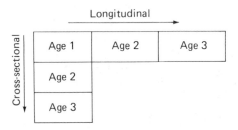

FIGURE 1.1. Cross-sectional and longitudinal comparisons of age effects. A cross-sectional study would be conducted by comparing age groups along the vertical dimension. A longitudinal study would be conducted by comparing the same people from ages 1 to 2 to 3.

functioning. As an explanation of the aging process, though, age alone is not sufficient. In the first place, using age as a criterion places too much emphasis on discrete, arbitrarily defined units of time to explain a continuous evolution of the individual's bodily structures and functions. The second problem with using age as a means of classifying people for research purposes is that in doing so variations are ignored among people of the same age in bodily capacities.

An even more basic problem in regarding age as a means of explaining something about how individuals change as a result of the aging process is inherent in the way that age is calculated. Age is measured in units of time. However, time is an index of an event occurring externally to and independent of the individual, that is, the movement of earth around the sun. Such an externally based chronometric reference system can never provide an explanation of events taking place within the individual's body as a result of the aging process. To overcome this problem, indices of biological age are suggested as alternatives to chronological age (Birren, 1959; Borkan & Norris, 1980; Damon, 1972; Ludwig & Smoke, 1980), as has distance in years from death (Lieberman & Coplan, 1970). One group of researchers even suggesting using subjective (self-perceived) age as a substitute for chronological or functional objective indices (Kastenbaum, Derbin, Sabatini, & Artt, 1972). None of these alternative indices has received sufficient testing to establish its validity and all are relatively difficult to use in practice. However, if a non-age-based index could be validated and efficiently used, it would provide an excellent alternative to the present age-based classification system.

One additional problem encountered with the use of chronological age that applies specifically to research on the aging process is related to the attempt by developmental psychologists to separate the effects of bodily aging (internal processes) from changes that occur in the outside world (external processes). Of major theoretical interest is the question of whether particular functions are affected more by the unfolding of the genetic program for development within the cells of the body, or if these functions are affected more by historical events, the evolution of social and political systems, interactions with other people on a day-to-day basis, and exposure to chemical and mechanical forces in the physical environment. It is considered important to be able to separate the influences of the two sets of processes; internal and external. However, because people age within the context of their surroundings, such a separation becomes difficult to achieve in practice. Various strategies for making this separation exist, but these methods are all plagued by the inevitable problem that age cannot be separated from external calendar time either experimentally or statistically (Schaie & Baltes, 1975). The best that can be done is to estimate the relative effects of both.

While these difficulties having to do with using age as a basis for studying the aging process are not easily resolved, it still remains the case that age is the most convenient and easily understood basis for describing the effects of the aging process on the individual. Age is the most frequently used reference point for comparing adults on a variety of bodily functions. As will be seen in the chapters that follow, virtually all researchers base their work on aging on comparisons of the people they study according to age. Age is therefore the most useful way to organize the findings of these studies for the purposes of discussion. An understanding of why changes occur in a particular function must go beyond descriptions of people who were born at different times, though, and descriptions made in terms of age should not be the final products of analyses of the aging process. While it is helpful to use age for describing what appear to be age-related phenomena, its limitations as a factor that can explain the aging process must be recognized.

Because of the arbitrariness of applying categories based on chronological age, no attempt will be made in this book to distinguish a set point at which a person moves from "adulthood" to "old age." The most common conventional demarcation for the onset of old age is 65 years, and to a certain extent, this book will follow that convention, but with the proviso that this age should be regarded as only a rough marker. When the term "older adult," "aged," "elderly," or "old person" is used in subsequent pages, it is intended to apply very generally to the later years of life, after 65 to 70.

Special Concerns in Research on Physical Functioning and Age

The material presented in this book on the physiological functioning of the body's major organ systems is based on research that has associated with it special problems in addition to those already mentioned concerning the use of age as a basis for categorizing people into groups for the purpose of studying the aging process.

For the most part, only cross-sectional studies are available on which to base the descriptions of the effects of aging. This situation is partially due to the impossibility of conducting longitudinal research on the effects of aging on the body's structures where organs must be totally dissected from the rest of the body in order to weigh them and study their cells. In order to study normal aging, diseased organs removed for surgical purposes would not be appropriate. Investigations of nonpathological age effects must be done on organs that are not diseased. The only way to obtain these organs for study is to conduct an autopsy. Tissues from normally functioning organs can be obtained by surgical biopsy, and this procedure is occasionally found in the research on aging. However, it would be extremely unlikely that a researcher could obtain a repeated series of biopsies from very many people, since the procedures are usually quite invasive. Cross-sectional studies of organ functioning are far less complicated to conduct, since they involve tests of the functions of organ systems using physiological measures from living human beings. While the procedures used to obtain such measures can sometimes cause discomfort to the subjects, they are at least amenable to longitudinal analysis as well as cross-sectional comparisons, and there are at least some data from repeated testings available with which to compare results obtained from different age groups.

Cross-sectional research, both on living persons and on tissue obtained from autopsy, has the possibility of confounding differences due to age with differences due to other factors. One of the major confounding factors in this type of research is that there are huge differences among generations in health practices, which have a major impact on the intactness of bodily organs. The material included in the following chapters on age effects in bodily systems was screened so that it has as few documented confounding factors as possible. In some cases, though, confounding factors are present, and there are few other studies that are less confounded on which to rely for better information. As a basis for selecting topics to include in each of the chapters, preference was given to information that is based on at least several investigations, longitudinal if possible, and where available, on both men and women. Since a number of major studies, particularly longitudinal ones, were on samples composed entirely of males, there are quite a few areas of physical functioning for which data are available only for men. The reader should be especially attentive to points raised in the discussion of the studies about possible factors other than age that could account for the findings and other factors, such as sex, which limit the generality of the conclusions.

It will also become evident to the reader that there is virtually no reference to studies that are not on humans. This is because the approach adopted in this book generally precludes the use of data from nonhumans on which to base inferences about the effects of aging on self-esteem and bodily self-concept as components of psychological adaptation. Cross-species comparisons would make such speculations even more tenuous. In addition, there is very little material in this book on the diseases of aging, except to distinguish them from the normal aging process. This is because it is important to describe the effects of aging independently of the effects that characterize the diseases that are more prevalent in later life. This distinction is important from the theoretical standpoints of psychological and physiological adaptation outlined in this chapter, for the purpose of making valid conclusions from research, and for the purpose of applying concepts of normal aging in working with older persons. The conclusions made in all three cases are very different for normal aging than they are for the chronic diseases that occur in middle and old age.

For the sake of simplifying the language used in this text to describe research findings, differences attributed to age through cross-sectional studies are referred to as "cross-sectional age differences," or in some cases as age-related "increases" or "decreases" across groups of adults, even though the people themselves did not change on that given attribute. It is important to remember that unless otherwise noted, these age effects were not observed in the same people studied as they aged, even though the title of a cross-sectional study may be "Age Changes in . . . ." and so lead the reader to incorrectly conclude that the study was longitudinal.

An attempt was made in the material on age effects to describe age differences in terms of the ages at which changes were first observed either by clinicians or researchers working with adult populations. However, the reader should not regard these ages as if they were engraved in stone, but rather view them as relative. The reason for not taking exact chronological age too seriously is partly related to the nature of cross-sectional research, and its tendency to capitalize on peculiarities of a sample (particularly when only a few studies are available). For instance, a researcher may have selected age groups constituted of people from 20 to 25 and 65 to 70. To say that an age effect is observed "at 65" is not necessarily a valid inference to draw from the existence of age differences, since the change reflected in the 65-year-olds may have occurred sometime earlier in the interval betwen 26 and 64 (if in fact the age difference reflects a true age change). Even where a more complete range of ages forms the basis for the conclusion, it is more realistic to regard the findings in terms of whether the age effect reported appears early or late in adulthood. Another reason to be cautious about drawing firm conclusions from the research on aging is that even relatively well-established age effects show wide interindividual variations. Some of these differences among people in the rate of aging are due to cultural influences. Other differences are the product of hereditary inputs to cellular structure and function, which may affect the rate of genetically programmed change during adulthood. Rather than reflecting the effects of age alone, some age differences may be due to these non-age-related factors.

CHAPTER 2

Appearance and Movement

The way the adult appears to other people is determined by the total picture the body presents at rest and in motion. The individual's own image of his or her body is at least partly a function of its static and dynamic appearance (Schilder, 1935/1950). The body's movement efficiency also has a significance, of course, that is more basic than the social impression the person makes. The ability to move the body so that the physical environment can be successfully negotiated is a primary component of adaptation and well-being in adulthood (Lawton, 1977; Lawton & Nahemow, 1973). Without mobility, adults are forced into being dependent on others to provide the basic essentials of life, and thereby a major contributor to feelings of personal competence is lost. A smoothly and effectively moving body also makes it possible for the adult to experience the enjoyment derived from physical activities, ranging from a long, relaxing walk to active, invigorating sports such as skiing, bicycling, and swimming.

Age Effects on Appearance in Adulthood

It is useful to discuss adult age differences in the face separately from those in the rest of the body, since the face and the rest of the body make somewhat distinct contributions to the individual's self-concept. However, in the case of the skin, since it is common to both, the effects of age will be described in terms of skin as a total organ system.

Skin

The most visible changes in the skin during adulthood are the development of creases, furrows, and sagging as well as a loss of firmness and resiliency. These effects of age, which occur in all adults to differing degrees and according to varying timetables, are due to a combination of changes in skin cells and the cellular elements found throughout the skin structure (Rossman, 1977, 1979; Selmanowitz, Rizer, & Orentreich, 1977) as well as to damage from exposure to the sun's rays (Robinson, 1983).

Wrinkling

Wrinkling of the skin is due to several causes. First, it is generally accepted that the outer, epidermal, layer of the skin becomes thinner as its cells are depleted in number at a faster rate than they are replaced (although Whitton, 1973, found no correlation between age and epidermal thickness in a sample of adult women). The greater fragility of the epidermal layer with age that would result from cell loss contributes to the development of wrinkles. In addition, as epidermal cells are replaced they form less organized patterns, becoming more irregular in their arrangement. The surface pattern of geometric furrows visible on the surface of the skin becomes less orderly in areas of the skin unprotected by clothing (Lavker, Kwong, & Kligman, 1980). In general, adults whose occupations or leisure interests take them outdoors are particularly likely to experience these changes, which are more prevalent in parts of the body exposed to the sun and wind. These exposed body parts include the backs of the hands, forearms, face, and "V"-shaped area at the junction of the neck and chest.

Changes in the middle, dermal, layer of the skin contribute further to its wrinkling. Collagen fibers, which make up a large proportion of the connective tissue of the body, and run throughout this layer of skin, exhibit systematic changes with age that reduce their flexibility. The strands of the collagen molecule form cross-links with other collagen molecules with increasing frequency throughout adulthood. These cross-links are very stable, since the replacement rate of collagen molecules is very slow in later life (Strehler, 1977). As a result, the flexibility is reduced with age of collagenous tissue in the skin (Thorne, 1981) and also in other organs of the body. Elastin molecules are another component of the dermal layer of the skin, and of other tissues of the body frequently deformed by movement (such as the blood vessels). These molecules have the "elastic" capacity to return to their original shape after they have been pushed or pulled by forces acting upon them. With increasing age, the elastin fibers become more brittle and in the case of the skin, its ability to conform to the moving limbs is greatly reduced. The skin is more likely to sag and form wrinkles, since when it is stretched out, it does not return to its original shape. Wrinkling is also contributed to by a reduction in the activity of the sebaceous glands located in the dermis. These glands lubricate the skin with the oil they secrete. As less oil is produced, the skin becomes dryer, rougher, and more likely to undergo mechanical damage when rubbed by other skin surfaces, clothing, and exposure to wind, sun, and water.

The fat in the innermost layer of skin normally provides an underlying padding to smooth out the contours of the arms, legs, and face and also contributes to the opacity of the color of the skin. With increasing age, the subcutaneous fat on the arms and legs diminishes, instead collecting in areas of fatty deposits. Since there is less underlying padding, in addition to lowered elasticity, the skin can be pulled downward more easily by the force of gravity and so it sags. A diminution of muscle mass contributes further to loss of firmness with aging of the skin. These changes are most apparent on the face and arms, and to a lesser extent, the upper torso.

Other Changes on the Skin Surface

The coloring of light-skinned people undergoes some age changes due to alterations in the melanocytes, the pigment-containing cells in the epidermal layer.
These cells decrease in number, and those that remain have fewer melanin pigment granules (Montagna, 1965) so that less deepening of skin color occurs as the
result of sun exposure. In addition, irregular areas of dark pigmentation that look
like large freckles ("age spots") appear, as do pigmented outgrowths ("moles")
and "senile" angiomas (elevations of small blood vessels on the skin surface).

Some capillaries and small arteries may become dilated on exposed portions of
the skin surface, creating small irregular lines. On the legs may appear varicose
veins, which are knotty, bluish, and cordlike irregularities of the blood vessels
which show through the skin. The coloring of fair-skinned older adults may lose
some of its pinkish flesh tone, due to a reduction in blood supply stemming from
capillary loss. In general, the blood vessels and bones become more visible, particularly on the arms, as subcutaneous fat layer is diminished. The skin on the
face, for the same reason, appears more translucent in color.

Some of the changes in the skin cells combine with age-related changes in the
blood vessels to have the effect of reducing the adaptability of older adults to
extremely hot and cold temperatures. The skin normally serves to insulate the
individual against wide variations in the environmental temperature. This insu-
lating function of the skin gradually becomes less efficient throughout adulthood.
The adult sweats less in hot weather, and in cold weather, less heat is conserved
due to thinning of the epidermal and especially the subcutaneous fat layers.
(Other age effects on body temperature regulation which compound the changes
in the skin are discussed in chapter 6.)

Face

Adults rarely cover their faces to protect them from the elements except in cases
of extreme cold, and at times when sun-blocking lotions are applied prior to
exposure to the rays of the sun. Because of this virtually continuous contact with
the environmental forces of sun and wind, the face is particularly likely to
undergo the changes described in the previous section in the epidermal layer of
the skin. In addition, the skin on the face is constantly being stretched and fur-
rowed in the course of everyday life as the individual smiles, frowns, wrinkles the
brow in concentration, and squints the eyes or crinkles the skin around them
when laughing. These facial expressions involve the use of characteristic muscle
groups, which lead to particular wrinkling patterns as the muscle pulls on the
skin. Talking and eating, which involve the movement of the jaw, result in the
development of horizontal rings and vertical lines in the neck (Rossman, 1979).

The reduced flexibility of the skin due to changes in the composition and
amount of elastin in the dermis with age means that as the skin is stretched as the
facial muscles move, it becomes less likely to return to its original shape and to
sag and wrinkle. The loss of subcutaneous fat in the deepest layer of skin on the
face and its shifting to areas of fat deposition further accentuates the sagging of

the skin. One area where this accumulation is particularly noticeable is under the chin, where the skin is weighted down and forms the proverbial "double chin" of middle age.

Changes in the underlying structure of facial features also take place with increasing age in adulthood. The nose and ears tend to become longer and broader (Damon, Seltzer, Stoudt, & Bell, 1972). A reduction of the amount of bone in the jaw due to resorption of the bone matrix can alter the shape of the lower portion of the face, which has the visual effect of shrinking the apparent size of the jaw. This process is compounded by loss and deterioration of teeth (Tonna, 1977), which probably relates some to oral hygiene practices than age (Kenney, 1982), but is more likely to have a cumulative impact in later life. Replacement of diseased teeth with dentures may also alter the shape of the face, even though they restor the size of the lower jaw and chin. There are also changes in the appearance of the remaining teeth, which may be stained yellow by accumulation of fluoride, be cracked and chipped, and be surrounded by recessed gums (Wantz & Gay, 1981). As the hard enamel surface of the tooth wears down with age, the tooth is more likely to become stained brown by food, tobacco, coffee, and tea. These changes would be most apparent while eating and during social interaction, when the person is talking and smiling.

The need to wear eyeglasses becomes increasingly common for adults over 40 years of age (see chapter 9). Even for adults who have worn corrective lenses since youth, the kind of glasses they must wear are usually changed from single vision lenses to bifocals, which have different prescriptions for the top and bottom halves. Bifocals are prescribed due to the need for separate corrections for near and far vision, which would otherwise require the person to switch glasses for reading and for distance vision. If the individual develops cataracts (a pathological development of opacities in the lens of the eye) which require surgical removal, special eyeglasses may be prescribed to compensate for the loss of the lens. The lenses in these eyeglasses greatly magnify the size of the eye, and give them a strange, distorted look to an observer. If contact lenses are used to replace the lens, this problem is avoided, but the person must learn to adapt to the somewhat complicated regimen of caring for them, and the new sensation of having an artificial substance in the eye.

The eyes change their appearance in several other ways during the years of adulthood. One is the previously mentioned development of wrinkles around the eyes, and possibly also sagging due to fat and fluid accumulation underneath the eyes, and puffiness of the upper lid, for the same reasons. Added to these changes is the accumulation of dark pigment in the skin around the eyes and eyelids, which can make the eyes appear sunken. Other changes in the eyes include a loss of brightness and translucency of the cornea (see chapter 9), and late in adulthood, the development of arcus senilis. This condition, which is virtually universal in white persons after the age of 80 (Rossman, 1979), is characterized by the accumulation of lipid (fatty substances) around the outer edge of the cornea. The result is the formation of a white circle around the cornea. Although this effect of age on the eye may not interfere with vision, it has a marked impact on the individual's appearance.

Hair

The major changes with age in the hair of the head are graying and thinning. The appearance of gray hair results from the interspersing of pigmented hair and the white hair produced by the aging hair follicle after melanin production has diminished. The exact shade of gray that emerges on the individual's head during the period of transition from totally pigmented to totally white depends on the original color of the pigmented hair and how it appears when mixed with white. Eventually, all the pigmented hairs are replaced by nonpigmented ones, and the overall hair color turns pure white. The rate of this process of hair color change varies both in the timing of when it begins in adulthood and the rate at which melanin production decreases in individual hair follicles across the scalp. While gray hair is regarded as a virtually universal phenomenon associated with growing old, the degree of grayness of the hair on the head is not as reliable an indicator of age as is the grayness of the axillary hair (Kenney, 1982).

A gradual and general thinning of the hair on the head occurs in both men and women over the course of adulthood, although hair loss is commonly thought of as a process reserved for men. Hair loss in both men and women results from the destruction or regression of the germ centers that produce the hair follicles. It is true, however, that pattern baldness, which is the most common form of baldness, is more prevalent in men who are genetically predisposed. In this case, the hair follicles are not actually destroyed, but change from producing coarse terminal hair which is normally present on the scalp to producing fine, almost invisible, vellus hair. Men generally do not lose facial hair as they age; in fact the hair of the eyebrows and on the inside of the external canal of the ear may grow longer and coarser. Women may develop patches of coarse terminal hair on their face, particularly the chin. This process seems to be related to hormonal changes associated with menopause.

Voice

The sound made by one's voice contributes in an important sense to the total impression the adult makes on others. Changes with age in the larynx, respiratory system, and the muscles controlling speech and hearing may contribute to alterations in voice (Benjamin, 1982; Ptacek, Sander, Maloney, & Jackson, 1966). These alterations include lower pitch, breathiness, tremorousness, slowness, stronger inflection patterns, imprecise articulation, and lowered intensity and vitality (Benjamin, 1981, 1982; Hartman & Danhauer, 1976; Ptacek & Sander, 1966). However, there is some indication that poor physical condition rather than advanced age is the cause of changes in vocal production, particularly under circumstances where the individual's vocal production abilities are stressed (Ramig & Ringel, 1983).

Body Build

Height

Most of what is known about the effects of aging on total body height comes from cross-sectional studies of adults of various ages. Cross-sectional studies always have the potential to present an exaggerated picture of age effects in adulthood because they involve the comparison of people who not only differ in age but also in exposure to differing health and nutritional conditions, which may affect their development. In the case of height, this problem exists in an extreme form, because height is so sensitive to the individual's opportunities to receive adequate vitamins and minerals during the years of bone growth in childhood. Since the general level of nutrition in the population has improved in successive generations, younger cohorts of people are likely to be taller than older ones, thus producing an apparent decrease of height with age when the cross-sectional results are examined. A cross-sectional reduction of standing height is what is generally reported in the literature (e.g., Damon et al., 1972; Durin & Womersley, 1974), and it seems to occur at a greater rate after the 50s (Shephard, 1978). The decrease in height after the 50s was confirmed in a small ($n = 170$) but randomly selected sample of men and women ranging from 55 to 64 who were followed up when their ages ranged from 66 to 75 (Adams, Davies, & Sweetname, 1970). The men in this study lost a little over one-half of an inch, and the women lost close to an inch during the 11-year period in between the two measurements. According to Garn (1975) the decrease of height in adulthood is a fairly universal concomitant of the aging process, and it seems to be related to the collapse of vertebrae resulting from loss of bone strength, a process described in greater detail in the following section.

Weight

Cross-sectional studies on total body weight in adulthood are subject to the same criticism made above regarding studies on height. In the case of weight, though, the problem takes on a somewhat different character because the generational differences in nutritional status apply throughout the period of adulthood, rather than only in the period of growth. The generations of adults who are currently considered aged are more likely to be carrying through the nutritional habits characteristic of the early and middle 20th century, before Americans were as diet-conscious as younger adults who formed their dietary patterns more recently and are more likely to consider the fat and carbohydrate content of the foods they eat. Conversely, older adults may alter their food intake such that they consume a smaller amount than middle-aged and younger persons due to social and psychological factors associated with eating habits in later life (see chapter 5).

These potential confounds with age notwithstanding, the experience of weight gain in middle adulthood is one that is well-known by many men and women. The cross-sectional studies provide documentation of this phenomenon, revealing weight gain from the 20s until the mid- to late-50s after which there is weight loss throughout old age (e.g., Pařizková, 1974; Shephard, 1978).

Body Composition

It is a generally accepted conclusion (albeit based on cross-sectional studies almost entirely) that the pattern of age differences in body weight reflects an increase in body fat and a loss of lean body mass composed of muscle and bone mineral (Chien et al., 1975; Cotes, Hall, Johnson, Jones, & Knibbs, 1973; Cunningham, Rechnitzer, Pearce, & Donner, 1976; Dill, Yousef, Vitez, Goldman, & Patzer, 1982; Ellis, Shukla, Cohn, & Pierson, 1974; Forbes & Reina, 1970; McArdle, Katch, & Katch, 1981; Novak, 1972; Tzankoff & Norris, 1977). The loss of total body weight in old age reflects the greater weight of muscle and bone compared to fat, yielding a net effect of loss once these components of lean body mass diminish in relative and absolute amounts in later adulthood.

The increase in body fat is quite consistently found to be more pronounced in the torso of the body, while the subcutaneous fat in the extremities is reduced in quantity in progressively older adults (Chien et al., 1975; Damon et al., 1972; Durin & Womersley, 1974). The abdominal girth shows a gain of 6%-16% in men and 25%-35% in women across the adult years (Shephard, 1978), confirming the existence of what many adults colloquially refer to as "middle-age spread." Other sex differences involve the distribution of bodily fat, with men accumulating more on the back, and women around the waist and also the upper arms (Malina, 1969).

Psychological Consequences of Age Changes in Appearance

The appearance of the face and body is determined by the particular configuration of the bones, muscle, and fat. Appearance is also affected by the texture and color of the skin, particularly that on the face, as well as hair on the head, and eye color, size, and the shape of the lids. Variation in all these components accounts for the highly individualized nature of personal appearance. This variation is so large that a picture of the face is used by governments and other large institutions as virtually the only nonforgeable piece of identification a person can have. The uniqueness of appearance, particularly of the face, contributes to the individual's sense of identity as a person separate and distinct from others. One's appearance is also intimately tied to the person's sense of continuity of self over time, another aspect of identity or self-concept (Guardo, 1968). Because of these two aspects of appearance, the individual is recognizable from day to day, month to month, and year to year, to him- or herself and to others. However, it has just been shown that aging has the potential to alter the individual's appearance in a number of ways. Although the adult's sense of uniqueness is unlikely to be particularly affected since variations in appearance are retained throughout adulthood, aging can threaten the person's sense of continuity over time. The older adult may still "feel" young and therefore be confused about his or her identity when looking in the mirror and seeing wrinkles and gray hair, or when hearing others refer to him or her as that "old man" or "old lady."

Appearance also serves a variety of social functions. One is by its effect on interpersonal perception. Physically attractive individuals are regarded with more favor than their less physically appealing peers, partly because people believe there is more to physical beauty than "meets the eye." The individual's physical prowess is one quality that is judged on the basis of appearance (Dowell, Badgett, & Landiss, 1968). Physical appearance also provides the basis for inferring "internal" qualities about others (Schneider, Hastorf, & Ellsworth, 1979). People who are more attractive are regarded by others as having more favorable psychological qualities as well and so are perceived to be more desirable to be with (Huston & Levinger, 1978). A pleasing appearance is also regarded as a positively valued attribute that one might use in "exchange" for other valued resources (money, possessions, services) in establishing long-term dyadic relationships (Burgess & Huston, 1979). The stereotype of this phenomenon is the pragmatic "arrangement" between the attractive young woman and the wealthy and powerful old man.

Attractiveness has an importance in and of itself on interpersonal relationships regardless of the effects of aging on appearance. However, most of the research on interpersonal attractiveness and appearance as a factor in exchange relationships has been conducted on young people, mostly college students, for whom age changes are not yet an issue. Because of the emphasis in American society on youth and beauty, it may be speculated that the older individual, particularly the woman (Sontag, 1972), is likely to "lose points" in the exchange equation, and to be regarded as possessing fewer desirable mental and physical capacities (Connor, Walsh, Litzelman, & Alvarez, 1978). Studies on physical self-concept, which includes self-perceptions of attractiveness, have also been conducted to determine the relationship to overall self-concept and attitudes toward the self, but again, these have been for the most part conducted on young people. Moreover, in these studies the distinction is usually not made between physical appearance and physical abilities as components of the physical self-concept.

There is, however, some evidence that individuals evaluate their own bodies on the basis of possession of qualities regarding appearance that are influenced by age and that the judgments so arrived at can influence self-esteem. Berscheid, Walster, and Bohrnstedt (1973) reported on the results of a survey of 2000 *Psychology Today* readers (randomly selected from the over 62,000 who responded to the questionnaire published in the magazine) on what they called "body image." These authors defined body image as satisfaction with one's body and although they did not find age differences in overall body image, some of the apparent effects of age on specific body parts were of concern to many of the respondents. Almost one-half of the women expressed dissatisfaction with the size of their hips, and 36% of the men were not satisfied with the size of their abdomens. In general, body image was positively related to self-esteem, and of the parts of the body, satisfaction with the face was the most important in terms of self-esteem.

Given the nature of aging effects on the face and body, it would seem to be the case that adults must more or less concede that as they grow older, their appearance will change in ways that they themselves and others will regard as unattractive. However, there are measures that are quite readily available for either concealing or even offsetting some of the effects of age on the outward physical characteristics of the body. Whether the adult takes advantage of these measures may depend on the seriousness with which age effects are regarded, which quite possibly stems from the importance the individual places on physical appearance. Given sufficient motivation, however, the effects of the aging process on the face and figure can quite successfully be modified (see chapter 7 for a discussion of how changes in appearance relate to sexuality in later life).

Compensation for Age Effects on Appearance

Up to a point, many of the changes in facial appearance associated with the aging process in adulthood can be disguised by commercially sold preparations that cover up wrinkles, discolorations, and sagging of the skin on the face. Men tend not to use these products, although hair dyes "for men" are available. Hair loss is a problem more easily corrected by women, also, since the wearing of hairpieces is considered less socially acceptable for men. More extreme steps are taken by men and women who elect to have cosmetic surgery to replace and tighten sagging and wrinkled skin or to transplant hair onto the scalp. Even with these ameliorative techniques, though, eventually most adults reach a point where there face and hair take on a distinctly "old" appearance recognizable if not by themselves, then by other people. Short of keeping one's face continually covered to protect it from the outside elements (and it is not clear whether even this would be effective), there is little that can be done to reverse the effects of aging on the face and hair.

The effects of age on body build, particularly weight and body fat, are not inevitable, however. Continued participation in active sports and exercise activities tends to offset the deleterious effects of aging on the accumulation of body fat, as shown by the fact that endurance athletes do not gain weight (Suominen, Heikkinen, Parkatti, Forsberg, & Kiiskinen, 1980) and do maintain their muscular physiques throughout adulthood (Kavanagh & Shephard, 1978; Shephard, 1978) for as long as they continue to train (Grimby & Saltin, 1966; Robinson et al., 1973). The amount of training an adult athlete engages in also seems to be an important factor determining body composition. Middle-aged (40- to 60-year-old) marathon runners compared to nonmarathon runners were found to have lower body fat than regularly exercising joggers, in one study (Hartley & Farge, 1977). In addition, participation in exercise training programs by previously sedentary middle-aged and elderly adults, although it may not result in actual weight loss, can achieve the desired effects of reducing bodily fat and increasing lean tissue. The exercise regimens in studies on this topic have involved from 30 to 60 minutes of dynamic exercise (vigorous walking, jogging, cycling) for from 3 to 4 days a week, and from 10 weeks to one year (but typically no more than 20 weeks). Some of the research that has yielded evidence to support this conclusion is methodologically flawed because there was no control group to compare to the exercise training group's improvement (Buccola & Stone, 1975; Getchell & Moore, 1975; Massie & Shephard, 1971; Sidney, Shephard, & Harrison, 1977).

In cases where a control group was used, however, the results were quite comparable in that there was a greater reduction of body fat in the exercise treatment group (Pollock, Dawson, Miller, Ward, Cooper, Headley, Linnerud, & Nomeir, 1976; Pollock, Miller, Janeway, Linnerud, Robertson, & Valentine, 1971).

Given the concern expressed by adults toward their weight and body fat distribution (Berscheid et al., 1973), it might be expected that weight loss would have favorable effects on bodily identity and self-esteem. However, there are no studies in which these relationships were directly assessed. The only existing relevant information concerns a study on a very small group ($n = 5$) of obese teenagers, who demonstrated an increased self-esteem following weight loss as the result of an exercise program (Collingwood & Willett, 1971). The many other investigations of the psychological benefits of exercise training (see chapter 3 regarding aerobic exercise training programs) have generally not examined the relationship between weight loss and gains in well-being. However, it has been suggested that one of the most salient psychological variables influenced by exercise training is self-concept (Folkins & Sime, 1981), which must surely include concepts of the appearance of one's body (O'Brien & Epstein, 1982).

Movement

Smooth coordination of the joints and muscles and their supporting tissues (tendons and ligaments) is essential if the person is going to be able to move effectively through the physical environment. At the same time, the strength and resiliency of the bones will influence the extent to which the individual is able to move around securely without risk of injury. The ability to carry out desired actions involving the moving, lifting, and manipulating of objects depends also on the integrity of the muscles, joints, and bones.

Muscles

Age Differences in Strength and Muscle Mass

The muscles of interest in the study of aging are the skeletal muscles, which are under the control of motoneurons in the central nervous system and hence respond to external stimulation and voluntary efforts of control (see chapter 8). The skeletal muscles make movement possible because when they contract they exert a force on the bones to which they are attached via the tendons. The structural variables of interest in research on the aging process are the number and size of the muscle fibers. The functional variables studied are strength and endurance.

Muscle strength is the functional variable having the greatest direct applicability in regard to age. There is no single pattern of age differences in this function across adulthood. There is variation in the effects of aging according to whether men or women are tested, the general level of activity (particularly occupational) typically engaged in by the respondents, and the muscle group being tested. Another factor affecting the pattern of aging demonstrated is the type of muscular strength being assessed, that is, whether it is static (isometric) or dynamic. Despite these potentially counteracting tendencies, the following general summary statement can be made: There seems to be little reduction in strength at least until age 40 or 50, and the loss up thereafter until 60 or 70 seems to be minimal, on the order of 10% to 20% total (Anderson & Cowan, 1966; deVries, 1980; Katsuki & Masuda, 1969; Larsson & Karlsson, 1978; Montoye & Lamphiear, 1977; Petrofsky, Burse, & Lind, 1975; Petrofsky & Lind, 1975). Persons in their 70s and 80s suffer a more severe loss (30% to 40%) (Asmussen, 1981; Grimby & Saltin, 1983) and this loss is more pronounced in the muscles of the legs than in the hands and arms (Shephard, 1981).

This description of the effects of age on muscle function is of interest in its own right, and the pattern of loss of strength has many implications for the functioning of the adult in performing daily activities (discussed in the following). However, the basic issue that intrigues researchers is to account for the underlying cause of these age effects. On the basis of this research, it seems safe to conclude that what appears to be the source of age differences in muscular strength is the loss of muscle mass that accompanies the aging process in adulthood (Asmussen, 1981; Grimby, Danneskiold-Samsoe, Hvid, & Saltin, 1982; Shephard, 1981; Young, Stokes, & Crowe, 1982). This reduction of mass is claimed to be due to atrophy of the muscle fibers, and in particular in later adulthood the loss of number (Larsson, Grimby, & Karlsson, 1979; Larsson & Karlsson, 1978; Larsson, Sjödin, & Karlsson, 1978; Örlander, Kiessling, Larsson, Karlsson, & Aniansson, 1978) or diameter (Aniansson, Grimby, Nygaard, & Salton, 1980; Grimby & Saltin, 1983; Tomonaga, 1977) of the "fast twitch" muscle fibers. These muscle fibers are important in developing the rapidly accelerating powerful contractions normally associated with "strength." Remaining constant through at least 70 years are the "slow twitch" muscle fibers, which are involved in postural adjustments and in maintaining a contraction over extended time and so are responsible for endurance during prolonged muscular exertion. Fast-twitch fibers are thought to atrophy because the motoneurons that activate the muscles themselves atrophy, and the muscle fibers die because they lose their innervation (Campbell, McComas & Petito, 1973; Petrofsky et al., 1975; Shephard, 1981; Tomonaga, 1977). The consequence of muscle atrophy is deterioration of the muscle fibril, which is replaced initially by connective tissue, and ultimately by fat (Inokuchi, Ishikawa, Iwamato, & Kimura, 1975).

The muscle fibers that do not atrophy seem to undergo some major changes in their reactivity and in their metabolism of energy. There are various findings reported in the literature that pertain to this issue of the characteristics of the muscles in later adulthood. One scenario is that the fast- and slow-twitch fibers dedifferentiate, the fast-twitch losing their quick and intense response to stimulation, becoming more like the slow-twitch muscles (Gutmann, 1977; Petrofsky et al., 1975; Sidney, 1981). Another view, which is not entirely incompatible with the dedifferentiation proposal, is that the muscles that do not atrophy fatigue more quickly and have impaired or deficient energy metabolism due to degenerative changes in the mitochondria (Tomonaga, 1977). These cellular organelles, it is reasoned, because of the deterioration they suffer, are less efficient in producing the ATP that provides the energy for muscular contraction (Brooks & Fahey, 1984; Shephard, 1981). It should be noted, though, that this view is not universally held. Örlander et al. (1978) argued that energy production in the muscle cell remains intact up to 65 years of age, and that other changes are responsible for reduced muscle strength such as deafferation from atrophying neurons. However, since the more significant age reductions in strength do not occur until after 60 to 70 years, this position need not be incompatible with the proposal that impaired energy metabolism in later adulthood contributes to loss of strength.

Another hypothesized change in the existing motoneurons is one with more favorable implications for the daily functioning of older adults. This is the suggestion that the intact motoneurons may reinnervate other muscle fibers to make up for the fibers that have atrophied. At the same time, the remaining muscle fibers may hypertrophy as they are called into action on a more frequent basis by the intact motoneurons (Campbell et al., 1973). As a consequence there may be a "dynamic reorganization" (Grimby & Saltin, 1983) of the neuromuscular system in later adulthood. There are also other ways through which muscle atrophy can be compensated that are more directly under the individual's control through exercise training. Research on the effectiveness of muscle strength training has the added feature of advancing knowledge of the cause of effects of aging on strength and muscle atrophy, particularly because much of this research is based on experimental designs rather than simple descriptions of age differences. By learning about which factors can be manipulated to halt or reverse age losses in strength, investigators may derive testable hypotheses about the factors leading to deterioration.

Training Effects on Muscle Function in Adults

Exercise training is the mode of compensating for muscular deterioration in adulthood that is most readily accessible to adults. This type of exercise need not be of the sort that weightlifters or bodybuilders perform. Instead the underlying principle of exercise training in adulthood is that the muscles simply need to be *used* in order to prevent atrophy, a thesis that has support from enzyme-histochemical analysis of the pattern and distribution of muscle fibers in whole muscle cross-section (Lexell, Henriksson-Larsson, Winblad, & Sjöstrom, 1983).

The basic question addressed by research on exercise training is whether it is possible for older individuals to prevent, minimize, or even reverse the pattern of age-related loss of strength. There are three types of evidence bearing on this question. One set pertains to the documented effects of enforced inactivity such as bed rest (Campbell et al., 1973) on muscle mass and strength. The rationale for this research is that lowered activity patterns in later life (Sidney & Shephard, 1977a) cause muscle atrophy (Bortz, 1982). The converse of this approach is to show that adults who maintain a high level of activity throughout their lifetimes as a result of work or leisure involvements (Shephard, 1978) show less deterioration of muscle mass and strength (Asmussen, 1981). Support for this line of reasoning comes from studies in which trained or highly active men are found to have larger and more powerful muscles than men who have led very sedentary lives (Gollnick, Armstrong, Saubert, Piehl, & Saltin, 1972; Kuta, Pařizková, & Dýcka, 1970; Montoye, 1975; Suominen et al., 1980). Within the same individual, in fact, differences in atrophy and strength may exist between the more- and less-frequently used muscles (Grimby et al., 1982; Gutmann & Hanzlikova, 1976).

Experimental studies in which previously sedentary middle-aged and older adults receive systematic exposure to exercise training supply the most clear-cut answers to the question of whether muscle atrophy can be compensated for by activity in old age. In one such investigation, gains of up to 22% in strength have been reported in 69 to 74-year-old men (Aniansson & Gustafsson, 1981). Although the amount of strength acquired is not always this impressive (e.g., Larsson, 1982), it is believed that older persons seem to gain as much relative to their initial levels of ability as do younger adults given the same training regimen (Sidney, 1981). Perhaps even more important than the sheer gain in power (although that may itself have favorable effects on everyday life) is the potential of the muscles of older people to develop their metabolic efficiency and so increase their work capacity. Some of the mechanisms that appear to be enhanced by exercise include increased respiratory activity in the mitchondria (Kiessling, Pilström, Bylund, Saltin, & Piehl, 1974; Örlander & Aniansson, 1980), greater capacity for aerobic metabolism and decreased accumulation of lactate (Cunningham & Hill, 1975; Suominen, Heikkinen, & Parkatti, 1977; Suominen, Heikkinen, Liesen, Michel, & Hollmann, 1977), greater fiber size, and more widespread recruitment of motor units (Aniansson & Gustafson, 1981). While there cannot, of course, be an actual regeneration of muscle fibers with exercise, it appears that exercise can make the remaining muscle become more "fit" (Grimby & Saltin, 1983). At the same time, these studies confirm the importance of the muscle fiber's efficiency of energy metabolism as a primary contributor to age effects in muscular strength above and beyond the process of muscle atrophy caused by neurological changes. However, in addition, there is some indication that neural activation of muscle fibers can be augmented by exercise in later adulthood. Moritani and deVries (1980) trained small ($n = 5$) groups of older and younger adult men with resistance and dumbbell exercises, focusing on the dominant elbow flexor muscle. The older adults only gained about one-half the strength of the young men (13.8 lb vs. 26.1 lb), but the main observation of the study was that the two age groups gained strength via different mechanisms. The effect for older adults was not due to muscle hypertrophy (as was true for the young), but to increased muscle activation due to "neural factors." The results of this research must be regarded as preliminary until supporting replication studies are conducted and there is greater specification of what is meant by "neural." Given this qualification, though, since denervation is seen as a primary cause of muscle atrophy in adulthood, the results of this research offer some encouragement for the possibility of offsetting a fundamental deterioration process in later adulthood through exercise.

Psychological Consequences

There are many activities in the course of a routine day that require the efficient functioning of the muscles, ranging from getting up from a chair to climbing the stairs. In general, adaptation to the environment is affected in a very basic way by the ability to have one's motions be well-placed and effective. The muscles make it possible for the adult to perform actions required at home, at work, and during recreational activities. Some of these actions involve strength, others endurance, and many especially require muscular coordination. The individual must be able to complete these actions within a small range of error, and also without feeling undue fatigue. Although most of these actions are automatic in nature because they are performed so frequently, it may be speculated that their successful completion on a continuous basis is a pervasive, if not conscious, contribution to feelings of personal competence and efficacy. Age losses in strength, endurance, and coordination would have the effect of communicating to the individual the sense that his or her body is no longer adapting effectively to environmental demands. Consequently, the older adult may lose a very basic sense of reliance on the smooth operation of the skeletal musculature.

In addition to performing actions requiring muscular work in the course of completing daily activities for the sake of one's own occupational and personal demands, adults also must perform muscular work in order to fulfill their responsibilities to others. This is especially true in the family, where even the most sedentary adult must at times perform strenuous muscular work, such as lifting children, boxes, furniture, and tools. In addition, muscular coordination is needed for many household tasks, such a sewing, repairing walls and furniture, gardening, cooking, and countless other manual tasks involved in work and leisure, such as typing, making handcrafts, playing a musical instrument, and dancing. The adult whose coordination and/or strength is diminishing may begin to engage in serious self-questioning when even these routine and taken-for-granted activities become more difficult to carry out.

Another aspect of age effects on muscular abilities in adulthood concerns the impact of the adult's changed capabilities on how that adult is evaluated by the others in his or her social environment. Those adults who value physical fitness and consider a muscular-looking appearance to be important may feel inadequate when their muscles atrophy and convert to fat. In order to retain their youthful physique, they will have to work much harder at controlling their diet and in faithfully completing their exercises. Loss of coordination may be more difficult to control. Even adults who do not place a great deal of significance on being strong may be concerned and upset about age changes they undergo in their ability to coordinate their movements. A lack of coordination in conducting such activities as walking, running, moving objects about, or doing delicate handwork is usually a grave source of embarrassment to adults in public settings. Missing a step and falling, being unable to thread a needle or connect a broken wire, and failing to button up a coat can all create distress when other people are present and watching. While reductions in muscular efficiency are not inevitable and indeed may be reversed or slowed down with exercise, it would seem that the potential is great for the adult's self-concept to be challenged by these changes in the body's ability to move and work.

Bone

Age Effects in Bone Strength

Bone development in childhood is marked by two processes. One is the growth of the length and thickness of the bones. The second process is ossification, the replacement of the cartilage that is the main skeletal component in early childhood with the mature bony matrix made up of collagen impregnated with calcium and other inorganic salts. During adulthood, new bone continues to be laid down, but at the same time, resorption occurs, so that there is a continuous remodeling of the skeletal structure. The reasons that remodeling occurs appear to be related, ultimately, to vascular and mechanical pressures that are placed upon the bones. New Haversian systems, structures within the bone matrix in which osteocytes (bone cells) are situated, develop in response to vascular growth, which may be related to circulatory demands. Muscular contractions and gravitational forces also stimulate bone remodeling, resulting in a net increase in density and/or cross-sectional transverse area of the bone. However, the overall thrust of development of bone in adulthood is loss of the ability to withstand mechanical pressure and consequently the bones become more vulnerable to fracture. Since resistance to pressure is one of the most critical functional requirements that bone must fulfill to support locomotion and other muscular activities, there is a great deal of practical value in being able to explain why this function is impaired with increasing age.

It appears that the progressive loss of bone strength in adulthood is due primarily to a loss of bone mineral content, which weakens the bone so that it cannot support the loads it must bear. Most of the evidence used to support this argument is based on x-ray or photon-beam absorption techniques that assess bone mineral content or density indirectly in living subjects. Either of these methods has the obvious advantage of being usable in vivo. The explanation of the underlying process that accounts for loss of bone mineral content is that the rate of resorption exceeds that of new bone apposition in later maturity so the net result is a reduction of bone mass (Adams, deVries, Girandola, & Birren, 1976; Avioli, 1982; Cohn, Vaswani, Zanzi, & Ellis, 1976). Presumably, such an increase in the resorption rate is due to the heightened activity of the osteoclasts, whose activity destroys bone matrix, while the bone-producing osteoblasts show a relative diminution of their activity.

According to this account of adult bone loss, osteoporosis is not really a separate disease entity from aging; instead it is a more extreme version of the universal process of adult bone loss. One of the reasons for this conclusion is that persons with clinical symptoms of osteoporosis (frequent bone fractures, back pain) are generally not distinguishable on the basis of in vivo measures of bone density from persons without these symptoms, and conversely severe bone loss is not always associated with the presence of reported clinical symptoms (Adams et al., 1976; Garn, 1975; Morgan, Spiers, Pulverloft, & Fourman, 1967; Riggs et al., 1981).

Because of the variety of measures used to describe bone density, and hence, adult bone loss, there are varying estimates made of the quantity of mineral content by different researchers. For instance, as percent of cortical bone, the decrease per year is given as 0.11% in the 40s and 0.26% after 55 in men (Beausoleil, Sparrow, Rowe, & Silbert, 1980); as grams of calcium, it is a loss of 3.8 g/year between the ages of 30 and 54, and 7.6 g/year after 55 in women (Cohn et al., 1976); as skeletal mass it is estimated as amounting to 8%/decade in women and 3%/decade in men (Garn, Rohmann, & Wagner, 1967); and as bone mineral density, the estimated total reduction is about 20% to 30% in women over the adult age span (Avioli, 1982; Riggs et al., 1981; Smith, 1971) and about one-half that amount in men (Mazess, 1982).

Another complicating factor in describing the age of onset and degree of loss of bone material is the type of bone studied and whether it is trabecular (also called cancellous) or cortical (also called compact). Trabecular bone is less dense than cortical bone, and is described as spongy because of its larger open spaces and weblike structure. It is found in relatively high proportion in the vertebrae, at the ends of long bones, and in flat bones. This type of bone seems to be particularly vulnerable to loss of density over the adult years (Adams et al., 1976; Avioli, 1982; Sparrow, Beausoleil, Garvey, Rosner, & Silbert, 1982), with a decrease beginning at 30-35, and amounting to at least a 6% to 8% decrease per decade (Mazess, 1982). The cortical bone, which forms the outer ring of the shafts of the long bones and bones of the hands and feet, is less vulnerable to bone loss, but nevertheless still shows a reduction in estimated mass over the adult age span. There is variation also within different cortical bones. The individual bones in the hand and foot lose less bone than the tibia and ulna (Garn et al., 1967), and the distal radius loses more than the midradius (Avioli, 1982).

Despite the variations across bone and method among these studies on aging and bone loss, the general statement is made that the loss of bone mineral content begins in the late 30s, according to Garn (1975) at very close to the "Jack Benny" age of 39. The rate of bone loss is reported to begin slowly and then accelerate, particular for women, in the 50s, possibly slowing down by the 70s (Avioli, 1982; Beausoleil et al., 1980; Cohn et al., 1976; Mazess, 1982; Riggs et al., 1981; Sparrow et al., 1982). Although both men and women suffer from bone loss, the rate is quite a bit (perhaps two times) higher for women (Garn, 1975; Garn et al., 1967; Morgan et al., 1967). The sex difference in rate and amount of bone loss is accounted for by the lesser bone content that women have at their peak in adulthood, and by the depletion of estrogen after the menopause, which might have the effect of increasing the bone resorption rate (Heaney, 1982). Other hypotheses concerning bone loss are based on age changes in calcium absorption in the small intestine due to a depletion of Vitamin D (see chapter 5).

Another line of research in the area of age effects on bone strength concerns the main functional property of bone: Its actual resistance to direct mechanical pressure. This research has led to a better understanding of not only the extent of loss of bone tissue but also the microscopic changes in bone structure that can affect its ability to withstand impact. In this type of work, it is necessary to subject a bone to enough stress to make it break, and this can obviously only be done on bone specimens taken from cadavers. Researchers working in this area base their findings on principles of bone mechanics, whereby a stress-strain curve is constructed to determine quantitatively the extent of decrease in adulthood. The results of this research confirm the clinical picture of heightened vulnerability to fracture in the bones of older persons in that the elastic and plastic qualities of bone are gradually reduced over the adult years (Burstein, Reilly, & Martens, 1976; Evans, 1973, 1976). By correlating tensile strength with structural changes at the cellular level, it is then possible to determine how structural changes in bone can account for loss of function in a way that could not be accomplished by in vivo methods.

On the basis of such research, the decreased resistance of bone to stress has been attributed to two distinct but nonmutually exclusive processes. One is an increase in the degree of mineralization with age (Currey, 1979), which has the effect of reducing the energy the bone can absorb before it fractures (Currey, 1984). Thus, there is a loss of total bone mineral content, which reflects the loss of bone tissue responsible for much of the loss of resistance to impact of older bone (Horsman & Currey, 1983). But in addition, there is also an increased mineralization of the remaining bone tissue, and this would have the effect of compounding the weakness caused by the total reduction in skeletal mass. This idea that there may be increased mineralization, which causes bone brittleness in osteoporosis, is not a new one to the gerontological literature, but it has not been incorporated into it (cf. Smith, 1971).

The second possible contributing structural cause of reduced tensile strength in cortical bone is an increase in porosity (Stein & Granick, 1980). This increased porosity may be due to an increased number of osteons as the result of the continued remodeling of bone over adulthood. The Haversian canals in these osteons create more open spaces in bone, making it more porous. In addition, there is an increase over the adult years in the cement lines that represent the accumulation of old sites where remodeling took place (Evans, 1976). In trabecular bone, structural changes may also take place that weaken the system of trabeculae so that the loss of tensile or compressive strength is more severe than would be predicted based on mineral density alone (Bell, Dunbar, & Beck, 1967; Twomey, Taylor, & Furniss, 1983; Weaver & Chalmers, 1966). As a result of one or both of these processes, when the bone of an older person is subjected to pressure, it is more likely to snap and cause a "clean" fracture that is difficult to heal. The bone of younger persons instead fractures in such a way that there are many cracks and indeed, complete fracture is less likely to occur because of the impedance offered by the tissue with its higher proportion of organic fibrous material. The metaphor used to represent the difference between how old and young bones respond to pressure is that of tree branches. A young bone is like a green twig, in that it takes a great deal of bending before it breaks. When it finally breaks, it creates many fracture lines that absorb the energy applied to it, so it is less likely to snap altogether. Repair can be very effectively accomplished because the fracture lines can be easily and completely rewoven. In contrast, old bone is likened to a dry stick, which easily snaps when bent. The kind of fracture that will probably occur is more complete because there is less impedance to the force causing the crack. The fracture is harder to repair because it will be smooth and straight across the two surfaces at the site of the injury.

Effects of Exercise Training on Bone

The main implication of the research on structural and functional qualities of bone is that the reason for loss of strength with adult age is a decrease in skeletal mass, reflected by a loss of total mineral content. Training studies aimed at determining the possibilities of compensating for loss of bone mineral content have focused on exercise as a way of reversing or slowing the aging process. The underlying assumption of this research is that reduced activity in the aged is one cause of loss of bone strength, because there is less muscular and vascular stimulation for bone remodeling. Prevention or remediation would therefore be accomplished by increasing the older adult's activity level.

The results of these exercise training studies have been mixed, however, with some demonstrating "positive" results in terms of increased mineral content (Smith, 1971, 1981; Smith & Reddan, 1977) and others showing no effect (Aloia, Cohn, Ostuni, Cane, & Ellis, 1978; Montoye, 1975). Nevertheless, the principle behind this research seems valid and given more extensive testing, may hold promise as a means of alleviating some of the more debilitating effects of bone loss on mobility in later life.

Psychological Consequences of Aging of the Bone

Since there is absolutely no research on this crucial topic of how aging of the bones affects people psychologically, the implications for environmental adaptation and feelings of competence can only be speculated upon. Given the importance of bone strength to every movement that is made, large and small, it would seem to follow that aging of the bones would have the potential to have a particularly significant impact on the quality of the older adult's daily life, a suggestion also made by Barzel (1983) in connection with clinical implications of bone loss. However, these effects are probably difficult for the older person to discern and differentiate from other sources of musculoskeletal losses. What is probably of greater psychological significance to the adult is the threat or actuality of bone factors due to the reduced resistance of the bones to mechanical stresses or pressure.

Anyone who has experienced a broken limb is aware of the real physical discomfort and inconvenience it creates. In addition, there is a loss of freedom to move independently through the environment. The physical pain and restriction mean that even simple everyday tasks cannot be carried out without difficulty and conscious effort. Work activities may have to be temporarily abandoned, and extra assistance required to perform self- and home-maintenance tasks. Depending on the site of the injury, walking and perhaps driving will not be possible, so that the person must depend on others for transportation. The outcome of this inconvenience and reduced solubility would probably be a temporary reduction in the person's sense of competence at being able to manage the affairs of everyday life.

For the older adult, particularly a woman, the experience of breaking a limb is quite likely to occur, based on the fairly high reported prevalence of serious adult bone loss (Avioli, 1982; Twomey et al., 1983). Like a younger adult, the older one will suffer a concomitant loss of ability to complete everyday tasks following a bone fracture. However, unlike a younger person, a number of serious, permanent consequences that have broader implications for her or his life-style are likely to follow from bone fracture in the aged individual. These include hospitalization followed usually by transfer to a nursing home for rehabilitation, and if the period of institutionalization is long enough, a permanent loss of independence. Obviously, if these sequellae follow from bone fracture, the psychological consequences will be very pervasive. But even if they do not occur to the older person him- or herself, the probability is high that they will occur to a friend or relative. Seeing these age peers in this unhappy circumstance of recovering from bone fracture and becoming institutionalized, the older person would naturally want to avoid this outcome. He or she may therefore curtail activities that carry the risk of bone injury and become more cautious about performing the relatively simple act of walking (Costa & McCrae, 1980). Even without actual bone fracture, then, the older adult may experience some of the psychological consequences that lessened mobility carries with it. Furthermore, necessary daily activities that previously were conducted with little concern, such as descending a steep flight of stairs or walking down icy pavement, may create fear and hence avoidance. It might well be imagined that these feelings would detract from the aged adult's independence and sense of competence.

Joints

Effects of Age on the Functioning of the Joints

The smooth functioning of the body's joints is made possible by the strength and elasticity of the tendons and ligaments and the synovial fluid, which makes possible frictionless movement of the bones as they rub against each other within the protective encasement of the joint capsule. Although aging of the joints is usually associated with the later years of life, degenerative processes that reduce the functional efficiency of the joints actually begin to take effect even before skeletal maturity is reached. Discomfort and restricted movement can therefore be a potential problem for the individual at any point in adulthood, and with increasing frequency as age progresses. Although there is surprisingly little empirical

information in the literature on joint mobility, flexibility, and restriction of movement, the findings that exist are consistent in reporting a peak in joint function
as assessed by these variables in the 20s and a constant decrease thereafter
(Boone & Azen, 1979; Greey, cited in Munn, 1981; Jarvey, cited in Munn, 1981).

Age losses have been documented, however, in virtually every structural component of the joints. The connective tissue of the tendons and ligaments is
described as becoming less resilient and less able to transmit the tensile forces
that act upon it as the collagen and elastin molecules within the tissues degenerate, and the fibrous structures become increasingly fragmented (Adrian, 1981;
Brewer, 1979). In addition, scar tissue (Brooks & Fahey, 1984) and areas of
calcification form within the connective tissue and joint capsule (Chung, 1966b;
Hass, 1956) threatening flexibility and elasticity of the fibrous structures. The
synovial membranes also develop areas of hypertrophied fibers, which make
them more fibrous and hence less flexible. The synovial fluid within the joint
capsule becomes less viscous (Balasz, 1977). Thus, the structures and substances
that are intended to serve as cushions for the forces of movement of and on bones
and to ensure that the joint can be maximally flexed or extended all become less
efficient throughout adulthood.

Complicating the serious consequences of these changes is a series of
deteriorative processes that compromise the function of the arterial cartilage.
The arterial cartilage is the clear and hard protective coating over the ends of
bone that minimizes friction and protects the bone from being worn away by
constant rubbing against the opposing bone within the joint. With increasing age
in adulthood, starting in the 20s, the arterial cartilage undergoes a series of harmful alterations. The color changes from a translucent bluish to an opaque yellow,
which might suggest some loss of superficial protective hardness. More importantly, there are damaging alterations throughout the cartilage layer (Adrian,
1981; Chung, 1966a; Hass, 1943). The surfce becomes thinner, less resilient,
pitted, and there is extensive fraying, cracking, and shredding. There are also
outgrowths of cartilage where the cells grow in clusters in response to the continuous damage that is suffered in the course of daily activities. Over time,
as a result of these processes the bones underneath the cartilage gradually
become eroded.

The degenerative form of osteoarthritis resembles in many ways the effects just
described attributable to normal aging processes. Distinction between normal
aging and pathology in the case of the joints has been notoriously difficult to
make both for clinicians and researchers. One set of investigators observed a
cross-sectional increase in the rate of osteoarthritis in the hand among the Baltimore longitudinal sample study of men (Plato & Norris, 1979a). On the basis of
longitudinal analysis, they went on to suggest that osteoarthritis of the distal
joints of the interphalangeal bones represents a true aging effect and that the same
condition in the proximal joints represents extraneous factors, not intrinsic to
aging (Plato & Norris, 1979b). This fine a distinction has generally not been
made in the literature, and represents a useful effort to separate intrinsic changes
in the body associated with aging from disease.

In trying to account for the cause of deteriorative changes in the joints, there does not seem to be one factor that can be singled out as the main source. There are probably changes at the cellular level in the structure of collagen and elastin molecules comparable to those that occur in the dermal layer of the skin, and which contribute to loss of flexibility, strength, and resiliency of connective tissue. In addition, hypovascularization due to diminished efficiency of circulation may contribute to some deteriorative changes (Brewer, 1979; Hass, 1956). Since the cartilage receives little vascular supply to begin with, any reductions in adulthood (due to aging or arterial disease) will further reduce its capacity for cellular repair.

What is probably the most commonly accepted implicit model for the aging of the joints, however, is the "wear and tear" theory (Brewer, 1979; Brooks & Fahey, 1984) which, although generally a poor explanation of the aging process (Strehler, 1977) appears to have some heuristic if not scientific value in the case of the joints. The joints are subjected to an extreme amount of trauma during life, due partly to constant stresses placed upon them during movement due to weight load in the legs and feet and constant small motor movements and stretching of the hands and arms. Over the course of the adult years, the individual is also likely to suffer from numerous major and minor strains and sprains encountered during everyday activities and during strenuous exercise. Unlike the muscles, the joints do not seem to benefit from their continued and heavy use. Instead, the reparative processes the cellular structures enlist seem to be in and of themselves detrimental. For instance, many of the changes noted by Chung (1966b) and Hass (1956) in the increase with age of calcified and fibrous material in the joints are abortive and inefficient fibrogenetic and osteogenetic repair processes, which result in an overgrowth of material that interferes with the effective working of the joints. Indeed, rather than a simple wear-and-tear model alone, it seems that in the case of the joints, a more useful model incorporates the reparative processes of living organisms: a wear, tear, and "faulty repair" model.

Exercise and Joint Function

If continued use of the joints is one of the factors responsible for age losses in function, then it would seem to follow that the aging individual would be well advised to refrain from engaging in most physical activities. Exercise training nevertheless seems to hold some promise as a way of alleviating restriction of movement and pain caused by joint deterioration. In order for positive outcomes to result from exercise training, though, there must be progressive increments built into the program so that at no one point is a joint ever overextended or stressed (Shephard, 1978). There are other ways to reduce muscle and joint problems in exercise training that has the purpose of improving cardiovascular functioning (e.g., deVries, 1980), but if the purpose of the training is to enhance joint flexibility and range of movement, then these are the problems that must be addressed by the program.

On the basis of several investigations (most of which involve some form of control group), it appears that exercise can promote joint functioning through jogging (Buccola & Stone, 1975), finger lifting (Chapman, deVries, & Swezey, 1972), flexibility exercises (Frekany & Leslie, 1975), rhythmic exercise (Lesser, 1978), and dance (Munn, 1981). Although these exercises may not be able to undo severe damage to the joints that has accumulated over a lifetime of use, they may have the effect of ameliorating restrictions in the individual's life caused by lack of flexibility and pain during movement (Brooks & Fahey, 1984). Thus, it is unlikely that the amount of exercise involved in these programs is sufficient to compound the effects of aging on the joints due to accumulation of trauma over the adult years, especially if the participant does not suddenly begin to place undue stress on a paticularly sensitive area. Instead, exercise probably has one of its strongest effects in that it strengthens the muscles that support the joints so that less stress is placed upon the tendons and ligaments. Secondarily, activity stimulates the cardiovascular system and can enhance the vascular supply to promote reparative processes in the joint capsules that are exercised.

Psychological Consequences of Aging of the Joints

Degenerative changes in the joints, whether due to osteoarthritis or normal aging, have many pervasive effects in the individual's life (Shephard, 1978). Restriction of movement in the upper limbs rules out many enjoyable leisure activities such as tennis or handcrafts, and makes it difficult for the adult to perform mandatory occupational activities that require finely tuned motor skills and frequent movement of the hand and arm. Pain and reduced flexibility in the lower legs and feet can hamper the individual's ability to walk at his or her desired pace. From restricted movement in the hip follow such effects as limping, trouble climbing stairs, and rising from low seats and a supine position. If the knee is involved, it is also more difficult to climb stairs, and in addition, the legs may lose their stability. Degenerative changes in the spine often result in back pain which, if not restrictive in and of itself, certainly detracts from the individual's enjoyment of even otherwise pleasurable activities and may make mandatory routine ones almost unbearable.

Since joint movement is an unavoidable aspect of daily life in maintaining personal hygiene, in performing work activities, in caring for family, and in participating in sports and hobbies, it is difficult for the individual to avoid situations in which pain or restriction might be encountered. Moreover, joint pain has the quality of being impossible to ignore, forget, or disguise. Changes in the body's appearance are important psychologically, but they are only self-evident upon deliberate inspection. They can also be disguised cosmetically so that other people and oneself are made less aware of them. In contrast, the ache of a sore shoulder, elbow, or knee is not very easily dismissed by the individual who experiences it. As a result, even the adult who would rather not think about his or her physical aging will be hard-pressed to overcome feelings of pain and restriction resulting from age losses in the functions of joint flexibility and range of movement. In addition, such limitations are inevitably apparent to other people and therefore may contribute in a negative sense to how the adult's competence is socially judged, and in turn, one's own self-evaluation of competence.

In contrast, when freedom of movement is restored as the result of a well-prescribed and adhered to exercise program, the individual can gain an improved sense of self-sufficiency and hence, real and perceived competence. Such activities as reaching for zippers and buttons while dressing, picking up objects from the floor, or putting on one's shoes are simple tasks, yet are very basic to a sense of competence in everyday life. If an adult has lost the ability to perform these tasks and then through exercise training regains these skills, an enhanced sense of well-being may ensue (Folkins & Sime, 1980; Frekany & Leslie, 1975). Individuals who improved their objectively measured range of movement following a 12-week program of dance and movement training in Munn's (1981) study also reported that they felt an enhanced sense of independence as well as improved balance and flexibility. It would seem that exercise can help offset the basic psychological processes initiated by loss of joint function to the extent that these feelings contribute to a sense of competence regarding one's body.

CHAPTER 3

Cardiovascular System

The sole functional requirement of the heart is to pump blood through the circulatory system on a continuous basis at a rate that adequately perfuses the cells of the body during rest, physical exertion, and mental stress. The aging process involves some rather significant limitations in this function that can have the effect of reducing the adult's enjoyment of and participation in a range of physically demanding activities. Growing older is also associated with the increased probability of developing fatal cardiovascular diseases. There is a rapidly growing body of evidence bearing on the question of whether participation in exercise can play a role in increasing the individual's chances of maintaining healthy cardiac functioning throughout the adult years. This research is perhaps stimulated in part by widespread concern that adults have over their developing age- and disease-related cardiovascular problems.

The exercise equipment used in this research is usually either a stationary bicycle ergometer, in which work load can be quantified in terms of the tension applied against the pedals, or a treadmill, which varies in the speed and grade at which the belt passes under the feet of the walking exerciser. For submaximal determinations, the oxygen uptake at given work loads is set at a fixed value. When maximum oxygen consumption is assessed based on direct performance observations (it can also be predicted from heart rate at submaximal levels), the method basically involves measuring the highest oxygen consumption the person can achieve. Determination of this measure depends on the subject's and experimenter's willingness and confidence in pushing the levels of performance to a point where the subject is completely exhausted, or physiologically the subject's maximum oxygen consumption has plateaued. This procedure may create problems with very old respondents, or those aged respondents who are not used to exerting themselves (Bassey, 1978; deVries, 1980; McArdle et al., 1981; Shephard, 1978). Direct measures of cardiac output and related variables require, in addition, arterial puncture and right atrial catheterization, obviously a very invasive procedure. For these reasons, the effect of exercise training is often measured by predicting maximum oxygen consumption from a person's submaximal heart rate.

Another difference between submaximal and maximal exercise concerns the criteria for successful performance. Submaximal exercise is generally reported to be indicative of the individual's ability to perform at fixed levels of work intensity that are preset by the equipment or by the maximum allowable performance on such parameters as heart rate and oxygen consumption. Because there is an upper limit to the work load that is being demanded, it is to the worker's advantage to accomplish this level of work with as little exertion or strain on the cardiovascular system as possible. Optimal performance on submaximal tests is, then, reflected in a low heart rate and also a low cardiac output (hence a low stroke volume). If the work can be performed without requiring a rapid heart rate or left ventricular effort, then there is an economy of work on the part of the heart. The oxygen uptake necessary to complete the work is achieved via enhanced extraction of oxygen by the skeletal muscles, which are making the most efficient use of the oxygen in circulation.

The situation is reversed for maximum levels of performance. In this case there is no fixed amount of work to perform; instead what is demanded is the highest level that the cardiovascular system is able to support. Optimal performance is achieved by enhancing oxygen extraction, of course, but in order for very high maximum work levels to be reached, the individual must pump more blood per minute. This demand is met by raising the stroke volume as much as possible via increased contractility of the left ventricle. In addition, the heart rate must be able to rise to very high levels through heightened responsivity to sympathetic stimulation and through decreased time necessary between beats for diastolic filling.

Age Effects on Cardiovascular Functioning

The two most serious primary effects of aging on heart functioning are evident when the individual is exercising dynamically (rhythmically contracting large groups of muscles) at a steady state of maximal activity and relying on aerobic muscle metabolism. One is a decreased aerobic power, or maximum oxygen consumption. This effect has been demonstrated in many studies, both cross-sectionally (e.g., Hossack & Bruce, 1982; Strandell, 1964) and longitudinally (Åstrand, Åstrand, Hallback, & Kilbom, 1973; Robinson, Dill, Tzankoff, Wagner, & Robinson, 1975). The decrease in maximum oxygen consumption is virtually linear throughout the adult years (McArdle et al., 1981) and is estimated to amount to a total loss of 30%-40% in the 65-year-old compared to the young adult (Shephard, 1978). The second widely observed age effect is a reduction in the heart rate attained during maximum levels of exertion. Again, this finding is demonstrated across a variety of samples and with both cross-sectional and longitudinal designs, including those cited above regarding maximum oxygen consumption (since the two variables are often measured simultaneously).

Taken together, these findings signify that throughout the adult years the heart continually loses its efficacy as a pumping device. The question of interest regarding the causal mechanisms that underlie this process is the extent to which this loss is the outcome of changes in the central (the heart) or peripheral (the veins and arteries) cardiovascular system. This question does not necessarily have one answer. Indeed, there seems to be a multiplicity of causes for the reported age losses in maximum heart rate and oxygen consumption.

The main centrally based mechanism that seems to lead to diminished cardiovascular functioning is the reduced capacity of the left ventricle of the heart to propel a large volume of blood into the aorta when it empties at systole. The stroke volume (amount of blood pumped at each beat) is, in turn, reduced to a commensurate degree. Consequently, the total cardiac output per minute (measured by multiplying the stroke volume by the heart rate) is also lower. This explanation is suggested by the series of studies in which one or more of these variables are directly or indirectly measured (Brandfonbrener, Landowne, & Shock, 1955; Hossack & Bruce, 1982; Julius, Amery, Whitlock, & Conway, 1967; Port, Cobb, Coleman, & Jones, 1980; Robinson et al., 1975). Maximum oxygen consumption is thereby reduced, since there is less blood flowing through the arteries from which oxygen could be extracted (based on the formula that maximum oxygen consumption is equal to cardiac output multiplied by the rate at which oxygen is extracted by the body's tissues). The anatomical changes that appear to underlie this set of consequences are an increase in the thickness of the left ventricle's wall and the total ventricular mass (Gerstenblith, 1980; Lakatta, 1979; Landowne, Brandfonbrener, & Shock, 1955), which reduces its potential to contract completely. As a result of these changes, more time is required to complete the cardiac cycle of diastolic filling, contraction, and systolic emptying of the left ventricle into the aorta (Granath, Jonsson, & Strandell, 1970; Kino, Lance, Shamatpour, & Spodick, 1975; McArdle et al., 1981; Port et al., 1980; Weisfeldt, 1980). The reduced contractility of the myocardium would also make it less easily stimulated by the activities of the sympathetic nervous system (see chapter 6). Electrocardiographic abnormalities during exercise may result from this reduction in the heart's efficiency, most notably S-T segment depression (Bengsston, Vedin, Grimby, & Tibblin, 1978; Montoye, 1975; Profant et al., 1972; Silver & Landowne, 1953; Strandell, 1963). This pattern signifies that with increasing age in adulthood, there are more individuals who will suffer from temporary ischemia during exercise, meaning that their muscles become more likely to receive inadequate oxygen to support aerobic metabolism.

The suggestion that peripheral cardiovascular factors play a role in causing reduced cardiac functioning is based on several observations. One is an increased resistance to blood flow, which is offered by the peripheral vascular system (Brooks & Fahey, 1984; Julius et al., 1967; Raven & Mitchell, 1980; Shephard & Sidney, 1978; Sutton, Reichek, Levett, Kastor, & Giuliani, 1980; Weisfeldt, 1980; Yin, 1980). The explanation that is usually offered for this phenomenon is that the arterial walls become more rigid with age and so are less able to dilate to accommodate increases in blood flow during systole, and especially during exercise, when there is more blood flow to the skeletal muscles. Some of the changes with age in the arteries are illustrated in Figure 3.1.

A similar explanation would probably account for higher levels of systolic and diastolic blood pressure observed in the older age groups within the population

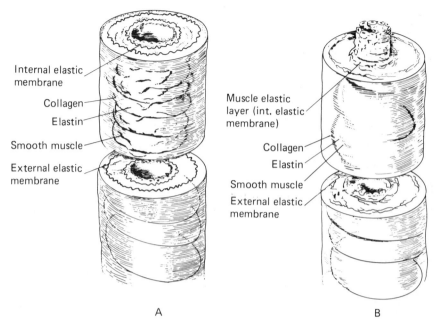

Internal elastic
membrane

Collagen

Elastin

Smooth muscle

External elastic
membrane

Muscle elastic
layer (int. elastic
membrane)

Collagen

Elastin

Smooth muscle

External elastic
membrane

A   B

FIGURE 3.1. Differences in structure between artery in young adult and artery in older adult. (A) Young adult, with normal diameter of walls and configuration of layers. (B) Old adult with thickening of diameter due to increase in collagen and calcium and abnormal innermost layer of epithelial cells. *Note*. From *Aging: Vol. 12. The Aging Heart: Its Function and Response to Stress* (p. 141, 143), edited by M.L. Weisfeldt, 1980, New York, Raven Press. Copyright 1980 by Raven Press. Reprinted by permission.

at rest (Sato et al., 1981) and during submaximal (Bengsston et al., 1978; Denolin, Messin, Degre, Vandermoten, & deCoster, 1970; deVries & Adams, 1972a; deVries & Adams, 1977; Julius et al., 1967; Norris, Shock, & Yiengst, 1953) and maximal (Bengsston et al., 1978) exercise. However, this pattern of results is not consistently reported (Asmussen, Fruensgaard, & Norgaard, 1975; Ordway & Wekstein, 1979; Proper & Wall, 1972; Tzankoff, Robinson, Pyke, & Brown, 1972), and it is possible that the lack of consistency across different studies is due to the confounding of the normal aging process with the disease process of hypertension, which is more prevalent in increasingly older age groups (Sleight, 1979). Smulyan, Csermely, Mookherjee, and Warner (1983) found that when blood pressure levels were controlled, the walls of the arteries in the forearm (between the brachial and radial arteries) were more rather than less compliant as a function of age, a finding that agreed with Learoyd and Taylor (1966) who studied arterial wall distensibility in iliac and femoral vessels in older and younger samples.

Another mechanism that may account for lowered maximum oxygen consumption in older groups of adults is that there is lowered demand for oxygen by the tissues due to the reduction in muscle mass that occurs with age (see chapter 2). In this case, less oxygen is extracted by the muscles because there are fewer skeletal muscles requiring a supply of oxygen during exercise (Brooks & Fahey, 1984; deVries, 1980; Julius et al., 1967; McArdle et al., 1981; Montoye, 1975).

Effects of Exercise on Aging of the Cardiovascular System

The question of whether the effects of aging on the cardiovascular system can be offset or reversed by habitual physical activity has stimulated two decades of intense research. Much of the early research was conducted in Sweden, where a large percentage of the population participates in endurance types of sports such as cross-country skiing and long-distance running. Investigators such as Grimby, Åstrand, Saltin, and Asmussen have been able to capitalize on the popularity of these sports to determine the effects of regular exercise on cardiovascular functioning in fairly good-sized samples of middle-aged and older adults. More recently, researchers in this country are finding more and more willing volunteers on whom to conduct short-term exercise training studies. This research has been made possible on a widespread basis by the greater emphasis in American society on being, becoming, and staying "fit," as well as a greater emphasis on certain participatory sports. Jogging, especially, has taken hold as a major part of life for many adults, ranging from the business executive to the homemaker. To be sure, part of what is meant by the use of the term "fit" among the general public also implies feeling and staying "young." Determination of the validity of this belief in exercise as a means of offsetting the effects of aging therefore has both practical and theoretical importance.

Attempts to quantify whether involvement in endurance types of sports actually does retard or ameliorate the loss of aerobic power in adulthood has taken three separate but complementary routes. The first is based on observations of functional decrements that adults (including some relatively young people) suffer when forced to avoid any physical activity altogether (Bassey, 1978; Bortz, 1982; Kenney, 1982). Most of the empirical basis for inferring the effects of activity is based on more direct research evaluating the way that training alters the normal age pattern of losses. As in studies on the effect of exercise on muscular strength in later adulthood, there is a group of investigations (some of which include muscular function variables) in which highly trained athletes perform tasks requiring maximum exertion while their cardiac functioning is monitored in the laboratory. The short-term exercise training experiment, in which pre- and posttraining cardiac functioning is compared, is the other means through which the effects of exercise on the aging process are evaluated.

The major dependent variable of interest in studies in which the effect of training is evaluated is aerobic capacity (maximum oxygen consumption) and secondarily, heart rate under conditions of maximum exercise (i.e., when no further increment in maximum oxygen consumption is observed despite increases in work load intensity). Some investigators also include indices taken at submaximal exercise levels. The latter type of research is based on the principle that a more efficient cardiovascular system will require less effort (measured in terms of heart rate and also stroke volume and cardiac output) to perform successfully at a given level of oxygen consumption. In all these studies, it is important not only to examine the exercise training approach, but also the type of people who serve as subjects, and, especially, their motivations. (The topic of motivation will be covered in the context of psychological consequences of cardiovascular aging.)

Long-Term Training Effects

The studies concerning long-term endurance exercise participation fall into several subcategories, based on the nature of their samples. One set includes those conducted on former physical education majors in college, many of whom had gone into the field of teaching as well as groups of physical education teachers (Asmussen et al., 1975; Asmussen & Mathiasen, 1962; Kanstrup & Ekblom, 1978). This type of sample is of interest for two reasons: First, because it would be expected that such individuals would have a long-term interest in the development of the body, and secondly, because they may be assumed to have an above-average level of daily physical exertion because they enjoy the sensations caused by bodily movement. All three of the studies in this subcategory are longitudinal, spanning from 13 (Kanstrup & Ekblom, 1978) to 40 years (Asmussen et al., 1975).

Contrary to what might be expected, though, individuals previously or currently in the field of physical education did not compare particularly favorably with others of their age in cardiac functioning. However, this result does not necessarily constitute a challenge of the value of training. Averaging across the entire sample disguises the differences that existed in their actual current levels of physical activity, because not all were still performing their conditioning exercises. In fact, in Kanstrup and Ekblom's (1978) study, because some had been so well-trained when young, they were especially likely to show a decrease in aerobic capacity after they had stopped exercising; what is called the "detraining" effect. Those who did continue hard physical training demonstrated improved cardiovascular fitness. Since the respondents in the Asmussen studies were not subdivided according to current training involvement, it is not possible to make a similar determination in this case. However, fewer of the women seemed to be physical education teachers currently and they showed a greater decrease than the men, most of whom were teaching this subject. It seems likely, then, that aerobic capacity in middle adulthood was, among this sample, related to current levels of training.

Another type of individual whose levels in adulthood of cardiovascular functioning have been studied is the former athlete who trained in youth but who is no longer actively engaging in strenuous activities. It appears that past athletic training does not protect the person from the effects of aging to any significant degree (Robinson et al., 1973), although there might be a slight advantage when the older former athlete is compared to an older adult who has never participated in sports (Saltin & Grimby, 1968). More importantly, it seems to be continued exercise throughout the middle adult years that makes it possible to avoid age losses in aerobic power and perhaps even reverse them. Gains in maximum oxygen consumption in the 13 years between testings in the Robinson et al. (1975) study were demonstrated only for those men who continued to exercise regularly throughout the duration of the period, whereas the man who had trained in youth and then stopped were found to have the same rate of loss of aerobic power as the others in the longitudinal sample (Robinson et al., 1973).

The athlete who continues to stay in competitive endurance sports throughout adulthood into old age is a special case of great interest to exercise physiologists. Because of his or her unique capabilities and unusually high level of continuous practice, this individual should, it is reasoned, provide the most extreme test case for the favorable effects of active sports participation on aerobic capacity. The orienteer is one such individual whose cardiovascular function has been measured, and the other is the long-distance cross-country skier (in some cases the athlete may engage in more than one of these activities). The orienteer is a particularly interesting case, and one that is found fairly specifically in the Scandinavian countries (although it is recently becoming more popular in America). Orienteering is a sport in which the competitors run from 4 to 10 miles (depending on their age category) in a wooded area, in which are found various artificial and natural landmarks (fences, hills, swamps, streams). The orienteer must find his or her way through this course with the aid of a map. Training for this sport involves at least 1-1.5 h of running from two to six times a week, at a varying running speed (Grimby & Saltin, 1966). Orienteers, skiers, and other long-distance endurance athletes have proved to have much larger aerobic capacities than their sedentary counterparts, even those who are considerably younger (Anderson & Hermansen, 1965; Cumming, 1967; Gollnick et al., 1972; Grimby, Nilsson, & Saltin, 1966; Grimby & Saltin, 1966; Suominen et al., 1980).

Another kind of unusually athletic older person whose aerobic capacity might be predicted to be especially large is the champion or master's athlete who continues to compete in special meets, including track and field events. The available literature from this group of persons is supportive of the research on the orienteers, that is, they have higher levels of maximum oxygen consumption than sedentary adults of the same age and younger (Heath, Hagberg, Ehseni, & Holloszy, 1981; Kavanagh & Shephard, 1978; Pollock, 1974). Nevertheless, even the middle-aged and older-adult master's athlete experiences an age-related reduction in this index of cardiovascular efficiency.

There is an extensive literature on the short-term effects of aerobic exercise training on the cardiovascular functioning of middle-aged and older adults, as will be shown in the next section. The average duration of the training programs that these adults complete is 10 to 12 weeks, although some continue for periods of up to 1 year. There is a smaller number of cross-sectional and longitudinal studies, in which "ordinary," that is nonathletic, adults participate in an intense training program for an extended period, or in which adults who have maintained an active life-style are compared to age-matched sedentary adults. Kasch and his co-workers followed a group of 15 men for 15 years, from the ages of 45 to 60, who had kept up their involvement in long-distance running and swimming, three days a week, for 1 h a session (Kasch, 1976; Kasch & Kulberg, 1981; Kasch & Wallace, 1976). Although the intensity, frequency, and duration of the training sessions had dropped somewhat in the last 5 years of the investigation, and there was a loss of maximum oxygen consumption, Kasch's subjects retained greater aerobic power than the average man their age who was not in training.

Other studies on adults who remain active throughout their middle years provide further support for the advantage offered by continued exercise participation in that these active men and women function at the level of sedentary people who are 10 to 20 years younger (McDonough, Kosumi, & Bruce, 1970; Plowman, Drinkwater, & Horvath, 1979; Wright, Zauner, & Cade, 1982). There have also been suggestions made that occupational activity may have some role in minimizing age-related losses in aerobic capacity. Evidence that would support such an interpretation comes from Proper and Wall's (1972) analysis of the cardiac functioning of pilots, who showed none of the usual decrements in any of the variables measured. It might be reasoned that these individuals have to meet such exacting health standards that they engage in health and physical activity practices which keep them "in shape," and that the active life-style engendered by their work prevents them from experiencing the negative effects of aging. However, the alternative interpretation is equally plausible—that these men were pilots because they were in such good health, and so no age decrements were observed. Moreover, based on other research comparing persons with different levels of activity in their occupations, it does not appear that the amount of exercise received during the working day makes a critical difference in aerobic capacity. Instead, activity during leisure hours is more strongly related to longevity (Rose & Cohen, 1977) and fitness (Brunner & Meshulum, 1970).

From this research on the relative benefits of continued exercise participation for cardiovascular aging in later adulthood, it appears that there is a favorable effect of aerobic training on the otherwise inevitable consequences of the aging process on the ability of the heart to supply blood to the body's tissues. One of the reasons for this favorable impact appears to be that continued involvement in endurance sports helps offset the specific sources of age-related functional losses that are based on the diminished capacity of the left ventricle to pump large amounts of blood into the aorta. On the one hand, athletes and highly trained older people have a slower resting heart rate and heart rate during submaximal exercise, but they also have an ability to increase their rate of contraction to very high levels during exhausting exercise. The highly trained older person in addition has a left ventricle that achieves high levels of contractility so that a maximal stroke volume and hence cardiac output can be attained during exhausting work (deVries, 1980). Finally, the efficiency of energy metabolism in skeletal muscles may be increased, so that oxygen from the blood is more effectively extracted and used (Kiessling et al., 1974).

Although athletes do show signs of aging, then, the aging process appears to be slower in the rate at which it causes the athlete to experience limitations in his or her endurance and strength (Shephard, 1978) and to suffer from pain and serious damage when vigorous activity becomes necessary (Shephard & Sidney, 1978).

Short-Term Training Effects

The research in which athletes are compared to their sedentary age peers has the obvious disadvantage of being sensitive to challenges to inferences about cause and effect. This is because it is possible that when favorable results are observed, it is due to the fact that the athletes were in better health to begin with and that this factor is what accounts for their maintenance of aerobic capacity compared to sedentary individuals. In addition, the researcher has little control over the athlete's exercise patterns, and it is not known whether variations in performance at the time of test are due to aging, variations in training intensity, or to individual differences. When lifelong physical activity is assessed in a retrospective fashion, it is also difficult for the researcher to know how accurate the respondent's recollection is of the amount and duration of exercise participation. Moreover, rarely, if ever, is a control group compared to the "experimental" group so that the effects of training can be compared to a no-treatment condition. It is also difficult to find extrinsic incentives that will stimulate a subject to participate in a training program for perhaps as long as 15 years. The subjects who remain in a study after that length of time almost certainly (although this has not been determined) differ in important ways from those who drop out, in that they probably have very strong desires to stave off the effects of age on their bodies or at least to stay in good health. Many of these limitations are, if not eradicable, at least possible to hold to manageable proportions in a short-term exercise evaluation study.

Nevertheless, short-term exercise training studies are not without their drawbacks. One is that there can still be selection effects due to the fact that the participants in any study in which an experimental training procedure is being evaluated are, by definition, volunteers, and so may have unusually high motivation to do well on the exercise measures. By the same token, these volunteers are probably in better shape than the people who stay away from situations in which their physical abilities will be tested. This limitation is particularly critical in the case of some of the experiments that will be examined later, in which motivational and adaptational correlates of exercise training are studied.

In order to control for the volunteer selection effect, it is necessary to have not only a no-treatment condition, but a waiting-list control group whose members are comparable to those who receive the treatment on factors such as motivation, physical status, and involvement in physical activity. Another problem in evaluating short-term training studies is that there is considerable variation across investigations in the training modality—treadmill, bicycle ergometer, ball games (such as racquetball or handball), swimming, or jogging. Moreover, there are extreme differences in frequency, duration, and especially intensity of the training sessions. Finally, there is the phenomenon of attrition, which depends on the length of the study and how much interest can be generated among the subjects by the training modality or the morale of the training group. If the investigator fails to take into account such motivational concerns or makes the training regimen so strenuous that unfit sedentary subjects injure themselves before they have had a chance to become conditioned, the drop-out rate may reach 50% or even higher.

Given these qualifications, the short-term exercise training study appears to hold great potential. It is an efficient way of evaluating the actions to protect against cardiovascular aging that can be taken by relatively nonathletic individuals employed in sedentary occupations and with correspondingly inactive lifestyles. In describing the results of the numerous studies on middle-aged and elderly adults, the major dependent variable that will be focused on here is maximum oxygen consumption.

If an exercise training program increases a subject's maximum oxygen consumption, it means that the oxygen transport system is better able to support the maximum amount of work being performed by the muscles. The effects of training on cardiac functioning under submaximal conditions are also of interest in that these performance situations are closer to what individuals actually do when they make demands on the cardiovascular system in daily life. Some researchers have used both invasive and indirect measures of hemodynamic functioning to provide estimates of cardiac output and oxygen extraction. The findings from these studies provide clarification of possible mechanisms underlying the improvement of aerobic capacity.

It appears that the otherwise negative effect of aging on maximum oxygen consumption can be offset to a large degree by aerobic exercise training which large groups of muscles are made to contract dynamically. If the normal loss of maximum oxygen consumption over the adult age span is figured at 1%/year (Brandfonbrener et al., 1955), or a total of 40% between 25 and 65, the loss can be reduced by up to one-half in any given 2- to 3-month training study in which the participants meet at least 3 h/week and exercise at 60% of their aerobic capacity or more (Hodgson & Buskirk, 1977). Most of the research in which this effect is demonstrated is conducted on persons up to but usually not over 65 (e.g., Pollock et al., 1976; Siegel, Blomquist, & Mitchell, 1970; Strauzenberg, 1978; Tzankoff et al., 1972). However, when the exercise training is sufficiently strenuous (unlike Benestad's, 1965, study), men and women in their 70s are capable of increasing their aerobic power in a relative sense that is equivalent to that achieved by sedentary younger adults who undertake training (Barry et al., 1966; Buccola & Stone, 1975; deVries, 1970, 1980; Heikkenen, 1978; Niinimaa & Shephard, 1978a; Pollock et al., 1976; Rost, Dreisback, & Hollmann, 1978; Sidney & Shephard, 1978; Suominen, Heikkinen, Liesen, Michel, & Hollman, 1977).

The mechanics through which aerobic exercise normally has its favorable effects on maximum oxygen consumption are twofold: one on the heart itself as a pumping device, and the other on the peripheral system through which blood is distributed to and used by the muscles. The effects on the heart are addressed by examining the variables of heart rate, cardiac output, and stroke volume. Although exercise training studies usually do not demonstrate a favorable effect of training on maximum heart rate in middle-aged and older adults (e.g., Hartley et al., 1969), an almost universal finding is that of a decrease in heart rate during submaximal exercise (e.g., Blumenthal, Schocken, Needels, & Hindle, 1982). This effect means that less stress is placed upon the heart in terms of the number of times per minute it must contract to satisfy the needs of the tissues at a fixed work load (deVries, 1980).

It does not appear that cardiac output at submaximal work levels is improved in older adults by training but instead remains constant (e.g., Kilbom, 1971b; Pollock et al., 1976; Rost et al., 1978). This finding, in combination with the decrease of the submaximum heart rate, suggests an increased efficiency of the left ventricle during submaximum exercise as the result of training (based on the formula: Cardiac Output = Stroke Volume × Heart Rate). Maximum cardiac output is increased by training in middle-aged and older adults (Adams, deVries, Girandola, & Birren, 1977; Fardy, 1971; Skinner, 1970). Since the maximum heart rate is not enhanced by training, the enhanced cardiac output reflects improved functioning of the left ventricle during maximum work, that is, a higher stroke volume.

Secondly, in young persons, training seems to have the effect of enhancing the peripheral circulation and the extraction and use of oxygen by the muscles during dynamic exercise. One possible mechanism for this effect is an increase in the number of mitochondria in the muscle cells. However, older adults do not seem to improve their rate of oxygen extraction as the result of training (Hartley et al., 1969), nor is there an increase in the mitochondrial volume as there is with younger adults (Kiessling et al., 1974). Moreover, due to reductions in the blood supply serving the muscles (McArdle et al., 1981) there is less opportunity for the muscles of an older adult to increase their uptake of oxygen. Instead, the effects of training that would improve the peripheral circulation are those that ultimately pertain to the resistance to blood flow offered by the arteries (Cureton, 1969).

The accumulation of lipids in the arteries over the individual's lifetime has the effect of narrowing the internal diameter of the vessels through which the blood must be pumped, hence increasing the work of the heart. Middle-aged and elderly adults seem to be able to benefit from the favorable effect that exercise has on enhancing lipid metabolism. This effect is due to the increase in the fraction of high-density lipoproteins, the plasma lipid transport mechanism responsible for carrying lipids from the peripheral tissues to the liver for excretion or synthesis into bile acids (Haskell, 1984). Training thereby reverses the normal effect of aging, which is an increase in concentrations of total plasma cholesterol, triglycerides, and the low and very low density lipoproteins that transport these substances throughout the circulation (Heiss et al., 1980). The more favorable rate of lipid and triglyceride catabolism that results from training therefore diminishes the chances for lipid accumulation in the arteries.

Although atherosclerosis (the rigidification of the arteries) is a disease process and not a normal consequence of aging, it does become more prevalent among adults beginning in the fourth decade of life. To the extent that training enhances lipid metabolism, it should lower peripheral resistance caused by the normal changes in the arteries and those associated with atherosclerosis. A similar process may account for the effect that exercise training has on reducing blood pressure during or immediately after maximum exertion (Buccola & Stone, 1975; Kilbom, 1971b; Tzankoff et al., 1972). Hypertension (chronic high blood pressure) is, like atherosclerosis, a disease that increases in prevalence throughout the middle and latter decades of adulthood. A reduction of blood pressure during dynamic exercise would have the impact of decreasing the load on the heart caused by heightened resistance to flow in the arteries.

It must be pointed out here that several decades of research have not established that there are long-term benefits of exercise for increasing longevity, despite what would seem to be cogent arguments, even before the relatively recent lipid metabolism research was conducted. However, the reduced risk of developing chronic circulatory disorders such as hypertension and atherosclerosis that would appear to result from exercise training via the mechanism of lipid metabolism would probably be a prime candidate for a factor that does offer the possibility of increasing longevity, or at least reducing the individual's risk of becoming disabled by these chronic diseases in the later years of life (Clarke, 1977; Shephard & Sidney, 1978).

Psychological Consequences of Aging of the Cardiovascular System

Whereas in the other physiological systems covered in this book, discussion of psychological consequences has been based on a minimal amount of empirical work, in the area of cardiovascular functioning there is an extensive amount of research on adult samples from which to draw. Although in many of these studies the subjects are in their 40s and 50s and so are not particularly "old," the results are applicable to the psychological consequences of the aging process because it is in this period of middle age that the risk of developing cardiovascular disease becomes high and that the changes associated with the normal aging process first become noticeable. Thus, this research is pertinent to the psychological consequences of aging, but it is not directly addressed to the question of how aging of the cardiovascular system affects the individual's adaptation and well-being. Instead, this research is oriented toward the psychological correlates of participation in aerobic exercise training programs, and to a lesser extent, the psychological attributes of athletes who specialize in endurance activities. By describing the motivational characteristics of middle-aged and older people who seek aerobic training, on the one hand, and the psychological effects of exercise participation, on the other, inferences can be made about what is important to people about the aging of their cardiovascular systems and what the psychological consequences are when aerobic capacity is reduced as a function of aging.

Motivation to Participate in Aerobic Training

The basis for the desire to participate in aerobic exercise training can be regarded as indicative of the concerns that are salient to adults regarding the aging of the cardiovascular system. There are two dimensions along which motivation for exercise training can be analyzed. The first is whether the motivation is intrinsic or extrinsic, that is, whether the interest is either in the activity itself, or in the external rewards received from conducting that activity (Wankel, 1980). An intrinsic reward would be the feeling of thrill or exhilaration (sometimes called "flow") that can accompany a steady state of physically intense but nonexhausting exertion. Another type of intrinsic motivation would be a sense of challenge to succeed at a difficult task, and subsequently, the satisfaction derived from a feeling of competence or mastery. In contrast, extrinsic motivation is oriented toward tangible rewards such as pay, or to satisfaction of goals not inherent in the movement of the body or successful completion of a challenge. Examples of these sorts of extrinsic rewards in relation to physical activity are conducting exercise in order to conserve one's health, a desire to satisfy one's family members, and the enjoyment of the social aspects of participation in group activities (which many training programs are).

The second dimension of motivation for exercise participation is whether the motivation is for initially joining the program or for continued adherence to the exercise activity, because these two phenomena are very different. One primary initial motivation, particularly characteristic of men in their 50s (on whom much of the aerobic exercise training research is conducted), is to reduce the risk of developing cardiovascular disease, to get into better "shape," or to improve their work capacity (Heinzelman & Bagley, 1970; Teräslinna, Partanen, Oja, & Koskela, 1970). Another reason is a desire for companionship offered by group exercise experiences (Stiles, 1967) or a desire for recreation or a change in routine (Heinzelman, 1973). It would appear that all of these are extrinsic motivations, and it makes sense that this should be the case, since it is not possible for a person to know that an activity will be inherently enjoyable until it has been tried, and also until a certain level of competence in performing that activity is achieved. Indeed, in the literature on exercise adherence it appears that involvement in aerobic exercise training over extended periods of time is enhanced by intrinsic motivation (deVries, 1980; McPherson, 1980), although social motivators may continue to operate (Heinzelman & Bagley, 1970). It would also follow, furthermore, that if a person is to maintain exercise participation, the effects must be positively reinforcing, either in terms of an enjoyable sensation derived from aerobic exercise (Hartley & Farge, 1977; Morgan & Pollock, 1977), or in terms of demonstration of positive effects on health (Shephard, 1978).

Another set of motivational concerns pertains to the type of exercise involved in the training program. If the activity is individually prescribed and monitored according to an individual's specific physiological needs and current abilities, then an individually based training modality is what is most advisable. The two methods available for this purpose are the bicycle ergometer and the treadmill. Each of these has advantages and disadvantages from a motivational standpoint. The bicycle may not be as familiar to an older adult as is walking (on a treadmill), and in addition, it places a great deal of strain on the quadriceps muscles (in the thighs), making it more likely that the exerciser will feel cramps and muscle strain (Montoye, 1975; Shephard, 1978). While admitting that it is boring, others

(Siegel et al., 1970) have argued that the bicycle has the advantage over the tread-mill of being easy to use and moreover that the treadmill causes muscular strain (in the lower back and calves) (McArdle et al., 1981). Running outside is less bor-ing than either method, but it presents other difficulties, partly because it depends on environmental factors (the temperature and the conditions of the ground) (Shephard, 1978), and partly because of the fact that the runner may feel "winded" (McArdle et al., 1981) or develop neuromuscular problems (Tzankoff et al., 1972). All three methods are potentially less stimulating than team sports or competitive ball games, methods that tend to lead to very high adherence rates (Hanson et al., 1968). Another advantage to competitive exercise is the constant challenge and immediate feedback it presents as well as the social companion-ship. However, if walking and running are conducted in group sessions, the camaraderie that develops among the participants may stimulate adherence rates almost equally as high (Pollock et al., 1976).

The conclusion that can be drawn from this discussion of motivational factors is that the changes in the cardiovascular system that start to be experienced in middle age, even in the absence of significant debilitative changes but with the perception of increased risk of cardiovascular disease, seem to have a strong enough effect, in many adults, to lead them to seek aerobic training. However, this initial motivation is not sufficient to overcome the potential boredom, muscular pain, and inconvenience that interfere with continued involvement. Instead, an intrinsic sense of satisfaction derived from mastery and a sense of competence must be stimulated by the activity. The next question to be addressed is whether exercise training can satisfy this desire and in so doing provide a source of compensation for age effects on the adult's basic sense of reliance in the effective functioning of the body.

Psychological Effects of Aerobic Exercise Training

As in short-term exercise training studies in which cardiovascular variables are evaluated, the studies in which psychological effects are assessed have the same weakness of generally not including a control group. Most of the research to be described in this section must, therefore, be regarded with a certain degree of skepticism. This is even more of a concern because of the enthusiasm that the authors in the area of aerobic training convey about the value of physical activity to psychological health, which might present a serious source of experimental bias. This attitude is evident in the following quotation:

Physical fitness is characterized by a positive outlook on work and health. It is character-ized by an ambition to work and succeed, a willingness to strive and to minimize ailments, fatigue, frustrations, and the hazards of life . . . . We fully believe that physical fitness is related to developing and maintaining integration between mind and body. (Cureton, 1963, pp. 17, 23)

Given these qualifications, the effects of exercise training on how adults feel about themselves and their bodies nevertheless have some important implications for the normal consequences of cardiovascular aging on the adult's feelings about his or her body. The most general finding across studies is the report that athletes and former sedentary adults alike say that they "feel better" after exercise training, both immediately afterward and for the period of weeks or months over which they train. This is a common but elusive phenomenon to document empirically, since it has been notoriously difficult to quantify (Griest et al., 1979). Some of the components of the "feeling better" sensation seem to be related to a decrease in anxiety, depression, and tension, and an uplifting of mood (e.g., Folkins, Lynch, & Gardner, 1973; McPherson et al., 1967). However, this result is not consistently observed (Bahrke, 1981; Kilbom, 1971a; Morgan, Roberts, & Feinerman, 1971). Although it is not known whether anxiety disorders can be effectively relieved by exercise as compared to a more traditional form of psychotherapy, it has been suggested that endurance exercise such as running provides a regular discharge of tension that can keep anxiety from becoming a chronic condition (Morgan, 1981).

One explanation for the decrease in anxiety and increase in mood that may be a factor in the "feeling good" report is that aerobic exercise training does provide an intrinsically rewarding sense of mastery and control and therefore leads to an enhanced self-concept (Blumenthal et al., 1982; Folkins & Sime, 1981; Hanson & Nedde, 1974; Heikkenen, 1978; Hilyer & Mitchell, 1979; Morris & Husman, 1978). In particular, it seems that the aspect of self-concept that is reinforced by exercise training is that which pertains to one's image of one's own body (Collingwood, 1972; Hammett, 1967; Sidney & Shephard, 1976). Conversely, if the individual is made to feel inadequate as the result of exercise training, there may be a negative effect on the bodily self-concept (Shephard, 1978).

The results concerning the effects of training on these self-concept variables have been far more consistent than those regarding personality traits (e.g., Hammer & Wilmore, 1973; Ismail & Young, 1977; Tillman, 1965) mainly because there is no reason to expect that a change in total personality structure would occur as the result of exercise training and also because personality trait measures are not sensitive to this type of manipulation. The effects of exercise also seem to be favorable in other domains, such as work attitudes and performance, improved sleep and less need for sleep, and even enhanced sexual relations (Donohue, 1977; Hanson, Tabalon, Levy, & Nedde, 1968; Hanson & Nedde, 1974; Heinzelman & Bagley, 1970; Hellerstein, 1973; Petrushevskii, 1966). These situational improvements extend the range of experiences in which the trained individual can experience a sense of competence in ways important to total self-concept and self-esteem (Morris & Husman, 1978).

Turning around the finding that exercise increases feelings of competence, especially the sense of bodily well-being and effectiveness, it may then be inferred that one of the primary effects of the aging of the cardiovascular system is to detract from the individual's sense of competence in situations demanding physical exertion. Moreover, all adults know that the efficiency of the cardiovascular system is essential to life, supporting the activities of all other bodily systems. Threats to the integrity of this system caused by aging or disease are threats to life itself. Age changes in the cardiovascular system that lead to impairment of the heart's pumping capacity are harmful to the extent that they bring the adult closer to death. They also serve as reminders of mortality to the individual who is made aware of such changes within himself or herself and can lead to serious concerns about personal life expectancy. These conclusions are based on the finding that interest in health, which is an extrinsic source of motivation, is a primary reason cited by middle-aged adults for seeking exercise training.

The cardiovascular response to activity is also an important contributor to the individual's adaptation to the environment. There are times of danger when even the most sedentary adult must run, climb, or swim to avoid some dreadful outcome. Living in a dwelling with stairs but no elevator requires constant exertion. There are numerous occasions during the course of the adult's occupational, family, and leisure activities when exertion is required. The aging of the cardiovascular system may become apparent to the individual when fatigue and discomfort are experienced while performing tasks in these realms which had at one time been performed with ease. In the case of activity regularly engaged in involving aerobic exercise, the adult may gradually come to feel as if he or she is working harder but accomplishing less. An unexpected demand for maximal exertion may fail to be met adequately. The interpretation that follows from the psychology of exercise literature is that a lack of ability to perform tasks placing demands on the cardiovascular system may make the adult feel less competent and this feeling can lead to a reduction of self-esteem.

Through exercise, the middle-aged and older adult can again complete more strenuous everyday activities and also have the opportunity to receive, on a regular basis, assurance that his or her body is capable of working effectively in response to the demands placed upon it. Although Folkins and Sime (1981) consider such a model of mastery and competence to be too simplistic to account for the psychological benefits of exercise to self-concept, it is consistent with the majority of interpretations offered by researchers in this area (e.g., Sonstroem, 1984) and in particular provides a useful framework for understanding the psychological effects of aging on this aspect of the body's functioning and a model for the effects of aging on other bodily systems.

CHAPTER 4

Respiratory System

When the respiratory system is operating efficiently, oxygen and carbon dioxide are exchanged in the lungs in such a way that the blood is adequately perfused with oxygen and rid of carbon dioxide. In order for this exchange to take place, the lungs must perform their bellowslike function of sequentially filling and emptying air into and out of the alveoli (air sacs) through the airways at a rate that matches the rate at which blood flows through the pulmonary capillaries that interface with the alveoli. It has been shown that in later adulthood, there is a reduction of the amount of oxygen that is taken up from the blood during aerobic exercise. The aging of the respiratory system structures has the additional effect of lowering the efficiency of gas exchange in the lungs. The effect of aging in the respiratory system therefore compounds the limitations on the ability to perform muscular work caused by the changes in the cardiovascular and muscular systems.

The effects of aging on the lungs are readily confused by the general public and easily confounded by researchers with other processes that impair respiratory functioning, notably diseases such as emphysema, and lifetime habits of cigarette smoking and exposure to environmental pollutants. These processes, which are extrinsic to the aging process per se, have cumulative effects and so are more likely to be apparent in older adult populations. Another factor that tends to obscure age effects is that some measures of lung functioning are related to bodily height. Since there are cohort-related and possibly true age differences in height in later adulthood, age effects may be exaggerated when cross-sectional adult samples are not equated on this variable. Apart from these additional factors, there seem to be some independent contributions of aging to the tissues in the respiratory system that cause these structures to be less efficient in accomplishing their function of regulating gas exchange in accordance with the metabolic needs of the body.

Age Effects on Gas Exchange and Ventilation

The main outcome of the aging process in the lung in functional terms is the reduction of the amount of oxygen delivered from the outside air to the arterial blood. The functional measure that represents the efficiency of diffusion across the interface of the alveolus and the capillary is the difference between the oxygen pressure in the arterial blood (after it has been oxygenated in the lungs) compared to the oxygen pressure that reaches the alveoli from the outside air. There are numerous cross-sectional studies that have documented the drop in arterial oxygen pressure and/or the increase across adult age groups in the alveolar-arterial oxygen difference (e.g., Begin, Renzetti, Bigler, & Watanable, 1975; Marshall & Wycke, 1972; Melmgaard, 1966; Sorbini, Grassi, Solinas, & Muiesan, 1968).

The implications of reduced oxygen transport to the blood in the older adult are very serious because of the difficulties that are created for the individual's ability to carry out everyday activities requiring exertion. Because of the importance of this age effect in pulmonary functioning, there is a great deal of interest in discovering its precise cause. The simplest explanation of this phenomenon would be that there is some type of heightened resistance with age to oxygen diffusion across the membranes in the alveoli, the capillaries, or both (Anderson & Shephard, 1969; Cohn, Carroll, Armstrong, Shephard, & Riley, 1954; Donevan, Palmer, Varvis, & Bates, 1955; Mauderly, 1978; McGrath & Thompson, 1959). However, it appears that the major cause of the reduction with age in the amount of oxygen that is diffused into the arterial blood is far more complex, and depends upon anatomical changes in a variety of pulmonary structures. Understanding of how these changes might cause reduced oxygenation of the blood in the aged lung requires an appreciation of the functional variables that describe the matching of ventilation and perfusion in the lungs. These values of these functional variables are derived from the performance of the lung during inspiration and expiration.

Compliance and Airway Closure

The primary cause of reduced oxygenation of the arterial blood in older adults appears to be a nonuniform distribution of air through the lungs, which lowers the efficiency of gas exchange. The result of the nonuniform distribution of air is that there are numerous zones where there is a mismatch between the rates of ventilation and perfusion (as observed by, e.g., Edelman, Mittman, Norris, & Shock, 1968; Greifenstein, King, Latch, & Comroe, 1952; Sandqvist & Kjellmer, 1960). A certain amount of disparity between ventilation and perfusion occurs in the lungs of healthy young adults. This because blood is more freely distributed through the lower part of the lung due to gravitational forces. However, the extent of this inequality is not great enough to have serious consequences for young adults (West, 1977). In the lungs of older persons, though, there are more areas of inequality between ventilation and perfusion so that there is a measurable effect on the oxygenation of the blood leaving the lungs. The differences between ventilation/perfusion spreads in the lungs of a young and a middle-aged adult male are illustrated in Figure 4.1.

The lower part of the lungs is one area in particular where there is a greater likelihood of a disparity being present in the older adult between the flows of air and blood. At all ages, the total quantity of blood leaving the lungs contains a

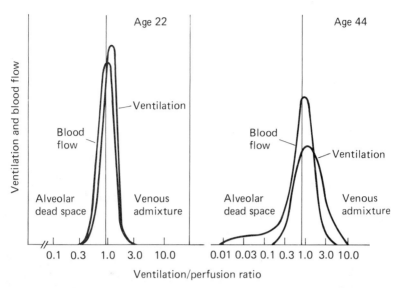

FIGURE 4.1. Spread of ventilation/perfusion ratios in the lungs of a 22- and a 44-year-old male. Increased alveolar dead space in the 44-year-old is shown by the relatively greater blood flow than ventilation. *Note*. Adapted from *Applied Respiratory Physiology* (2nd ed., p. 299) by J.F. Nunn, 1977, London: Butterworths. Copyright 1977 by Butterworths. Adapted by permission.

disproportionately large volume of blood from the lower lungs. Since this blood is especially likely not to be well-oxygenated in the lungs of older adults, it contributes an unusually large volume of poorly oxygenated blood to the total quantity leaving the lungs. The reason that there is greater inequality between ventilation and perfusion in the lungs of the older adult is that the lower, dependent, portions of the lungs are much less likely to be ventilated than is true for the younger adult. This reduction of ventilation in the lower part of the lungs is due to age changes in elastic recoil.

Elastic recoil of the lung is the tendency of lung tissue to resist expansion as it becomes filled with air. Its value is estimated from a plot showing the increase or decrease in lung volume that corresponds to an increase or decrease in the pressure across the lung surface during inspiration or expiration. This plot is referred to as a pressure-volume curve. Lung compliance is another way of expressing this same quality of resistance to inflation, but it is a dynamic measure, indexing the rate of change of volume in relation to pressure based on the slope of the curve. Because of the elastic recoil that normally exists in lung tissue, there is a positive pressure across the lung surface during inspiration which holds open the airways. During expiration, the elastic recoil of the lung helps keep the airways open until the pressure placed upon them via the respiratory muscles forces them to collapse. If the airways close prematurely, air will be trapped inside them and the lungs will not be able to expire to their maximum capacity.

The pressure across the surface of the upper part of the lung is normally more positive (by about 7.5 cm H_2O) than the pressure across the surface of the lower part of the lung due to the effect of gravitational forces. As a result, the top portion of the lung has more elastic recoil than the bottom of the lung up until the point that the lung is completely full. As the lung tissue loses elastic recoil with age, the bottom portion of the lung is, then, relatively more affected than the top from a functional standpoint. This is because the elastic recoil of the lower lung is reduced to below the critical point needed to keep the airways open during the end of the expiration phase of the ventilatory cycle. Consequently, the airways close prematurely in the normal ventilatory cycle in the lower part of the lung relative to the point at which they close in a younger person, for whom they may not close until expiration is almost at its maximum point. This sequence of events has been demonstrated in several cross-sectional studies in which the "closing volume" (point of airway closure in the lower part of the lung) of adults of different ages is compared (Anthonisen et al., 1969-70; Begin et al., 1975; Bode et al., 1976; Buist & Ross, 1973; Holland et al., 1968; LeBlanc et al., 1970). The alveoli are therefore more likely to be underventilated in the lower portion of the older adult's lung relative to the rate at which blood flows through the pulmonary capillaries.

A mathematical function fitted to the pressure-volume curve from the data of normal nonsmoking men and women is the most recently developed technique for describing age differences in elastic recoil and compliance (Colebatch, Greaves, & Ng, 1979; Gibson, Pride, O'Cain, & Quagliato, 1976; Knudson & Kaltenborn, 1981). It is a more precise and objective technique than the visual inspection of curves, the previously used method (e.g., Pierce & Ebert, 1958). The lowered elastic recoil across the adult age span is signified by less pressure across the lung surface (less resistance to inflation) at a given lung volume. Secondly, lung compliance, the volume increase for each pressure increase, is greater in older adults. This increase in compliance reflects the loss of resistance with age to distension of the lung as it is filled with air.

The structural basis for the loss of elastic recoil with age in adulthood or, conversely, increased lung compliance, is at the level of the alveoli. Some of the age effects thought to account for the loss of recoil are alterations in the composition and structure of the elastin (Brandstetter & Kazemi, 1983) and collagen composition and structure (Lynne-Davies, 1977; Turner, Mead, & Wohl, 1968). In addition, there seems to be an increase in the size of the alveolar ducts and alveoli, which results in a decrease in the total number of alveoli and their corresponding total surface area (Reid, 1967). The effect of this decrease is the number of alveoli is to reduce the surface tension created by the gas-liquid interface and so contribute to the loss of elastic recoil (Campbell & Lefrak, 1978).

In addition to being influenced by the compliance of the lung, ventilatory dynamics are determined by the compliance of the ribs and connective tissue that together make up the chest wall. The chest wall's elastic recoil when the lungs are expanded to less than their maximum capacity is in a direction opposite to that of the lungs; that is, the chest wall tends to expand outward. When the lungs are fully expanded, the elastic recoil of the chest wall is in the opposite direction, such that the chest wall tends to push inward toward the center of the body. At rest, the elastic recoil of the lungs and chest wall balance each other exactly. With increasing age in adulthood, the chest wall structures become more rigid (Lynne-Davies, 1977; Shephard & Sidney, 1978). This increase in rigidity combines with loss of elastic recoil of the lungs to lower even further the ability of the lung to be fully compressed during expiration and fully expanded during inspiration. Since less than the maximal amount of air can be brought into and out of the lungs, gas exchange is therefore compromised. The greater rigidity of the chest wall also increases the amount of work that must be performed by the respiratory muscles during the inspiratory and expiratory phases of the ventilation cycle (Brooks & Fahey, 1984; deVries, 1980; Kannel & Hubert, 1982).

Static and Dynamic Ventilatory Volumes

The description of age effects on gas exchange just given is based on the use of arterial oxygen pressure as the indicator of the efficiency of lung functioning. Another way to index the efficiency of the lung's functioning other than oxygen pressure in the arterial blood is to quantify the extent to which the lung and its associated structures can move air in and out for the purpose of gas exchange between blood and air at the alveolus-capillary interface. The measures of lung functioning that address this aspect of functioning fall into two categories: those that describe various components of the total capacity of the lung to hold air, and those that describe the volumes of air that can be moved in and out during a fixed time period. Like the measures of cardiovascular functioning, these respiratory measures can be taken at rest and during dynamic exercise, either at fixed levels of submaximum work or at maximum work capacity. Again, as with cardiac functioning measures, lower values during submaximum work represent greater efficiency of the respiratory structures in that they are able to provide sufficient oxygen to meet the body's metabolic needs with minimal effort. Conversely, high levels of maximum performance on these measures indicate that the lungs are able to support the requirements of the muscles for oxygen under circumstances in which there is an intense demand to provide oxygen. The functional measures of the lung based on lung volumes are useful for describing age differences in the pulmonary response to exercise.

Vital capacity is the volume of air that is moved in and out of the lungs when the person is ventilating maximally. The volume of air that remains in the airways and air sacs within the lungs at the end of the maximal expiration that the individual can produce is the functional measure called residual volume. Together, these two indices equal the total lung capacity, the greatest amount of air that can be held by all the structures of the lung. Part of the residual volume includes the air that remains trapped in the small airways that collapse when there is insufficient pressure to keep them open at the end of a maximal expiration. The most consistent finding across cross-sectional and longitudinal studies based on measures of lung volumes is that there is a reduction in the vital capacity and an increase in residual volume (Asmussen et al., 1975; Drinkwater, Horvath, & Wells, 1975). This process seems to begin at around the age of 40 (Brody et al., 1974) and results in a total of 40% loss of vital capacity between 20 and 70 (Lynne-Davies, 1977). It has already been shown that age changes in compliance are largely responsible for an increase in the closing volume, and it follows that the same process would account for the age-related increase in the residual volume (and concomitant decrease in the vital capacity). There is also an increase with age in the amount of air remaining in the lungs after expiration during normal breathing, as indexed by the functional residual capacity (Muiesan, Sorbini, & Grassi, 1971).

Another set of measures of the efficiency of the lungs in their function of moving air that show age effects is based on the quantity of air that is brought into the lung per minute. One of these measures is the ventilatory rate (per minute), which is equal to the volume of air (in liters) inspired during a normal breath (the tidal volume) multiplied by the frequency of breaths per minute. At submaximal levels of work, it is desirable to have a low ventilatory rate, which means that the lungs do not have to work overly hard (hyperventilate) in order to provide enough air to oxygenate the blood. The results regarding age effects on this measure are not clear-cut, with some investigations revealing no change (Cotes, Hall, Johnson, Jones, & Knibbs, 1973; Denolin et al., 1970), a decrease (deVries & Adams, 1972b) or an increase (Robinson et al., 1975). In contrast, it seems to be fairly well-established that older individuals are less able to maintain a high ventilatory rate at maximum levels of exercise (Daly, Barry, & Birkhead, 1968; Saltin & Grimby, 1968). There are two possible sources for this age effect, and both appear to play a contributing role. One possible explanation is that younger adults are able to attain a higher ventilation rate because they increase their tidal volume, an efficient mechanism in terms of lung economy (Norris, Shock, Landowne, & Falzone, Jr., 1956). Another reason for this age effect is that older adults are less able to maintain a high breathing rate because of decreased chest wall compliance. The maximum breathing frequency that can be achieved on demand is a good index of the "bellows" function of the chest (Baldwin, Cournand, & Richards, 1948), that is, its ability to support air movement to meet the requirements of maximum exercise. This function seems to decrease in later adulthood, reaching as low as 50% of that achieved by a young person (Drinkwater et al., 1975; Griefenstein et al., 1952; Norris et al., 1956). Since age does not appear to affect tidal volume under conditions of maximum exertion (Montoye, 1982), it would seem that the decrease in maximum breathing frequency with age is the major factor responsible for the age loss in maximum ventilatory rate.

A measure of lung functioning that takes into account the body's metabolic needs is the ventilatory efficiency, the ventilation rate divided by the oxygen uptake. This measure indexes how many liters of lung ventilation are required for each liter of oxygen uptake, and so it reflects the efficiency of the oxygen transport mechanisms of the body (deVries, 1980). In contrast to the results on cross-sectional age differences in ventilation rate, there is a fairly consistent body of research indicating that at submaximal levels, the lungs of older adults are less effective in transporting oxygen. This decrease in effectiveness is indicated by a higher ventilatory efficiency index in older adults (Robinson et al., 1975; Skinner, 1970). At the same time, the maximum level of ventilatory efficiency that can be achieved by older persons is reduced (Norris et al., 1955). The meaning of this latter finding is that older adults are less able to provide sufficient ventilation to meet the body's oxygen demands at maximal levels of exertion.

The rate at which air flows into and out of the lung is also used to describe the inspiratory and expiratory capacity of the respiratory apparatus. There are many possible measures that can be derived based on flow rates, and these have been made under a wide variety of conditions and on a variety of age and sex groupings of subjects. Consequently, it is difficult to reach a general conclusion regarding age effects on air flow rates to complement the other findings on lung volumes and ventilatory rate, which are based on a more consistent set of measures and test conditions, if not subjects.

One flow rate measure that has been fairly widely used is the forced expiratory volume, which is the amount that can be expired in a specified amount of time (usually 1 s). It may also be expressed as the percent of vital capacity that can be forcibly expelled during this time. Theoretically, the flow rate should be reduced in older adults to the extent that it is influenced by the elastic properties of the lungs (Pierce & Ebert, 1958). Although not observed consistently, there does seem to be a diminution of the flow rate as expressed by forced expiratory volume across adulthood (Ericcson & Irnell, 1969; Gibson et al., 1976; Kannel & Hubert, 1982; Shephard & Sidney, 1978). As with other indices of lung functioning, though, there is a confound of body height with age. In the studies on vital capacity across the adult years, height is frequently correlated with age and also with lung volumes (e.g., Baldwin et al., 1948; Berglund et al., 1963; Mittman, Edelman, Norris, & Shock, 1965). In some cases, partialling out height substantially reduces the correlation between age and lung function (as can be seen from the data of Morris, Koski, & Johnson, 1971). It is also possible that there is an interaction between age and height, at least for forced expiratory volume. Cole (1974) observed that among the older group studied, it was the taller men who lost forced expiratory volume to a greater degree over a 9-year period than did the shorter men. This finding was attributed to the stronger effect of gravitational forces on the larger lung of the tall man, which creates more of a distortion of the lung due to its own weight. As a result, aging would have a more serious effect on elastic recoil of the taller individual, and so reduce the flow rate more than for the shorter man.

In addition to providing sufficient respiratory volume and rate to support exercise performance, another requirement of the respiratory system is to be able to increase the ventilation rate when the oxygen pressure in the arteries reaches too low a level to support adequate oxygenation of the muscles. It appears that there is a reduction in the responsivity of the respiratory response to hypoxia in older adults (Kronenberg & Drage, 1973; Petersen, Pack, Silage, & Fishman, 1981). This finding could be due to a number of structural and functional changes in either the respiratory system or the neuromuscular control over ventilation.

Based on the pattern of results they observed, though, in the compliance curves, Petersen et al. (1981) suggested that the source of this age effect is a diminished neural output to the respiratory muscles under conditions of hypoxia. According to this view, the lowered ability of the respiratory system to provide adequate oxygen to support maximum levels of exercise may be contributed to by the loss of sensitivity to oxygen deprivation so that there is less driving of the inspiratory and respiratory forces.

However, Petersen et al.'s (1981) interpretation is not consistent with the extensive data (described in chapter 6) showing increased sympathetic nervous system activity in the aged. Based on that evidence, it would be argued that the diminished response to hypoxia in old age is due to the inability of the respiratory structures (lung and chest wall) to respond to sympathetic stimulation. In contrast to the response to hypoxia, the respiratory control of breathing in response to carbon dioxide pressure variations seems to undergo little change with age (Rubin, Tack, & Chermak, 1982).

Effects of Exercise Training

The key requirement of the respiratory system for adequate functioning during aerobic exercise is to supply enough oxygen to the working muscles at a rate sufficient to support their metabolic activity. If the respiratory system is not functioning efficiently during exercise, the individual will feel "winded," or "out of breath," a state more technically referred too as dyspnea. Fatigue will rapidly develop as the muscles lose one of their primary sources of energy. The purpose of exercise training for the middle-aged and older adult is to reverse the deleterious effects of aging on the respiratory system functions of gas exchange and ventilation. The variables most commonly examined in relation to exercise training that index these functions are vital capacity and ventilatory efficiency. The assumption underlying the use of these variables is that enhanced movement of air into and out of the lungs will produce the desirable goal of enhancing oxygenation of the blood so that the individual is better prepared to meet the metabolic demands of the muscles during physical activity.

Because of the nature of the composition of the lung and chest wall structures, it would seem logical to postulate that exercise training would not have the same potential for reversal of age effects as is true for the cardiovascular system. The only muscular tissue in the respiratory system that can be strengthened by exercise training are the muscles that control inspiration and expiration. In contrast, the heart is a muscle, and the ability of the left ventricle to contract is a function considerably more amenable to exercise activity that has the effect of strengthening it than is true for the lungs, which are not composed of muscle. However, it would be expected that the respiratory muscles themselves can be strengthened so that they can move the chest wall structures more effectively and therefore overcome some of the problems created by altered compliance of the rib cage and chest wall in later adulthood.

The aerobic exercise training studies in the area of respiratory function generally parallel the lines of research in other physiological systems described so far. The major exception is that the long-term effects of total inactivity have not been investigated to test the possibility that lowered levels of participation in exercise in later life serve to contribute to age effects in respiratory functioning. The studies on the effects of activity on respiratory functioning all concern short- and long-term effects of participation in exercise. One set of studies concerns the long-term effects of exercise participated in through leisure- or work-related activity. The second group of studies is directed at evaluating the effects of short-term (an average of 3 months) intensive dynamic exercise training on groups of volunteers.

The number of training studies specifically focused on respiratory functioning alone is decidedly smaller, however, than the number concerning the cardiovascular system. There seem to be several possible reasons for this. One is the fact that among the general public, the connection between respiratory functioning and exercise has not been made to the same degree as is true for cardiovascular functioning. Instead, adults seem more likely to try to enhance their own respiratory functioning by changing their smoking habits, especially cutting down on or quitting cigarette smoking. Researchers tend to follow this pattern as well, using cigarette smoking as an independent variable to predict various health problems that can reduce longevity. Because of this lack of popular and professional interest, there may be fewer volunteers who wish to spend time in training programs where respiratory functioning is the principal area of concern and fewer investigators who manipulate exercise as an independent variable.

Moreover, researchers may be less interested in measuring respiratory functioning in the context of a training study in which cardiovascular variables are assessed, although these measurements would be relatively easy to make in such a study. Thus, even though there may be volunteers participating in an exercise training study, the data are not collected that would make it possible to determine, on a wide-scale basis, whether training can benefit the respiratory system in later adulthood. This may be because there is less optimism about whether training can offset the aging process in the structures of the respiratory system which, on a physiological basis, would seem to be less responsive to training. At the same time, there is much more social interest in finding the causes of lung cancer than in the normal aging process and its effects on respiratory functioning. This interest has stimulated little research on the effects of training and little research relevant to the normal aging process. In contrast, the studies on cardiovascular disease permit more generalizations to be made about the normal aging process, because there are more similarities between the effects of chronic illness and those that are due to normal age changes than between lung cancer and normal age changes in respiratory functioning. Although aging and pulmonary emphysema have more in common in terms of their effects on elastic recoil of the lung, public and professional concern about emphysema has not stimulated widespread research on prevention or treatment through exercise training.

The research that exists on the effects of training on respiratory functioning in middle-aged and older adults, perhaps because of the relatively small number of studies on a set of diverse samples, presents an unclear picture of the extent to which exercise training can enhance respiratory functioning. This is partly because the studies involve a wide variety of measures of respiratory functioning, and different types of samples in terms of age, sex, degree of current and past occupational and leisure physical activity, and especially smoking habits. While the frequency of smoking is an important variable to control for in any study on respiratory functioning, it is usually easier to accomplish this in a cross-sectional age sample of normal volunteers, where smokers can be eliminated from the sample. In a long- or short-term training study it is not so easy to control for the effects of smoking, because often the subjects "get religion" about health and change all their habits, especially smoking (Morgan & Pollock, 1978). Therefore, the effects of training cannot be separated from the effects of these sorts of changes in life-style and habits made in addition to undertaking an exercise regimen.

Given these qualifications, it may be said that very long-term participation in endurance-type sports appears to have somewhat of an ameliorative effect on respiratory functioning. Athletes and even former athletes have been found to show less of a decline in vital capacity with age than sedentary adults, or to show no decline over the period of a longitudinal investigation (Asmussen & Mathiasen, 1962; Grimby & Saltin, 1966; Kanstrup & Ekblom, 1978; Plowman et al., 1979; Robinson et al., 1973). However, this has not consistently been the case (Åstrand et al. 1973). In short-term studies, it has been more difficult to demonstrate a favorable effect on vital capacity of aerobic training (Barry et al., 1966; deVries, 1970; Getchell & Moore, 1975; Niinimaa & Shephard, 1978b; Pollock et al., 1976), although an increase in vital capacity at maximum levels of exertion was demonstrated in an 8-week training program in which young and middle-aged men and women participated (Heikkenen, 1978). If vital capacity can be enhanced by exercise, it therefore seems that, in general, it requires many years of continued participation to accomplish this, and probably also adherence to a set of health habits associated with an active life-style. Moreover, any beneficial effects of exercise are probably limited to middle-aged adults up to age 50 or 60 at most.

The deleterious effect of age on ventilatory rate (tidal volume multiplied by respiratory frequency) does seem to be modifiable through participation in long- or short-term exercise training programs, but this is true only for ventilatory rate measured at maximum, not submaximal, levels of exertion. Thus, although a "savings" of ventilation at submaximum levels does not seem to be made possible by training among middle-aged and older adults, it does appear that training raises the maximum rate of ventilation that can be attained during dynamic exercise (Åstrand et al., 1973; Barry et al., 1966; deVries, 1970; Kanstrup & Ekblom, 1978; Kasch & Wallace, 1976; Kilbom, 1971b; Saltin, Hartley, Kilbom, & Åstrand, 1969; Wessel & Van Huss, 1969). In only one exercise training study, in which middle-aged men were the participants, was there reported a training-related reduction in submaximal ventilatory rate (Tzankoff et al., 1972).

The improvement in maximum ventilatory rate is attributed by some authors to an increase in the maximum breathing frequency (Saltin et al., 1969; Kilbom, 1971b) and by others to an increase in tidal volume (Barry et al., 1966; deVries, 1970; Robinson et al., 1973). An increase in breathing frequency might imply that there is an enhancement of sensitivity of oxygen chemorecepteors to hypoxia, although this idea has not been investigated.

Measures of ventilatory flow also are enhanced by exercise training in middle-aged and older adults. Findings regarding these measures further implicate as a mechanism for the effect of exercise training the enhanced ability to move the chest in and out to increase the amount of air flow through the lungs. This enhanced air flow, in turn, makes it possible for a greater volume of air per unit of time to be available for gas exchange and hence provide more aerobic support for muscular activity. Thus, it has been observed that the amount of air that can be inhaled and exhaled during a maximum ventilatory effort is higher in athletes and former athletes than in sedentary adults of comparable ages (Grimby & Saltin, 1966; Saltin & Grimby, 1968). This measure of flow has also been found to be increased with training in men in their 40s (Hanson et al., 1968). Similarly, the amount of air that can be forcibly expired during a maximum respiratory attempt was found to be higher in atheletes and former athletes. For both of these measures, however, it must be pointed out that improvements with training have not been consistently demonstrated (Kanstrup & Ekblom, 1978; Pollock et al., 1976).

It would seem to follow that if the maximum ventilatory rate and ventilatory flow measures can be increased by exercise training, then the index of ventilatory efficiency (ventilatory rate relative to oxygen consumption) should also be enhanced. However, this has not been the case. There is generally no effect of training on this index of respiratory functioning either at submaximal or maximum levels of work (Benestad, 1965; deVries, 1970; Grimby & Saltin, 1966; Hartley et al., 1969; Kilbom, 1971b; Saltin et al., 1969; Shephard & Sidney, 1978; Stamford, 1973).

One explanation of the disparate results of training studies concerning ventilatory rate and ventilatory efficiency is that training's main value is that it makes it possible for the middle-aged and older adult to increase his or her rate of breathing to levels high enough to fulfill the demands posed by work at maximum levels of muscular exertion. However, training has no effect on the structures of the lung that underlie elastic recoil and hence lung volume. Measurements that depend on lung volumes (such as tidal volume, vital capacity, and ventilatory rate) therefore do not show positive effects of training.

Another possibility is that the effects of training on oxygen consumption (the denominator of the ventilatory efficiency equation) and ventilation rate (the numerator) proceed at different rates in older adults. A high ventilatory efficiency is indexed by a match between the rate of oxygen taken in by the respiratory system and that used by the muscles from the circulation. Exercise training seems to have no beneficial effect on either ventilatory rate or oxygen uptake in middle-aged and older samples, measured at submaximum levels, so only the case of maximum ventilatory efficiency is relevant to this discussion. The rates of improvement as the result of exercise training may differ such that increases in oxygen uptake (due to circulatory deficits or to less uptake by the muscles) do not keep pace with increments in the ventilatory rate. Consequently, there would be no apparent gain in maximum ventilatory efficiency as the result of exercise training.

This interpretation is consistent with the data of several exercise training studies in which the ventilatory rate seems to be more enhanced by activity or training than maximum oxygen consumption (Heath et al., 1981; Kasch & Kulberg, 1981; Kilbom, 1971b; Montoye, 1982; Plowman et al., 1979; Robinson et al., 1975; Saltin et al., 1969; Tzankoff et al., 1972). On the one hand, such an effect would signify lack of efficiency of the respiratory system, since the person who has undergone training is still ventilating at more than the required rate. The alternative interpretation is that the cardiovascular system or the muscle's aerobic metabolism rate is what sets the limit on the degree of respiratory efficiency that can be achieved.

Psychological Consequences of Aging of the Respiratory System

It may be proposed that the main outcomes perceivable by the individual of the aging of the respiratory system are dyspnea and fatigue during physically exerting activities that require more oxygen to be transported to the muscles than the lungs are capable of providing. Distress associated with dyspnea and fatigue is probably the major psychological consequence of reduced respiratory functioning in middle and later adulthood. This distress would, it may be reasoned, seem to be similar in quality to the perception of strain associated with impaired cardiovascular functioning during physical exertion. It can be an extremely alarming sensation for the individual to feel extreme shortness of breath as the result of exertion, since the ability to breathe is one of the physiological activities performed by the body that is probably one of the most taken for granted among healthy persons.

For most everyday activities, the loss of ventilatory efficiency and vital capacity associated with aging will not approach the point at which dyspnea is experienced. On more strenuous tasks, however, the older individual is more likely to perceive fatigue and dyspnea at lesser degrees of activity than was true at earlier ages. Heightened concern over loss of respiratory function may be expected to follow particularly serious or novel episodes of dyspnea. Such episodes may be so unpleasant and frightening that the individual vows not to engage in the activity that provoked it ever again. In some cases, this response is an adaptive decision, leading the individual away from participation in work, recreational, or family tasks that would have a detrimental effect on other aspects of physiological functioning. Another equally adaptive response would be to change one's habits so that dyspnea is less likely to occur in the future. One adaptive response of this kind would be giving up heavy cigarette smoking. However, avoidance of physical activity is a response that can also have a more damaging effect in the long-term on the efficiency of other bodily systems, in addition to the respiratory functions sensitive to activity levels in adulthood.

Unlike the area of cardiovascular functioning, where relationships have been quite well-established between more favorable psychological functioning and improvements in aerobic capacity, there have been no investigations directly exploring the effect of training of ventilatory functions and well-being. However, there is one short-term training study in which both cardiovascular and respiratory functions were evaluated as well a psychological measures. With men in their 40s who exercised for 3-4½ hr/week over 7 months, significant increments were observed in maximum ventilatory volume at maximum work levels (Hanson et al., 1968). Although no psychometric evaluation was made of changes associated with training in well-being, the results were consistent with the increase in feelings of well-being oberved in aerobic training studies focusing on cardiovascular variables. It would certainly not be unreasonable to speculate that training that reduces the likelihood of experiencing dyspnea during daily activities and also activities that are more strenuous would have a comparable effect on feelings of well-being and competence that accompany reductions in the perception of cardiovascular strain and perception of enhanced endurance. Moreover, even though the motivation that instigates a middle-aged or older adult to participate in exercise training may be the desire to enhance cardiac functioning, the result may be an improvement in respiratory efficiency that adds to that achieved in the cardiovascular system. The same argument can also be made as was put forth regarding cardiovascular functioning and feelings of bodily competence. That is, it may be hypothesized that reduced respiratory functioning has negative effects primarily because of its influence on feelings of bodily competence. The perception that one is out of "shape," which accompanies the sensation of dyspnea, may have the effect of reducing the individual's sense of well-being because of the implications that this experience has regarding the impaired effectiveness of one's aging body in carrying out its functions crucial for adaptation to the environment.

CHAPTER 5

Digestive and Excretory Systems

The digestive and excretory systems together function to process all nutrients that enter the bloodstream except the oxygen that is brought in by the respiratory system. Through the digestive system, nutritive substances are absorbed and any that are not are eliminated. The excretory system functions in parallel, eliminating the by-products of metabolism from the cells throughout the body, by its constant cleansing of the blood. Both systems, in addition to serving these life-support functions, also have associated with them numerous psychological meanings regarding the ingestion of food, and the elimination of wastes through the gastrointestinal system and the genitourinary tract.

Digestion

It is the function of the digestive system to extract sugars, proteins, fats, fluids, minerals, and vitamins from the various kinds of foods that enter the body and pass these nutrients into the bloodstream. The components of food that cannot be used by the body are passed through the digestive tract and eliminated. The critical features necessary for normal digestive functioning are, then, the transformation of food into usable form, and the transportation of food along the digestive tract.

Age Effects on Digestive Functioning

As with the effects of age on other physiological functions, it is extremely important to separate normal aging from the effects of diseases in describing digestive functioning. There are numerous gastrointestinal diseases that impair the efficiency of digestion and must be ruled out before the effects of age on an individual's functioning can be determined. Diseases affecting other physiological systems can also have indirect effects on the digestive system, and the contribution of these must be determined before attributing any abnormalities in functioning to the aging process. Also, lifetime habits of alcohol ingestion may influence the functioning of various digestive processes in old age. Isolation of aging effects in the case of the digestive system is made even more complicated by the contribution of individual differences in nutrition patterns. These differences are influenced by a host of psychological, cultural, and economic factors whose cumulative effects on food intake over a lifetime interact with any physiological changes due to the aging process.

The process of digestion occurs in phases, named to correspond to the location of the food within the system immediately after its ingestion.

Oral Digestive Phase

The oral phase begins when food enters the mouth, where the grinding and tearing actions of the teeth begin to break it into pieces. The presence of food in the oral cavity stimulates the salivary glands (parotid, submandible, and sublingual) to secrete saliva, which serves a variety of digestive and related functions. First, it protects the teeth and tongue from bacteria, and at the same time, lubricates the food, making it easier to chew. Secondly, by dissolving and washing away food particles, saliva allows the taste buds on the tongue to be maximally receptive to each incoming bite of food. Thirdly, the critical digestive function of saliva is served by the enzyme salivary alpha amylase, which begins to break down starches into digestible sugars, a process completed in the small intestine by pancreatic alpha amylase.

In later adulthood, these functions mediated by salivary secretion are compromised to a considerable degree. An increased incidence of the symptom of a dry mouth is reported to occur as a function of age in adulthood (Minaker & Rowe, 1982; Straus, 1979). The reduction in saliva, which causes this symptom, is linked to diminished secretion of saliva by the parotid glands after the age of 70 years (Kamocka, 1970). The result of a reduction in the volume of saliva is decreased protection of the teeth and tongue from bacteria, a reduction of taste sensations, greater difficulty in chewing, and additionally, discomfort when talking.

The second part of the oral phase is the propelling of the bolus of food through the pharynx and down the esophagus by peristalsis. The bolus then enters the stomach through the lower esophageal sphincter. This sphincter, which is normally closed, temporarily relaxes at the same time that swallowing begins and remains open until ingested material passes through it. In this manner, a one-way flow of food into the stomach is maintained. Occasionally, food that has already reached the stomach does flow back into the esophagus, due to some weakness of the lower esophageal sphincter. When this occurs, it is experienced by the individual as "heartburn." Another abnormal event that might occur in this part of the oral phase is when food becomes lodged in the esophagus. In this situation, peristaltic waves are initiated to move the bolus into the stomach. This type of peristalsis is called secondary, to distinguish it from the primary peristalsis that normally propels the food bolus down the esophagus.

An issue of controversy in the gerontological literature is whether there is dimished primary peristalsis in older adults, and an increase in secondary peristalsis, a condition caused by weakness of the esophageal muscles called "presbyesophagus" (literally, "old" esophagus). This condition would adversely affect the normal passage of the food bolus into the stomach. Description of this condition was first based on studies of elderly persons who were not at all in good health, some of whom suffered from esophageal disease. More recently, it has been established that esophageal motility is largely retained intact into old age in healthy individuals (Khan et al., 1977). There is a reduction in the healthy elderly in the intensity of the peristaltic wave, however (Hollis & Castell, 1974), and higher resting esophageal pressure, reflecting some weakness of the esophageal muscles (Khan et al., 1977). However, this age effect appears to be relatively small in functional terms, and so esophageal motility seems to be retained if not intact, then at least well preserved into later adulthood.

Gastric Phase

In this phase of digestion, the food bolus becomes transformed into chyme. In this liquid form, the food is then ready to enter the small intestine where it can be absorbed. The stomach accomplishes this processing of the food bolus into chyme by secreting gastric juice, and agitating the food and gastric juice together until the food bolus is totally dissolved and ready to be passed into the small intestine. The components of gastric juice include peptin, hydrochloric acid, mucus, and "intrinsic factor," a mucoprotein essential for the absorption of Vitamin B_{12} in the small intestine. Secretion of gastric juice occurs in the upper part of the stomach in response to stimulation by gastrin, a hormone produced by cells in the lining of the lower portion of the stomach and upper part of the small intestine in response to the presence of chyme. During a meal, some of the solid food is being acted upon by the gastric juice in the lower portion of the stomach, while the unprocessed food is held in the upper portion of the stomach. The gastric phase continues until gastric juice secretion ceases as the result of there being too little chyme left in the lower part of the stomach to stimulate gastrin production.

There is some indication that with increasing age in adulthood, the motility is reduced by the muscles in the stomach responsible for agitating the food with gastric juices (Minaker & Rowe, 1982). However, evidence that this is indeed an age-related phenomenon is incomplete, and it is not at all clear that reduced gastric motility is a function of aging (Bowman & Rosenberg, 1983). There does seem to be decline in the rate of gastric emptying of liquids after the age of 70 (Evans, Triggs, Cheung, Broe, & Creasey, 1981) but not solids (Moore et al., 1983) or carbohydrates (Webster & Leeming, 1975). That there is a reduction of the secretory function of the upper portion of the stomach to produce gastric juice is far more firmly established. It appears that gastric juice secretion drops off progressively across adult age groups to reach about 75% of the normal volume by the age of 60 years (Baron, 1963; Bernier, Vidon, & Mignon, 1973; Fikry, 1965).

There are several effects of the reduction in gastric juice secretion on the digestion of specific nutrients. The amount of activity by peptin is decreased by about 60% in persons over 60 years old (Fikry, 1965), and the digestion of protein is commensurately reduced in efficiency. A reduction in hydrochloric acid and Vitamin B_{12} intrinsic factor has been described as contributing to a series of changes in the intestinal absorption of iron, calcium, Vitamin B_{12}, folic acid, and protein, as well as increased bacterial growth in the intestinal tract (Bowman & Rosenberg, 1983). The mechanisms that might be responsible for these changes will be described in more detail in the following.

Intestinal Phase

Small Intestine

Digestion of all food substances contained in the chyme occurs as the result of processing by enzymes secreted from within the small intestine added to by other enzymes and digestive agents secreted into the small intestine from the liver and pancreas. Chyme is forced to move backward and forward throughout the small intestine by segmentation, the contraction of circular muscles in the wall of the small intestine. The process of segmentation ensures that all of the chyme is exposed to the lining of the intestinal wall, allowing for maximal absorption into the bloodstream to take place. Since segmentation occurs at a faster rate in the upper as compared to the lower segments of the small intestine, there is a net forward movement of chyme through this part of the digestive tract. Like gastric motility, segmentation is governed by neural and hormonal factors. There is no information concerning the adequacy of the segmentation process in the small intestine across adulthood, since research on motility throughout the digestive tract has focused on the stomach and the large intestine.

The absorption of nutrients from the chyme takes place in the villi, fingerlike structures projecting into the inside of the tract from alongside the wall of the small intestine. The existence of the villi and the folds between them greatly increases the surface area available for absorption. Columnlike epithelial cells, which make up the majority of cells along the outer surface of the villi, are the sites where nutrients begin their transport to the bloodstream, entering through the capillaries on the interior of the villi. The absorption process is facilitated by the sweeping action of the villi as chyme passes along them. The epithelial cells secrete some digestive enzymes and also hormones that stimulate the secretion of other digestive enzymes in the stomach and pancreas as well as the release of bile from the liver.

There are some structural changes with age in the small intestine that have the potential to impair the absorption process in the intestinal phase of digestion. The weight of the small intestine is reported to decrease in adulthood, and also there is a reduction in the mucosal surface area (Minaker & Rowe, 1982). This is probably because the villi tend to become broader and shorter, eventually becoming transformed into parallel ridges (Webster, 1978). The distribution of the lymphatic follicles on the surface of the intestinal wall also shows changes (Cornes, 1965).

These age effects appear to have only limited consequences for the efficiency with which absorption of most nutrients takes place. This is due, possibly, to the fact that the epithelial cells of the villi turn over so rapidly that they do not "age." There are approximately 25 million villi in the small intestine, and their epithelial cells replace themselves completely every 3-5 days. With this high turnover rate, aging would therefore pose little of a threat to the structural integrity of the epithelial cells (as contrasted to the neuron; see chapter 8). Moreover, due to the sheer quantity of villi, even though the villi themselves might flatten, the loss of surface area that would result causes little functional impairment.

Other information on the absorption of specific nutrients is suggestive of some selected but potentially significant decreases in later adulthood. There is reported to be some reduction in the absorption of fats (Minaker & Rowe, 1982; Webster, 1978), possibly related to changes in the distribution of lymphatic follicles across the intestinal surface. The absorption of the sugar D-xylose was reduced in one study after the age of 80 years (Guth, 1968), not at all in another on a somewhat younger sample (Kendall, 1970), and only in 26% of a sample of persons averaging 80 years in another (Webster & Leeming, 1975). In the latter study, however, some of the older sample (16%) showed more efficient intestinal absorption than the young. There is no direct information on protein absorption in the small intestine, which is surprising considering the importance of protein to overall bodily functioning. There is a fairly large and rapidly growing literature on the absorption of various vitamins and minerals. Much of this research has been stimulated by interest in improving the nutritional status of older adults to treat chronic health disorders sensitive to vitamin and mineral deficiencies.

Calcium and Vitamin D. One of the most significant and intriguing sets of studies concerns Vitamin D and calcium absorption. It is fairly well-established that calcium absorption in the intestine decreases after the age of 70 in men and women (Bowman & Rosenberg, 1983; Heaney et al., 1982). Calcium absorption is normally mediated by two processes: an active transport system and passive transport through simple diffusion across the intestinal membrane. The active transport of calcium is regulated by the active form of Vitamin D ($1,25\text{-}(OH)_2D$, called calcitriol). Dietary Vitamin D ($25\text{-}OH\text{-}D$) is hydroxylated at the 25 position in the liver and in the 1-alpha position in the kidney. The decrease in calcium absorption has been attributed to reduced serum levels of calcitriol (Bowman & Rosenberg, 1983; Heaney et al., 1982).

One hypothesis proposed to account for the bone loss that occurs in osteoporosis, and also adult bone loss, is that the diminished absorption of calcium in the intestine causes a rise in parathyroid hormone (the body's normal response, which serves to increase calcium levels when the absorption rate is low). The abnormally high levels of parathyroid hormone in turn leads to an increase in the rate of bone resorption, and hence bone loss. The alternative hypothesis involves a reversed direction of causality, and accounts for the decrease of calcitriol levels in the blood as the result of bone loss. According to this hypothesis, the flow of calcium from bone into the bloodstream that results from bone resorption exceeding bone growth leads to a decrease in parathyroid hormone, and hence a reduction in the synthesis of calcitriol.

Heaney et al. (1982), who presented these alternative hypotheses, suggested that both may be true, applying to different individuals with different kinds of osteoporosis. One the basis of available evidence on calcium alone, such a resolution is probably appropriate. However, it seems that the evidence on other causes for adult bone loss, specifically weakening due to accumulation of Haversian canals with a resulting increase in porosity and an increase of cement lines due to the accumulation of old remodeling sites (as described in chapter 2), is strong enough to support the second hypothesis, at least as an account of normal aging changes. This interpretation is consistent with the findings of a lack of relationship between calcium intake and bone loss in adulthood (Garn et al., 1967). It is possible, however, that other factors contribute to diminished serum levels of Vitamin D, especially lack of exposure to sunlight due to other chronic health limitations that require the individual to be housebound (Hodkinson, Rand, Stanton, & Morgen, 1973). In addition, it is possible that a decrease in the activity of 1 alpha-hydroxylase enzyme in the kidney contributes to reduced calcitriol levels (Gallagher et al., 1979).

Vitamin B_{12}, Folic Acid, and Iron.   Adequate absorption of iron, folic acid, and Vitamin B_{12} is essential to the normal development of red blood cells, and deficiencies in the intake, absorption, or excretion of any one of these can lead to anemia. There is some controversy in the literature over whether the elderly are especially likely to suffer from anemia. This applies to both pernicious anemia, which is due to Vitamin B_{12} deficiency (Hsu & Smith, 1984) and megaloblastic anemia, which is due to folic acid deficiency (Rosenberg, Bowman, Cooper, Halsted, & Lindenbaum, 1982). It seems that variations in the health and nutritional status of the samples, differences in levels of medication that can affect iron absorption, intake of vitamin supplements (as in Meindok & Dvorsky, 1970), and cooking and eating habits complicate survey-based assessments of the extent of anemia in the elderly population.

Despite some uncertainty about the incidence of anemia in the aged due to conditions that cause iron deficiency, there seems to be some basis for the position that there is impaired absorption of Vitamin B_{12} and folic acid. In the first place, a reduction of the production of the intrinsic factor by the parietal cells in the stomach lining in the aged would reduce Vitamin B_{12} absorption (Bowman & Rosenberg, 1983; Hsu & Smith, 1984). At the same time, there may be impaired intestinal transport of folic acid as the result of reduced acidity due to lowered hydrochloric acid production in the gastric juice (Bowman & Rosenberg, 1983). Although Marx (1979) found no reduction with age in iron absorption from soluble ferrous salts, this does not eliminate the possibility that lowered acidity of the intestinal environment impairs iron absorption from other sources of food iron (Lynch, Finch, Monsen, & Cook, 1982). Another contributor to folic acid deficiency may be the depletion of folate conjugase, which mediates the deconjugation of folates present as polyglutamates (their ingested form) to monoglutamates (their digestible form) (Baker, Jaslow, & Frank, 1978).

Vitamin C and Vitamin D.   Vitamin C (ascorbic acid) is required for the formation of intercellular substances, particularly the collagen that forms the bone matrices, cartilage, dentin, and the vascular epithelium. The elderly, especially women, have low Vitamin C levels, but this seems to be due to poor intake rather than absorption. Vitamin B_6 is involved in the synthesis and metabolism of amino acids, as well as formation of antibodies, and folic acid metabolism. There does appear to be an age-, not a diet-related deficiency in this vitamin in older persons (Hsu & Smith, 1984).

Liver

The normal digestion of fats is made possible by bile secreted by the liver, whose components serve to emulsify fat, facilitate the action of a fat-digesting enzyme produced by the small intestine, and put into solution the products of fat digestion so they can be transported into the villi. After passing through the small intestine, bile is returned to the liver through the portal vein, where it may be recycled again during a meal. In between meals, bile is stored in the gallbladder, from which it is expelled after eating begins as the result of hormonal stimulation brought about by the presence of chyme in the small intestine.

There are conflicting data about the extent of changes with advancing age in adulthood in the structure of the liver as indexed by total weight. In some investigations, decreases are demonstrated in adults over the age of 50 years (e.g., Thompson & Williams, 1965). In one very frequently cited study in the literature, there is no decline whatsoever in total liver weight throughout adulthood (Morgan & Feldman, 1957). It is possible to resolve this discrepancy by considering the possibility that when a decrease in liver weight is shown, it is a function of decreased total body weight, and not to an unusual degree of cell death in the liver.

There is a more uniform body of evidence regarding the effects of aging on liver cell size and structure. A variety of abnormalities have been reported to exist at the microscopic level. However, there is reason to doubt that these abnormalities in cell size and structure are either a direct result of aging or have serious functional implications for the digestion of fats in the healthy elderly. In the first place, alterations in cell size and structure are not restricted to the livers of aged persons, nor do they occur universally in the aged (Carr, Smith, & Keil, 1960). Moreover, some age effects may be nonspecific reactions to changes in other bodily systems (Schaffner & Popper, 1959). With respect to the functional implications for any age changes in liver weight or cell structure, it appears that in normally functioning elderly persons, there are no decrements shown on various measures of liver functions that reflect the adequacy of bile formation. In addition, there are no age effects on the levels of various enzymes in the blood, which would otherwise rise when liver cells are damaged (Calloway & Merrill, 1965; Cohen, Gitman, & Lipshutz, 1960; Kampmann, Sinding, & Møller-Jørgensen, 1975; Koff, Garvey, Burney, & Bell, 1973).

The lack of correspondence between anatomical and functional changes that occurs in the case of the liver appears to be due primarily to the large "safety" margin that is built into this organ. Up to 80% of the liver can be removed without adversely affecting its functioning, as has been shown in cases where surgical removal of the majority of the liver was performed. In addition, the liver has a tremendous regenerative capacity, and so can compensate for the normal losses it may experience throughout adulthood (Cohen, Gitman, & Lipshutz, 1960).

Disturbances in the gallbladder may arise at any point in adulthood as the result of various physiological disorders. Apart from these abnormalities, the gallbladder does not appear to undergo significant alterations in its functioning due to age (Hyams, 1978). Increases with age in gallbladder disease seems to be related to age in Western society (Bowman & Rosenberg, 1983), suggesting that there is a relationship to the high fat content of the diet rather than aging itself.

Pancreas

The exocrine glands of the pancreas produce two kinds of secretions that pass through the pancreatic duct into the small intestine. One is an aqueous alkaline fluid containing bicarbonate to neutralize the chyme's high acidity level, enabling the other pancreatic secretion to operate most effectively. This other pancreatic secretion contains enzymes to break down sugars, proteins, and fats. Hormonal stimulation from the small intestine stimulates secretion of these pancreatic fluids.

There appear to be some structural changes related to age in the pancreas that could interfere with its exocrine secretion (Webster, 1978). However, there are no reported age-related reductions in the volume of the aqueous pancreatic secretion, so that the acidity of the chyme can adequately be neutralized in the small intestine (Rosenberg, Friedland, Janowitz, & Dreiling, 1966). The only reduction of pancreatic function that has been associated with aging is in a component of the enzymatic secretion responsible for breaking down fat. However, the effect of this decrease does not appear to be substantial enough to have a noticeable impact on the older individual's everyday life (Webster, 1978).

Large Intestine

By the time the chyme has passed through the small intestine, digestion of nutrients is essentially complete. The major task remaining to be accomplished is the absorption of water to solidify the digestive waste products and to expel these waste products, in the form of feces, from the body. The large intestine completes the absorptive function by slowly agitating its contents back and forth, in a process similar to segmentation, in order to maximize the time available for water absorption. The intestinal contents are propelled forward through the large intestine when the segmental movements become more frequent and sequentially organized. This frequently occurs immediately after a meal or physical activity.

Material passing through the large intestine reaches the rectum, where it is stored until it can be expelled. The presence of feces in the rectum causes it to become distended, serving as a stimulus for the internal anal sphincter to relax. Defecation does not occur, however, unless the external anal sphincter is relaxed. This muscle is under voluntary control, and this control is what forms the basis for fecal continence. The natural frequency for defecation ranges from three times a day to once every 3 days.

Based on the media's presentation of the need the elderly have for laxatives, and the commonly held myth that constipation afflicts many older people, an automatic inference would be that motility in the large intestine is greatly impaired in the elderly. There is some evidence that there are cell abnormalities, reduced secretion of mucus, and atrophy of the muscles responsible for moving fecal material through the large intestine (Yamagata, 1965). However, despite these changes, and contrary to popular opinion, there do not seem to be any major functional losses in the large intestine due to the aging process. This conclusion is based on the evidence from studies on age differences in the transit time of radioactively labeled substances through the large intestine (Brauer, Slavin, & Marlett, 1981; Eastwood, 1972), sensitivity to distension of the rectum (Minaker & Rowe, 1982), bowel emptying habits, or control of the anal sphincters (Brocklehurst, 1978). When constipation occurs in an aged individual, it seems to be more likely the result of factors other than the physiological effects of aging.

Psychological Consequences of Aging of the Digestive System

The concern that many adults have over digestion is reflected in part in the social importance placed on the taste, appearance, and caloric content of food. With many choices available along these dimensions and more, many people allocate a considerable portion of their resources in time and energy to selecting food for themselves and their families. Digestion has another important influence on the individual's everyday life, and this has to do with the adequate functioning of the digestive system. Regulating one's food intake to maintain a sense of satiety without being overly full at the end of a meal is important to adequate functioning while awake and asleep. Intestinal gas, abdominal cramps, diarrhea, constipation, and nausea are conditions that make it difficult for the individual to carry out the activities of his or her normal daily routine.

It seems clear that healthy older individuals are generally comparable to their younger counterparts in the role that digestion plays in their lives. Compared to other physiological systems, age changes in digestion are generally much less marked. Despite the lack of any dramatic age effects on the digestive organs, many older persons nevertheless seek medical help for a variety of digestive ailments. Often, the symptoms for which the aged seek help cannot be traced to a physical cause, but seem to be the result of psychological factors.

An adult of any age can suffer from impaired digestive function when undergoing periods of intense emotional stress. Such digestive problems are due to physiological responses mediated by the autonomic nervous system. In the case of aged persons, stress-provoked responses of this nature can amplify what would otherwise be quite subtle alterations due to the aging process. For instance, anxiety or fear can inhibit salivary and gastric juice secretion, two functions shown to be impaired under normal conditions by aging. A more common situation is the development of functional constipation as the result of a variety of dietary and medicinal habits the older individual acquires (Minaker & Rowe,

1978). One such habit is to reduce fluid intake as a means of avoiding excessive frequency of urination due to age effects in the excretory system (Straus, 1979; see pages 84–85). Lowered fluid intake can lead not only to constipation, but also to an exacerbation of the problems of a dry mouth caused by diminished saliva production.

Another habit that can cause constipation in the older individual is the consumption of inadequate amounts of natural food fiber, a situation that results in increased transit time of the feces through the large intestine (Weisbrodt, 1981). Elderly people may reduce their fiber consumption because they are taking outdated medical advice that roughage is detrimental to the aging digestive system. The view that roughage is harmful rather than beneficial for the older person was generally held in the 1950s and 1960s as exemplified by the advice that only cooked fruits and vegetables be eaten by aged people with functional bowel disorders (Sklar et al., 1956). The physical inactivity that accompanies the sedentary lifestyle typical of many older people may also lead to constipation, by delaying the propulsive movements of feces in the large intestine (Brocklehurst, 1978).

There is general agreement in the literature that in addition to other contributory factors, perhaps the major factor leading to constipation in the elderly is overuse of laxatives (Bhanthumnavin & Shuster, 1977; Brocklehurst, 1978; Minaker & Rowe, 1982; Sklar, Kirsner, & Palmer, 1956). This situation may come about because of the following type of scenario. Older persons may be distressed by a frequency of defecation which is lower than what they think is "healthy," due to their adherence to medical opinions held earlier in this century that a daily bowel movement is essential to avoid building up toxic waste products in the body (Brocklehurst, 1978). This myth is reinforced by present-day laxative commercials, which stress the importance of being "regular." Habitual use of a laxative can have the opposite to the desired effect, ultimately detracting from the large intestine's intrinsic regulatory capacity.

At any age, an adult can be disturbed or embarrassed by various gastric symptoms such as heartburn, belching, flatulence, acid stomach, constipation, and diarrhea. However, these digestive disorders carry additional meanings to the elderly, because of the association between such symptoms and serious chronic diseases. Some symptoms, including constipation, are associated with cancer of the gastrointestinal tract, the most frequent form of malignant disease in the elderly (Brocklehurst, 1978). Elderly adults may also fear that they will develop loss of voluntary control over defecation. Fecal incontinence occurs in the later stages of senile dementia. Older persons who believe that "senility" is a normal part of aging may regard with alarm any indication that their defecation patterns are changing, interpreting what may be minor age-related alterations as evidence that their mental competence will soon be lost. Anxiety over such symptoms may exacerbate the effects of any other psychological and dietary factors leading to constipation and other digestive ailments. What originates as a temporary bout of indigestion or constipation can thereby come to have a more prolonged course.

Changes with age in other bodily systems can also have the effect of diminishing the older adult's food intake. One important set of changes is in mobility. Such tasks as pushing a shopping cart, getting groceries in and out of an automobile or having to carry groceries on a bus, and reaching for an item on a high shelf at home or in the supermarket may be complicated by losses in freedom of movement and strength. Sensory losses may reduce the aged person's ability to read shopping lists, grocery advertisements, and coupons, and take other steps involved in food preparation (Wantz & Gay, 1981). Other sensory losses in taste and smell have a more direct role in reducing the older person's enjoyment of food (see chapter 11). Finally, loss of teeth and problems with dentures may further interfere with pleasure derived from eating and so reduce intake (Bowman & Rosenberg, 1983).

Other stresses associated with aging may contribute to altered digestive patterns in old age. Adults typically adjust their food preparation habits in order to conform to the needs of the family, and these needs change with the ages of the children. The older adult who has no one else to cook for may feel a sense of futility in preparing elaborate or even adequate meals for him- or herself. A more radical change, perhaps, is when the older individual has to learn for the first time all about shopping and preparing meals, after his or her spouse is no longer there to do so. In either case, the transition from eating within a family to eating by oneself may, by virtue of serving as a reminder of past mealtimes and time spent together in the kitchen, create depression, lack of interest in food, and avoidance of the room where so many family experiences were shared. It is also expensive to cook for one person, and this can create a disincentive for planning and preparing adequate meals (Wantz & Gay, 1981). Poor nutrition will inevitably follow, unless the older person is able to find substitute sources of social interaction at mealtimes. Adverse changes in economic circumstances associated with retirement and widowhood will also have a depressing influence on the older adult's eating patterns, perhaps prohibiting him or her from enjoying the fuller range of choices made possible by the availability of discretionary income for satisfying one's gastronomic whims.

Thus, even though the digestive system itself seems to retain a high degree of stability into old age, there are threats to its functional integrity from the psychological and social complications that may arise in later life. These complications have the potential to make age changes in digestion take on a self-perpetuating quality and so reduce the enjoyment and comfort that the older person would otherwise be able to derive from a normally functioning system.

Excretion

The excretory system's function is to eliminate the chemical waste products of cellular metabolism on a regular, continuous basis in order to keep them from accumulating to harmful levels. After being excreted from the blood as it passes through the kidneys, the liquid wastes are stored within the bladder as urine until such time as they can be conveniently eliminated through the urethra. The kidneys play a critical role in ensuring the consistency of the fluid environment surrounding the cells of the body. If the body's fluid level varies outside a specified range of acidity and sodium content, the cells of the body will not be able to carry out their functions.

Age Effects on the Structure of the Kidneys

The nephron is the structure responsible for extracting chemical wastes from the body and regulating the amount and composition of the extra- and intracellular environment. With increasing age in adulthood, the volume and weight of the kidneys are reported to decrease. The loss of kidney mass is particularly pronounced in the cortex, the locus of the glomeruli. The number of glomeruli appears to be reduced with increasing age in adulthood, which conforms to the finding of decreased weight and size of the glomeruli-containing cortex (Dunnill & Halley, 1973; MaLachlan, Guthrie, Anderson, & Fulker, 1977). It is in the glomeruli that blood from the renal artery is cleansed of all substances in the blood except proteins and red blood cells. The remaining glomeruli show signs of partial degeneration, with scarlike tissue forming at the sites of degeneration (Sworn & Fox, 1972). In addition, age is associated with an alteration in the shape of the glomerulus from spherical but indented to smoothly spherical, so there is a loss of the surface area at which filtration of the blood can take place (Goyal, 1982; McLachlan et al., 1977).

After the blood is filtered in the glomerulus, the residual liquid (called the glomerular filtrate) passes through the Bowman's capsule, which encircles the glomerulus. From the Bowman's capsule, the glomerular filtrate passes through a set of complex convolutions into the tubule, down and through the loop of Henle (a long U-shaped tube), through another set of convolutions, and then into the collecting duct. It is in the tubule and collecting duct that the filtrate is converted to urine. In the process of being converted to urine, materials in the glomerular filtrate that are needed by the body (water, sodium and other minerals, glucose, amino acids) that have been removed from the blood are restored to it by being passed from the tubule, through osmosis into the surrounding capillaries, and from there to the veins. The waste products are retained in the urine and excreted through the renal pelvis, down the ureters, into the bladder, and ultimately expelled through the urethra.

Increasing abnormalities in the tubules have been associated with the aging process, including thickened membranes, fatty degeneration, shortening, the development of diverticuli that accumulate debris (Darmady, Offer, & Woodhouse, 1973), and also a loss of convoluted cells per unit area (Goyal, 1982). It would seem to be a likely possibility that such structural changes compromise the urine-concentrating capacity of the tubules, impairing the process of osmosis across the tubular membrane surface.

Abnormalities in the blood vessels in the kidneys are also reported to increase as a function of age in adulthood, beginning after 40 years of age (Davidson,

Talker, & Downs, 1969). These abnormalities, which are more pronounced in the larger blood vessels, include stiffening and narrowing, changes comparable to those reported in other parts of the circulatory system. Since the allowance of ample blood flow through the kidneys is essential to their capacity to cleanse a significant portion of the blood (enough so that all of the blood is thoroughly cleansed every 30 min), such age changes in the structure of the blood vessels would seem to have important functional implications, since they would have the effect of reducing renal blood flow.

Other vascular changes reported to rise with age in adulthood are localized at the site of the glomeruli. These fist-shaped clusters of blood vessels are of importance because they are the site at which the glomerular filtrate is formed, and if their structure is impaired, the rate at which this filtrate is produced will slow and the excretory process will be slowed. The glomeruli in the outermost portion of the cortex of the kidney degenerate completely in later adulthood so that they eventually disappear entirely. Other glomeruli, located deeper within the interior of the cortex near the medulla, also degenerate, but do not completely vanish. Instead, they leave behind a continuous link between the arterioles that formerly had passed into and out of the glomeruli. Thus, the connection from the renal artery to the renal veins is retained, but there is no glomerulus through which the filtrate is processed. Both types of glomerular degeneration are illustrated in Figure 5.1. Degeneration of glomeruli in the inner cortex of the kidney is

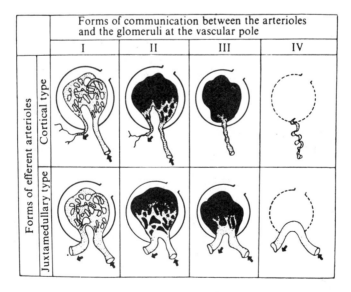

FIGURE 5.1. Diagram of the degenerative process in the cortical and juxtamedullary nephrons. *Note.* From "Intrarenal Vascular Changes with Age and Disease" by E. Taka-zakura et al., 1972. *Kidney International, 2,* p. 225. Copyright 1972 by Springer-Verlag, reprinted by permission.

reported to occur in substantial portions by the age of 45 years, increasing steadily to affect virtually all glomeruli in this region of the kidney by 80 years of age (Takazakura et al., 1972).

Age Effects on Kidney Functions

Renal Blood Flow

Age differences in the direction of greater prevalence of impaired efficiency in progressively older groups of adults have been reported on every measure of renal functioning representing diminished ability to excrete wastes at every step of the process. However, it should be noted that these descriptions of age effects are based on cross-sectional studies. Longitudinal analysis of the data from the Baltimore Longitudinal Study revealed that many subjects showed long periods of stable values, and in some adults, kidney function actually improved over a 10-year period (Shock, Andres, Norris, & Tobin, 1978).

The first step to consider in the excretory process is the movement of blood through the renal structures. The renal blood flow is the amount of blood that passes through the kidneys per unit of time (measured in milliliters per minute). The technique used to measure renal blood flow involves injecting a harmless weak organic acid into the bloodstream and recording how much of it passes through the kidneys in a minute. The amount of acid excreted in the urine is used to determine the rate of blood flow. Use of this procedure is based on the process of secretion that takes place in the tubules, whereby substances are transported from the capillaries that surround the tubule into the liquid passing through the tubules. Some of the organic acid injected into the bloodstream goes directly into the glomerular filtrate through the glomerulus. The rest of the acid remaining in the bloodstream is secreted into the filtrate when it reaches the tubules. The amount of acid excreted in the urine indicates how much blood must have passed from renal arteries to the renal veins during that same time interval in order for that amount of acid to have passed through the nephron.

A summary of 38 studies on a total of 634 men and women of different ages on renal blood flow using this method yielded a pattern of relative stability through to about 50 years of age, followed by a steady cross-sectional decrease into later adulthood (Wesson, 1969). More recent studies of renal blood flow using radiographic techniques have supported the earlier results and yielded the additional information than the rate of blood flow became much slower with increasing age in the cortical compared to the medullar nephrons (Hollenberg et al., 1974). This pattern of decline in later adulthood in renal blood flow signifies that vascular obstructions, such as structural changes in the arteries and glomerular degradation, impede the flow of blood through the kidneys of the older adult. Consequently, less blood is available per unit of time for cleansing by the excretory system, and so waste products are less efficiently eliminated in the bloodstream of the middle-aged and aged individual.

Glomerular Filtration Rate

The next step in the excretory process at which age differences are detected is in the rate at which the filtrate is formed from the portion of the bloodstream that passes through the glomerulus. The glomerular filtration rate, as it is called, is measured by injecting into the bloodstream a substance that passes through the glomerulus. Its rate of clearance, or excretion, as it passes through the kidneys is used as an indication of the amount of filtrate manufactured per unit of time. Inulin is one substance used to measure glomerular filtration rate because it is filtered as freely as blood plasma through the glomerular membrane. Consequently, the amount of inulin excreted in the urine per unit of time is a valid indication of how much blood was converted to filtrate. Another technique involves measuring the clearance of creatinine from blood to urine. Creatinine is a by-product of muscle metabolism, and since it is already in the bloodstream, it does not need to be injected. While this quality of creatinine is an advantage in terms of the ease of conducting research on renal functioning, creatinine has the disadvantage that some of it is normally added to the urine via the tubules through secretion. The amount of creatinine excreted in the urine is, therefore, slightly higher than that filtered at the glomerulus, and the actual glomerular filtration rate may thereby be overestimated.

In Wesson's (1969) analysis of age differences in renal functioning, there was a cross-sectional decrease after the age of 50 in the glomerular filtration rate (as measured via inulin injection) as well as in renal blood flow. A more recent study on glomerular filtration rate as indexed by creatinine rather than inulin clearance was conducted on a large cross-sectional sample of men by Rowe, Andres, Tobin, Norns, & Shock (1976). A gradual decrease across age groups was observed starting in the mid-30s and continuing through the mid-60s, after which there was an accelerating rate of decrease in creatinine clearance. These age differences in glomerular filtration rate were confirmed over five repeated testings conducted 2 years apart over a 10-year period on a subsample of men in the study. The diminution of glomerular filtration rate adds further to the loss of excretory efficiency implied by reduced renal blood flow.

Tubular Transport

Not only must sufficient blood pass through the kidneys and adequate filtrate be formed to cleanse the blood sufficiently, but substances must be secreted at a fast enough rate through to the urine passing through the tubule in order for excretion to proceed efficiently. Moreover, the urine concentrating ability of the kidney tubules requires adequate transport of materials across the tubular membrane. One measure of tubular transport involves using glucose as the transport substance to determine the efficiency of resorption from the urine in the tubules into the surrounding tissues and hence to the blood in the capillaries. The second measure involves the converse process, whereby weak organic acids are injected into the blood to measure the capcity of the tubular membrane for secretion from the capillaries surrounding the tubules into the urine. These measures are independent of the rates of blood flow and glomerular filtration.

Cross-sectional decreases have been reported on measures of tubular transport in samples of men spanning the adult age range. Resorption of glucose and secretion of weak organic acids show similar rates of decrease from about age 30 on. These decreases constitute proportions amounting to a total of about 35%-45% between the ages of 20 and 90 (Davies & Shock, 1950; Miller, McDonald, & Shock, 1951, 1952; Watkin & Shock, 1955). These cross-sectional reductions signify progressively less efficient functioning in adulthood in the capacity of the tubules of the nephrons to exchange materials with their surrounding capillaries.

Urine Concentrating Mechanisms

One of the most crucial functions taking place in the kidney's tubules is the regulation of the amount of water lost from the body in the urine. Almost all of the water (99%) in the glomerular filtrate is taken back into the body's tissues through resorption across the tubular membrane and as a result the urine becomes more concentrated than the glomerular filtrate. A concentrating mechanism in the loop of Henle is responsible for creating a high concentration of sodium in the tissues surrounding the tubule, so that water diffuses through osmosis from the highly dilute filtrate inside the tubules. Even more water reenters the tissues surrounding the tubule as the urine passes through the straight collecting duct neighboring the loop of Henle, around which there is a gradient of increasingly high sodium concentration.

The relative degree to which the urine becomes concentrated in the tubule is delicately regulated in accordance with the body's fluid supply. A dilute urine will be produced in large quantities when there is an excess of fluid in the body, a state called diuresis. When the fluid level in the body is low due to dehydration or excess water loss through evaporation, the kidneys will concentrate the urine, thereby returning as much water as possible to the body's tissues. Control over water loss and retention is exerted by two mechanisms, one under neural control and the other under control of local conditions of sodium concentration in the glomerular filtrate. The neural mechanism is the hypothalamus, which contains osmoreceptors sensitive to the water concentration of the blood. When the water concentration is low, these cells stimulate the pituitary gland to release antidiuretic hormone. This hormone increases the permeability to water of the tubule's surface so that the urine becomes more concentrated, conserving water for the body's tissues. The secretion of antidiuretic hormone is inhibited by the ingestion of caffeine and alcohol, and stimulated by exercise activity, pain, anxiety, and stress. Stimulation of the osmoreceptors produces the sensation of thirst which, if satiated, means that antidiuresis will not occur.

The local control of urine concentration is under the influence of the sodium level of the glomerular filtrate. When the filtrate has a high sodium content, it will retain water through osmosis, the urine will be dilute, and water will be lost through excretion. Cells in the nephron near the glomerulus are stimulated by the high sodium content of the filtrate to produce an enzyme, renin, that has the effect of reducing the water content of urine. It accomplishes this by indirectly causing the renal arterioles to constrict, thereby slowing down the rate of urine formation. In addition, renin stimulates the secretion of aldosterone, a hormone produced by the adrenal glands. Aldosterone in turn facilitates water loss from the tubule by increasing the movement of sodium out of the filtrate in the tubule. Water is then drawn out of the urine through osmosis.

When the kidney successfully concentrates urine, the proportion of sodium and other solids in solution increases as water is forced back into the bloodstream. Osmolality indexes urine concentration such that the higher the osmolality, the lower the proportion of water the urine contains. Osmolality measurements are usually taken to determine the kidney's responsiveness to various conditions in which diuresis or antidiuresis is induced.

The capacity of the kidneys of adults of different ages to concentrate urine or increase urine flow has been investigated under a variety of physiological states. Most investigations regarding the ability of the kidneys to concentrate the urine under conditions of water deprivation have shown there to be age-related decreases (e.g., Lindeman, van Buren, & Raisz, 1960; Rowe, Shock, & DeFronzo, 1976). Conservation of sodium under conditions of sodium deprivation also has been shown to be impaired in older adults (Epstein & Hollenberg, 1976).

Explanations of Age Effects on Renal Functioning

An observation made by Nathan Shock and his associates in some of the earliest research on age differences in renal functioning was that there were striking similarities in the rates of cross-sectional decrease of various renal functions over the adult years, amounting to about 6%/decade. This similarity was interpreted as evidence for the hypothesis that aging impairs the functioning of the nephron as a total unit to reduce simultaneously the efficiency of the glomerulus and the tubule (Davies & Shock, 1950; Miller et al., 1952; Shock, 1961). According to this view, nephrons are lost as total units, such that anatomical changes with age in the tubules are matched by anatomical changes in the glomeruli (McDonald, Solomon, & Shock, 1951). To the extent that the remaining nephrons are not impaired, compensation for the age-related reduction in number may occur (Lindemann, 1975). It has been observed that the glomerular filtration rate declines only slightly before the age of 50 to 60, after which it drops more precipitously. The pattern may reflect compensation for nephron loss before that age by surviving nephrons until the point is reached where so many nephrons are lost that the remaining ones cannot compensate (McLachlan, 1978).

This explanation of the effect of aging of the kidneys is actually more descriptive than explanatory, however. It does not specify what might be the primary cause of the nephron's loss of functioning. A second approach provides more of an explanation, attributing the primary effect of aging to changes in the vessels of the kidneys, which then cause deterioration of the nephrons. According to this view, the sites of these vascular changes are the arteries, which tend to become more occluded in later adulthood (Hollenberg et al., 1974). The effect of these changes is hypothesized to be a reduced blood supply to the nephrons, and particularly in the cortex in which the glomeruli are located. This has the effect, in turn, of reducing the amount of blood available to be converted into filtrate, thereby accounting for the reduced glomerular filtration rate with increasing age in adulthood (Hollenberg et al., 1974; Rowe, 1982).

There are several challenges to this position that vascular changes are the primary cause of the aging of the kidneys. One is that the events described as typical of aging are similar to those that occur as a function of cardiovascular disease, and so are not useful for understanding normal age effects in adulthood. Secondly, the associations between vascular and tubular changes have not been causally linked in direct experimental tests or through demonstration that vascular changes temporally precede renal ones. In support of this criticism, McLachlan (1978) argued that there is evidence of nephron loss without vascular deterioration in nonhuman species that do not contract cardiovascular disease. In his own reearch, McLachlan observed an association between vascular deterioration and ... number of glomeruli that was no greater than the association between age and the number of intact glomeruli (McLachlan et al., 1977). Thus, age was equally as good a predictor of the number of intact glomeruli as was vascular deterioration, and thus no explanatory value could be attributed to vascular changes alone.

A third challenge to the explanation of nephron loss based on vascular changes in later life is that no mechanism has been postulated to account for how restricted blood supply would actually cause nephrons to cease functioning normally. While a reduction in the glomerular filtration rate may be a logical consequence of a diminished blood flow to the glomerulus, there is no reason why the nephron should degenerate as a result. Moreover, as will be shown later in the case of cerebral blood flow, a diminished rate of flow through an organ may be the result, not the cause, of a reduced number of functioning cellular structures.

The effects of age on adaptation to altered physiological states that induce diuresis or antidiuresis has constituted another major line of inquiry into the cause of aging of the kidneys. In one of the first experiments following this approach, older men were found to have lower urine-concentrating ability than younger men in response to administration of antidiuretic hormone (Miller & Shock, 1953). Paradoxically, it was found in another study that older men spontaneously produced more antidiuretic hormone in response to injection of salty water. This result was interpreted by the investigators as signifying greater sensitivity to water deprivation in the osmoreceptors of older men (Helderman et al., 1978). As will be seen in chapter 6, such a finding may be accounted for by postulating heightened sensitivity of some of the body's control mechanisms as compensation for reduced functional capacity of the organs that these mechanisms are intended to regulate.

Loss of water-conservation ability has been attributed to diminished efficacy of the concentrating mechanism in the loop of Henle. At least part of this decline may be the result of an increase in blood flow in the medulla relative to the cortex. With greater blood flow, the tissues surrounding the tubules perfused by the capillaries in the medulla may be "washed out" so that they no longer possess the same power to attract water and hence concentrate the urine (Rowe, Shock, & DeFronzo, 1976). The shunting of blood away from the glomeruli in the area of the kidney bordering between the medulla and cortex may contribute to this effect (Takazakura et al., 1972).

A second possible explanation for decreased water and also sodium conservation in the aging kidney is a deficit in the mechanism whereby renin stimulates sodium and water retention (Epstein, 1979; Rowe, 1982). A decrease in renin activity from the age of 30 and up has been found in almost all investigations of this variable, including those in which respondents were tested in basal conditions, and also those designed specifically to stimulate renin production (Nakamura et al., 1981; Noth, Lassmon, Tan, Fernandez-Cruz, & Mulrow, 1977; Weidmann, DeMytteraerre-Bursztein, Mexwell, & Lime, 1975; Weidmann et al., 1978). This renin activity decrease is what appears to be responsible for a diminution in aldosterone secretion in the kidney rather than a deficit in the adrenal gland itself (Weidmann et al., 1975). In addition, the constriction of the renal arteries, which is indirectly stimulated by renin secretion, is not impaired with age (Hollenberg et al., 1974).

Thus, impaired renin activity with increasing age in adulthood would appear to be a primary cause of diminished salt- and water-conservation capacity in the kidneys of older persons. Some possible reasons for this decrement are that there are losses of cellular function at the site of renin production, and in addition, there is a decrease in the total number of nephrons throughout the kidney (Nakamura et al., 1981; Noth et al., 1977). It is also possible that some of the diminution in renin production attributed to aging is the result of hypertension, a disease known to inhibit renin activity (Schalekamp, Krauss, Schalekamp-Kuyken, Kolsters, & Birkenhager, 1971; Tuck, Williams, Cain, Sullivan, & Deuhy, 1973).

This discussion of the causal factors responsible for loss of kidney functioning in the aged has shed more light on the urine-concentrating mechanism than on the process of nephron loss. If there is some intrinsic process that causes nephrons to die, it has yet to be identified. If, instead, the vascular hypothesis is correct, then the means through which this would cause a reduction in kidney volume and weight requires further explanation.

Applications of Changes with Age in Renal Functioning

Despite age differences in measures of kidney functioning, it is nevertheless the case that the kidneys of older persons are still able to excrete harmful wastes and regulate the body's fluid composition and volume as long as no severe physiological stresses are imposed (Goldman, 1977; Lindeman, 1975; Rowe, 1982). It is only when the kidneys are required to function in extreme situations, placed under unusual demands, that age differences in renal efficiency become apparent.

One such stressful condition is dynamic muscular exercise. During strenuous exercise, blood flow is diverted away from the kidneys to the working skeletal muscles. The effects of an age-related loss in renal blood flow would, therefore, be magnified when the older adult is exercising to a greater extent than would be true for younger adults. Consequently, the excretory process would be relatively more impaired in the older person. This is not the case for the glomerular filtration rate, which is maintained at a constant level during exercise. Since there is an age reduction in the glomerular filtration rate, older adults would be producing less filtrate per unit of time, but not disproportionately less during exercise.

From the point of view of the individual's feelings of comfort, perhaps the most critical function during exercise to be adversely affected by age is the ability to conserve sodium and water. Both of these substances become depleted during exercise when the individual begins to perspire. Fatigue will occur more rapidly in the older exercising adult who cannot adequately conserve salt and water, so that strenuous physical activity would have to be discontinued at an earlier point than would be the case for a younger adult. No studies are available on the effects of aerobic exercise training on water or sodium conservation in middle-aged and older adults, so it is not known whether compensation is possible for age effects on renal functioning under the stress of exercise.

Impaired tubular functioning has consequences for the medical treatment of older adults with certain kinds of drugs. As described above, the tubules in the kidneys of the older adult are less efficient in secreting substances into the urine from the blood than is true for younger persons. However, drug dosage levels are usually determined on the basis of their secretion rates in the kidneys of adults without regard to their age. When the same dosage appropriate for a younger adult is administered to an older one, more of the drug will remain in the bloodstream over the period of time between doses. When repeated doses are given, the drug levels in the blood can then build to harmful levels. Unless the dose is adjusted to take into account the lower secretion rate in the aged individual, the drug may have an adverse rather than a beneficial effect.

Effects of Aging on the Bladder

The bladder is a baglike muscular structure whose function is to store urine that passes from kidneys via the ureters. It is operating efficiently when it can expand to hold urine between opportunities for voiding without discomfort, and when, upon voiding, it can empty completely. Both of these functions are compromised with increasing age. The total amount of urine that can be stored in the bladder is lower in adults older than 65 years, and the quantity of urine remaining in the bladder after voiding is greater. There appear to be age changes in the connective tissue in the bladder that could contribute these phenomena such that the bladder loses its expandability and compressibility. There is also greater variability in older adults of the perception of the need to empty the bladder. This sensation normally initiates a reflex controlling one of the spincter muscles around the opening of the urethra. Among adults younger than 65 or 70 years of age, this sensation occurs when the bladder is half-full. In the older person, this sensation to void may not occur until the bladder is almost completely filled, or there may not even be any sensation at all. This would mean that when the need to void is first perceived, the older person has less advance warning so that he or she can reach a lavatory in enough time so that no leakage or spillage occurs. Older persons also appear to be more likely than younger adults to awaken during the night with the need to urinate, but this age effect apparently occurs independently of any structural changes in the kidney or bladder (Goldman, 1977).

Psychological Consequences of Aging of the Excretory System

Changes with increasing age in the efficiency with which the kidneys function threaten the delicate regulation of blood chemistry and volume needed to maintain optimal celluar functioning throughout the body's organ systems. Most adults probably tend not to be concerned about how aging might affect the functioning of their kidneys, compared to the effects of aging on other systems, particularly the cardiovascular and nervous system. However, otherwise healthy adults of any age would seem to notice and show concern over any changes that might occur in their patterns of urination, whether these be due to age, the presumed effect of age, or to temporary disease states that cause heightened frequency or discomfort. Part of the reason for being concerned about greater urinary frequency might be that it is annoying to have to take time out of and interrupt daily activities to urinate. It is also very uncomfortable to have to hold one's urine when toilet facilities are not readily available. Finally, it might be speculated that the older adult is embarrassed at having to excuse himself or herself more than is considered "appropriate" in a social or public situation.

Another source of concern that the older adult might be affected by is that there is an association between "senility" and urinary incontinence. The loss of control over bladder function is greatly feared because of the threat that it poses to living independently in one's own home. When an older person becomes incontinent, institutionalization usually becomes necessary, and this is a condition that is greatly feared (as was pointed out in chapter 2). In addition, the aged individual may fear a loss of self-respect caused by return to the status of the infant or very young child who lacks control over urinary function. If an older adult perceives changes with age in his or her urinary patterns as leading ultimately to the state of incontinence, it might be anticipated that such a person will be very sensitive to what might otherwise be only moderately inconveniencing requirements for more frequent urination.

The general model used in other chapters to account for the psychological effects of aging on bodily functions by describing their impact on feelings of competence and mastery may be hypothesized to apply, then, in a negative way in the case of the excretory system. If this system fails to work properly, or threatens to undergo severe and detrimental losses, the older individual may experience feelings of depression due to the actual, concrete effects such changes have on daily life, as well as the foreboding of even worse changes in the future. Although there is a great deal of literature on urinary incontinence and its medical management, there is little information in the gerontological field on the potential psychological impact of this condition. The suggestions made here are speculative and require further testing, but are consistent with the model, which relates the sense of bodily competence to the effects of aging on emotional well-being and adaptation.

CHAPTER 6

Control Mechanisms: Endocrine System and Autonomic Nervous System

The body's cells require a stable internal environment in which to carry out their functions despite variations in the external environment to which the body is exposed. The maintenance of a stable internal environment is accomplished by complex regulatory devices involving the endocrine system and the autonomic subdivision of the nervous system. Not only do these systems function to maintain homeostasis, but they also permit the individual to respond quickly to the requirements for increased energy use when the situation demands mobilization of the body's resources.

The individual is not aware of these processes of internal regulation, since they occur independently of conscious control. Indeed, the person may at times wish that control were possible over these systems when they produce results that are uncomfortable, such as a racing heartbeat during a time of emotional unheaval. Other results may be socially awkward, such as increased activity of the sweat glands in response to either physical exertion or fear produced by a psychologically threatening situation (as in a job interview). In other cases, the individual is made unpleasantly aware that control mechanisms are not functioning adequately, since disorders of these systems can have massive effects on physical comfort, appearance, and diet.

Both the endocrine and the autonomic nervous systems also participate in regulation of reproductive behavior. The focus of the present chapter is on the control of the organ systems that support basic life-support functions. The role of these systems with regard to reproduction is treated in chapter 7.

Before proceeding, it should be pointed out that there is a great deal of interaction among the various components of the body's control mechanisms, with numerous overlapping functions and many opposing ones. In addition, the functions of the control mechanisms interact with the processes taking place within the organ systems that are being controlled. It is extremely difficult, for these reasons, to study control mechanisms in an intact, living, human being. Yet, it is within living, complex, and developing individuals that these systems operate, and in fact, contribute by their intricacy to the complex physiology of human aging.

Age Effects in the Endocrine System

The release of hormones into the bloodstream by the glands in the endocrine system is governed by the needs of particular tissues for such activities as protein synthesis, storage or release of glucose, sodium retention, and reduced inflammation at sites of injury. These needs are communicated to the endocrine glands by a variety of feedback loops, some of which are under control of the hypothalamus. The feedback loops of hormones released by the pancreas are regulated by glucose levels in the blood.

The endocrine system was once thought to play a critical role in the aging process by virtue of its dominant role in controlling the various organ systems of the body. When hormones were first discovered in the late 19th century, some scientists believed that they would eventually prove to be the "elixirs of youth ." Although that belief has not been substantiated, there remains a tremendous interest among present-day researchers in elucidating the role played by hormones in the aging process. Research in this area has focused on the hypothalamus-anterior pituitary system and the glands they control and the endocrine glands of the pancreas. Each of these systems has at one time or another been given special emphasis as a major influence on the aging process. The current approach is a more conservative one, describing and trying to explain changes in these systems over adulthood in their structure and function. Apart from any causative role these systems may play in the aging process, changes in the adult years in the endocrine glands are regarded as important because they influence the body's use of energy at rest and during exertion, and have tremendous significance for the individual's day-to-day feelings of well-being.

Hypothalamus and Pituitary Gland

The hormones produced by the hypothalamus enter the pituitary gland, and together these two organs form the core of a network that has an extremely wide range of control over the body's functions. These functions are best described in terms of the pituitary gland's two divisions, the anterior and posterior. There are no direct studies in humans on the effects of age on the quality or quantity of hypothalamic-releasing factor (HRF) secretion, the process that initiates the release of the anterior pituitary hormones. There are, however, fairly extensive investigations on the effects of age on pituitary gland functioning, particularly on the anterior pituitary hormones: thyroid-stimulating hormone (TSH), adrenocorticotropic hormone (ACTH), and growth hormone (GH). The other pituitary hormones, prolactin, follicle-stimulating hormone (FSH), and luteinizing hormone (LH), are primarily involved in changes in the reproductive system and are discussed in chapter 7.

Anterior Pituitary Gland

Since the anterior pituitary gland controls so many of the body's functions, changes in this gland and its functions were thought to play an especially key role in the aging process. This view now is modified, given the lack of overpowering evidence of losses in the anterior pituitary hormones. For instance, the age differences in structure (Verzar, 1966) and weight (Lockett, 1976) reported to exist in the anterior lobe of the pituitary gland do not appear to be reflected in any severe impairments with increasing age of hormone reserves or responsiveness to stimulation (Blichert-Toft, 1975; Lazarus & Eastman, 1976). Studies on the individual anterior pituitary hormones illustrate some age effects, but the overall picture is one of stability across adulthood.

TSH

There appear to be no age differences in the levels of TSH circulating in the blood in healthy men and women (Azizi et al., 1979; Tunbridge et al., 1977). The TSH content in the anterior pituitary, when studied at autopsy, does not show age effects (Bakke et al., 1964). A decreased TSH response to hypothalamic thyro-tropin-releasing factor was found to be present in men but not women in two series of studies (Azizi et al., 1979; Snyder & Utiger, 1972a, 1972b), but was not found to exist in either sex in another large-scale study (Blichert-Toft, Hummer, & Dige-Pedersen, 1975). However, not all evidence favors the interpretation of unaltered TSH basal levels and responsivity across adulthood (see Cole, Segall, & Timiras, 1982).

ACTH

Aging of the anterior pituitary's capacity to produce ACTH would have significance because of this hormone's mediating role in mobilizing the body's energy reserves in stressful situations. There are, however, apparently no age effects on the concentration of ACTH in the blood (Blichert-Toft, 1975; Jensen & Blichert-Toft, 1971) or the content of ACTH in the pituitary (Verzar, 1966). The normal responses appear to be maintained into old age of increasing ACTH levels following decreased blood levels of glucocorticoids or blood glucose, and lowered ACTH following steroid administration (Andres & Tobin, 1977; Lazarus & Eastman, 1976; Lockett, 1976).

GH

The function of growth hormone in adulthood and its role in the aging process have intrigued researchers for many years. Since the primary importance of this hormone is the regulation of growth during the maturation process, it was once thought that the onset of aging might also be under the control of GH (Everitt & Burgess, 1976). Thus, it was believed to be possible that maintenance of high GH levels would slow down the rate of aging. However, this hypothesis was not consistent with the information that excessive GH production in adulthood leads to premature death. Consequently, the effects of this hormone on adults, who are past the "growth" stage, and the effects of changes in GH levels with age on the body's functioning, have remained a puzzle.

There are functions of GH other than the promotion of skeletal growth in children that would have more obvious applictions to adult development. GH secretion stimulates the liver to increase the release of glucose into the blood from its stored form, glycogen. When blood glucose levels are low, GH secretion is increased. Exercise also stimulates GH release because it creates heavy demands for glucose delivery to the working muscles. The secretion of GH is an important part of the body's defense against hypoglycemia, a dangerous condition for the nerve cells, which are high consumers of glucose but have no capacity to store it. Protein synthesis is also promoted by GH secretion, as is energy release from fat tissue. The metabolic actions served by GH make a critical contribution, then, to the availability of energy to the body's tissues during strenuous physical work. Two factors that complicate the assessment of GH levels are its reactivity to stressful stimuli (pain, anxiety, temperature changes) and its cyclicity.

In many regards, the secretion of GH appears to be normally maintained into old age. In the first place, the content of GH in the cells of the anterior pituitary seems to remain constant across adulthood (Gershberg, 1957). There is no decrease with age in plasma GH under basal conditions (Blichert-Toft, 1975; Kalk, Vinik, Pimstone, & Jackson, 1973). Variations in adulthood in this measure of GH secretion have been related more to the relative degree of adiposity of the individual rather than to age (Dudl, Ensinck, Palmer, & Williams, 1973; Elahi, Muller, Tzankoff, Andres, & Tobin, 1982). The clearance rate of plasma GH is maintained to normal levels into old age (Taylor, Finster, & Mintz, 1969). Moreover, and particularly crucial in that a major function of GH is to mobilize the body's energy sources, the extent of increase in GH in response to physiological stimulation (arginine, hypoglycemia) seems to remain normal throughout later adulthood (Blichert-Toft, 1975; Cartlidge, Black, Hall, & Hall, 1970; Dudl et al., 1973; Kalk et al., 1973; Lazarus & Eastman, 1976; Root & Oski, 1969).

However, complicating this picture of stability are various results on different aspects of GH secretion that reveal selected areas of loss. In children and young adults, GH secretion peaks during the night, associated with periods of slow wave sleep. The number of peak periods of GH secretion and hence the total amount secreted during sleep decreases steadily from childhood through later adulthood (Carlson, Gillin, Gorden, & Snyder, 1972; Finkelstein, Roffwerg, Boyar, Kream, & Hellman, 1972; Prinz & Halter, 1983). Under conditions of moderate submaximal exercise, older persons seem to show a greater response of GH (Shephard & Sidney, 1975), a reaction that is not considered to be optimal (Galbo et al., 1977). When performing more strenuous, close to maximal, work the response of heightened GH secretion made by older persons is more adaptive.

The hormonal reaction to exercise at close to a maximum level of intensity has been shown to be sensitive to training effects in the elderly. Sidney and Shephard (1977b), in their work on the exercise and training response of persons in their 60s to a 10-week aerobics program, demonstrated that the GH response to exercise increased over the training period. Although no young control group was used for the purpose of direct comparison, it may be argued that such an increase constituted a beneficial effect. Along similar lines, Szanto (1975) demonstrated that extremely fit older men showed greater GH responsiveness to release from glucose suppression than age-matched sedentary individuals. One possible explanation for these findings of a training effect in the aged is that continued participation in endurance sports serves to preserve an otherwise age-impaired sensitivity of the tissues to GH (Pliska & Gilchrest, 1983) or of the hypothalamus to stimulate GH secretion (Sidney & Shephard, 1977b).

Posterior Pituitary

There is very little information on the effects of aging on this gland. The posterior pituitary cells are stimulated by the neuroendocrine cells in the hypothalamus to secrete the hormones vasopressin (antidiuretic hormone) and oxytocin (the hormone that stimulates the "let-down" reflex of the mammary glands during lactation). Several structural changes have been described as occurring as a function of age (Turkington & Everitt, 1976), but there are no data to indicate that there are functional losses resulting from these cellular alterations. If anything, the osmoreceptors of the hypothalamus may be more sensitive to low water concentrations in the blood in older persons, and so stimulate greater release of vasopressin (see chapter 5). Age effects on the response of the kidneys to vasopressin are described in chapter 5. Since oxytocin is produced specifically around the time of childbirth, it would not appear to have much relevance to the aging process.

Glands Controlled by the Anterior Pituitary

Thyroid Gland

Age changes in the thyroid gland have received a great deal of attention because of the instrumental role served by the thyroid hormones in controlling the body's expenditure of caloric energy. A major question in this research is whether the body's rate of caloric expenditure slows down with age, due to a diminution of thyroid hormones, necessitating a reduction in caloric intake if the adult wishes to maintain a constant weight.

Several structural changes in the thyroid gland observed in autopsy studies suggest a picture of a less active, "involuting" gland in later life (Cole et al., 1982) with some atrophy and detrimental changes observed at the cellular level (Everitt, 1976b; Gregerman & Bierman, 1974; Lasada & Roberts, 1974). It is also possible that apparent age differences in the degree of atrophy of the thyroid (as indexed by its volume) are due to a confound of age and body weight (Hegedus et al., 1983). Whatever the effects of age alone on the thyroid's structure, these are generally not regarded as serious threats to the integrity of this organ's functioning. The amount of free T_4 available in the blood for immediate use by the body's tissues was reported to be lower after 65 years on one study (Herrmann, Rusche, Kroll, Hilger, & Kruskemper, 1974) but to increase after this age in another (Evered et al., 1978). In general, though, there seems to be fairly good agreement that the amount of T_4 in the blood does not decrease across age groups of adults (Blichert-Toft et al., 1975; Snyder & Utiger, 1972a, 1972b). In addition, the responsiveness of the thyroid gland to externally administered TSH is not affected by aging (Azizi et al., 1979; Gregerman & Bierman, 1974; Melmed & Hershman, 1982).

This hypothesis of reduced tissue demand for T_4 receives support from the results of research on the basal metabolic rate (BMR), an index of oxygen consumption at rest. Because the BMR reflects the rate of energy use by the body's tissues, particularly by muscle, it can be considered an indirect measure of thyroid activity. The BMR decreases consistently across adulthood as indexed by cross-sectional studies, and this decrease is regarded as being due to a decreased oxygen uptake by individual muscle cells (Keys, Taylor, & Grande, 1973; Tzankoff & Norris, 1977). The diminished oxygen uptake is, in turn, accounted for by a reduction in the number of actively metabolizing muscle cells and hence less of a demand for the thyroid hormones. It is also possible that a diminution in daily energy expenditure because of lowered activity levels in middle-aged and older persons can contribute to a lowered BMR (McGandy et al., 1966), perhaps reducing even further the demands on the thyroid gland to secrete T_4.

From these results on the thyroid hormone T_4, it would seem evident that the thyroid gland itself is not affected by the aging process in any serious way, and that reductions in the secretion and use of T_4 are secondary to changes in the body's needs for caloric energy. There are indications, though, that T_4 levels in the blood do decrease across adulthood, particularly after the age of 60 years (Blichert-Toft et al., 1975; Herrmann et al., 1974; Hesch, Gatz, Juppner, & Stubbe, 1977; Rubenstein, Butler, & Werner, 1973). This decrease appears to be relatively greater for men (Evered et al., 1978; Sawin, Chopra, Azizi, Mannix & Bacharach, 1979; Snyder & Utiger, 1972a, 1972b). Although there were no age differences in serum levels of T_3 in one study, older adults were found to show diminished responsiveness of this thyroid hormone to thyrotropin-releasing hormone (Azizi et al., 1979). An age difference in baseline levels or responsiveness of T_3 would suggest a deficit in the thyroid gland's ability to meet the body's needs for this hormone. On the other hand, it is also possible that a reduction in T_3 occurs secondarily to a change in T_4 activity, since much of the T_3 produced by the thyroid is derived from T_4 (Gregerman & Bierman, 1974). It is also possible that the apparent age reduction that has been observed in T_3 is an artifact of disease, as was demonstrated in a well-controlled study by Olsen, Laurberg, & Weeke (1978).

Adrenal Cortex

Cells in the cortex of the adrenal glands each produce one of several types of corticosteroids: glucocorticoids (primarily cortisol), mineralocorticoids (primarily aldosterone), and sex steroids (primarily adrogens and estrogens). Secretion of all of these hormones is stimulated by release of ACTH by the anterior pituitary.

There appears to be a reduction with age, significant after 50 years, in aldosterone levels in the blood and urinary excretion rate, but this age effect is secondary to a decrease in plasma renin activity (Hegstad et al., 1983; Takeda, Morimoto, Uchida, Miyamori, & Hashiba, 1980; see chapter 5). The production of adrenal testosterone appears to decrease across age groups of adult men (Wolfsen, 1982), but since the role of the testosterone produced by the adrenals for male sexuality and reproductive functioning is not clear (Gregerman & Bierman, 1974), the significance of this age difference is not well-understood either.

Of the adrenal hormones, the glucocorticoids (cortisol) have received the greatest attention with regard to the effects of aging because of the consequences that these changes would have for the manufacture by the liver of glucose and proteins in response to bodily demands for these substances (a process stimulated by the glucocorticoids). As with the description of the effects of age on the thyroid, whatever structural changes there are in the adrenals (Cole et al., 1982) they do not seem to detract from the functional capacity of the adrenal cortex. The level of cortisol in the blood after an overnight fast or over a 24-hour period is stable across adulthood (Prinz & Halter, 1983). Like T_4, the excretion and disposal rates of cortisol show cross-sectional age reductions over the adult age span, and these decreases appear to be offset by a diminished secretion rate. However, unlike T_4, which shows an adequate response to TSH stimulation in older adults, the cortisol level does not increase as much in older adults as in younger adults when stimulated by ACTH, its corresponding anterior pituitary hormone. This pattern of results has been interpreted as being due to a diminution in muscle mass, so that less cortisol is required by the body's tissues (Andres & Tobin, 1977; Blichert-Toft, 1975; Wolfsen, 1982). The function of the adrenal cortex, then, appears to be well-maintained into old age. What changes are the body's demands for producing their hormones.

Pancreas

Outside the realm of hypothalamic control is the endocrine gland component of the pancreas, located in the islets of Langerhans. The endocrine pancreas is the primary regulator of glucose availability to the other cells of the body, via the secretion of insulin (by the beta cells) and glucagon (by the alpha cells). Insulin facilitates the conversion of glycogen to glucose in the liver, other energy-storing reactions in muscle and fat tissues, and the uptake of glucose from the blood by muscles. When the glucose level of the blood becomes low, glucagon is stimulated, and glucose-producing reactions are stimulated throughout the body, including the liver. Insulin and glucagon also regulate the metabolism of other nutrients by the body's tissues. Insulin inhibits the breakdown and promotes the synthesis of proteins, fats, nucleic acids, and glycogen, so that more energy is stored in muscle and fat tissue. The converse reactions are stimulated by glucagon, whose effects are to liberate stored energy from these tissues.

The high prevalence of diabetes mellitus (a disease characterized by abnormally low insulin levels in the blood) in populations of older adults has stimulated an enormous amount of research directed at understanding the functioning of the pancreatic endocrine gland in old age. In addition to those older adults who are diagnosed as having diabetes mellitus, there are many aged men and women who do not show clinical symptoms of diabetes, but whose scores on the glucose tolerance test are out of the normal range. A reduction in glucose tolerance (i.e., less reduction of glucose in the blood by insulin 1-3 h after glucose administration) that is progressive across the adult years is a virtually universal finding of numerous cross-sectional studies on glucose tolerance and age (reviewed by Andres, 1971; Andres & Tobin, 1977; Davidson, 1982; DeFronzo, 1982).

Investigations into the causes of diminished glucose tolerance in older adults have focused on whether there is lowered secretion of insulin by the beta cells in the pancreas or, conversely, whether the body's tissues are somehow resistant to the effects of insulin on glucose uptake and conversion to glycogen. Decreased secretion of insulin in response to glucose administration was found to occur in older adults in several studies (Dudl & Ensinck, 1972; O'Sullivan, Mahan, Freedlender, & Williams, 1971; Soerjodibroto, Heard, & Exton-Smith, 1979). However, the great majority of evidence offers support for the interpretation that beta cell secretion of insulin is maintained intact in the pancreas of the older person (Berger, Crowther, Floyd, Pek, & Fajans, 1978; Davidson, 1982; DeFronzo, 1979, 1982; Dudl & Ensinck, 1977; Kalant, Leiborici, Leibovici, & Fukushima, 1980; McConnell, Buchanan, Ardill, & Stout, 1982; Ratzmann, Witt, Heinke, & Shulz, 1982; Sherwin et al., 1972). A stronger case is made for the reduction with age in sensitivity of the body's cells to insulin (Davidson, 1979, 1982; DeFronzo, 1979; Jackson et al., 1982; Ratzmann et al., 1982). Insulin insensitivity is most clearly demonstrated using the "euglcyemic clamp" technique (DeFronzo, Tobin, & Andres, 1979), where by the amount of insulin is varied, and the glucose level is maintained at a steady state by continuous glucose infusion. The dependent variable is, then, the rate at which glucose is metabolized, as measured by the rate of infusion (it is assumed that hepatic glucose production is minimal). This method places the plasma glucose concentration under the control of the investigator and avoids the problem of having the subject become hypoglycemic, which is dangerous to the subject and creates problems of interpretation for the investigator because of the glucose-insulin feedback loop.

Lowered insulin sensitivity in the aged is attributed to a loss of permeability to insulin across cell membranes (Soerjodibroto et al., 1979), or possibly a decrease in the amount of metabolizing tissues in the body (Silverstone, Brandfonbrener, Shock, & Yiengst, 1957), but probably not to a diminished feedback system in the pancreatic beta cells (Feldman & Plonk, 1976), changes in the kinetics of insulin metabolism (McGuire et al., 1979), or insulin binding to receptor cells (Fink, Kotterman, Griffin, & Olefsky, 1983; Rowe, Minaker, Pallotta, & Flier, 1983). The reduced sensitivity, it is further postulated, is shown particularly in muscle (DeFronzo, 1982).

Another perspective on the issue of insulin sensitivity in adulthood is provided by a growing number of studies in which it is found that there is no age-related decrease in insulin sensitivity when the amount of body fat of the respondents is taken into account or controlled (Björntorp, Berchtold, & Tibblen, 1971; Kalant et al., 1980; Kimmerling, Javorski, & Reaven, 1977). Reduced physical activity in the aged is cited as another factor which can reduce insulin sensitivity in older persons, as was demonstrated by Seals, Hagberg, Hurley, Ehsani, & Holloszy (1984). These authors attributed the favorable effects of a 1-year exercise training program on lowering the amount of plasma insulin needed to maintain glucose tolerance at steady-state levels to adapatations in the pancreas, and reduction of adiposity. In addition, they suggested that the effects of exercise were due to increased sensitivity by the muscles to insulin and greater glucose uptake by the muscles during exercise. The results of this experimental study confirm the findings of Björntorp et al. (1972) on physically fit well-trained men in their 50s, in whom increased sensitivity to insulin seemed to be due to a greater uptake of glucose. This heightened glucose uptake was attributed, in turn, to a higher proportion of muscle tissue in these active men. Added to the list of possible reasons for reduced sensitivity to insulin (and hence reduced glucose tolerance), then, are the changes that many adults undergo in their levels of daily exercise and in their eating habits, as described in chapters 2 and 3.

Whether the loss in insulin sensitivity by the body's tissues is an inevitable consequence of aging, or a response to altered dietary and activity patterns, it seems reasonably clear that age has little, if any, effect on pancreatic functioning. This conclusion is reinforced by findings that other functions related to the pancreatic endocrine hormones do not show age effects. The glucose level in the blood under basal conditions does show a slight increase over the adult years (Davidson, 1982). However, this age difference may be related to the greater adiposity of older adults (Elahi et al., 1982). The level of glucagon in the blood under basal conditions and after physiological stimultion is maintained into old age (Dudl & Ensinck, 1972; Elahi et al., 1982). The conversion of glycogen to glucose, then, would not appear to be impaired by the aging process.

Implications of Age Effects in the Endocrine System

Studies on the functioning of the endocrine glands involved in the control of metabolic functions do not reveal a universal pattern of loss in old age in the endocrine glands or in the effectiveness of the hormones produced by these glands. While changes have been demonstrated in some endocrine functions, the predominant impression is one of stability. The glands that have been studied most intensively in relation to age are some of the most powerful regulators of total bodily functioning, because they control fundamental aspects of energy metabolism. Given the intricacy of these systems, and the sensitive balance that must be maintained among their numerous components, such stability in old age is indeed impressive. This stability is made possible, in part at least, by the adjustments noted in the discussion of the various hormones of decreased secretion in response to diminished disposal rates, which in turn, is a reflection of loss of lean tissue mass. In virtually all systems, it seems reasonable to take the position that the complex feedback loops involved in hormonal control are maintained intact into old age.

Apart from the lack of generalized age-related alterations in the endocrine glands and the hormones they produce, in almost every subsystem studied, there appear to be alterations in the hormone's effectiveness on the bodily cells on which it acts. In some cases, such as T_4 and cortisol, the alteration may be due to a loss of tissue mass, so that less of the hormone is required to achieve the necessary effect. In the case of insulin, it appears that the body's tissues are less responsive and so more insulin is required to achieve the adequate storage of glucose. In contrast, there appears to be loss either in tissue quantity or sensitivity to GH, although there is some reduction of GH production in during sleep.

Since the endocrine glands underlie so many aspects of functioning in daily life, this stability of the system would help to maintain feelings of well-being in the older individual. In particular, access to ample energy reserves, made possible by the pituitary as well as pancreatic hormones, should serve to ward against feelings of fatigue caused by exertion arising from this source. Instead, the feelings of discomfort the older person may experience during or after exhausting activity are probably much mor likely to originate from the cardiorespiratory adjustments to exercise, or muscle fatigue. Were the endocrine glands to become less efficient as well, the individual would have extreme difficulty at this kind of work.

It may be premature, at this point, to draw general conclusions about aging of the endocrine glands, though, given the diversity of subsystems within the larger totality. In a very global sense, it does appear likely that the stability rather than change does prevail. It also must be recognized that there is a tremendous diversity in the composition of the samples of adults who have participated in the research on aging in factors that may have profound relationships to endocrine functioning. One such factor that has already been addressed is adiposity, shown in several cases to be more strongly related than age to various aspects of metabolic functioning. Other factors include gender, nutritional status, activity level, and overall health status. Although attempts are usually made to solicit as subjects people who are in good health, in some cases hospitalized individuals have made up the older samples, and hospital staff the younger groups, so the groups may not have been comparable in some very important physiological characteristics that could affect endocrine functioning.

Another factor totally unaccounted for is the amount of stress associated by the individual with hospitals and the medical procedures involved in gathering the data. Since many of the studies are conducted in hospital settings, this may be an important threat to the validity of the conclusions (especially if the younger, hospital staff are more comfortable with these procedures). Stress can influence the levels of many of the hormones studied in relation to aging, and extraneous influences of the settings and procedures may throw off in differential ways across age groups some of the delicate balances in the systems under study.

Effects of Age on the Autonomic Nervous System

Many life-support functions involving the regulation of diverse bodily systems are served by the autonomic nervous system (ANS). Like the endocrine system, the ANS accomplishes its tasks continuously and without conscious control. However, instead of influencing other organ systems via the blood, as do the endocrine hormones, most of the neurons of the ANS are in close physical proximity with their target organs, permitting more direct and rapid communication than can be achieved with endocrine regulation. The subdivisions of the ANS, the parasympathetic nervous system (PNS), and sympathetic nervous system (SNS), operate in complementary fashion to take part in regulating features of the body's internal environment critical to life, such as respiration, digestion, excretion, and circulation. The hypothalamus plays a major role in this process, serving as the brain structure that regulates many autonomic activities independently, coordinates autonomic and endocrine information, and communicates with other parts of the brain regarding the state of the body.

Sympathetic Nervous System (SNS) Activity

The SNS is primarily responsible for controlling the body's response to stressful and energy-taxing activities. Stimulation by norepinephrine, the neurotransmitter released by most SNS neurons onto their effectors, results in energization of the body's resources, including a rise in oxygen consumption and cardiac output, glucose release from the muscles and liver, secretion of sweat and oil from the exocrine glands in the skin, and diversion of the blood from digestive and restorative activities toward the skeletal muscles. The secretion by the adrenal cortex of glucocorticoids in response to SNS activity further enhances this energization of the body's reserve.

The starting point for describing age effects on the SNS is with the almost universal finding of numerous recent studies that the level of norepinephrine in the blood is progressively higher across age groups of adults, when tested at rest in the recumbent position (Christensen, 1973; Coulombe, Dussault, & Walker, 1977, 1977; Esler et al., 1981; Franco-Morselli, Elghozi, Joly, DiGiulio, & Meyer, 1977; Lake, Ziegler, Coleman, & Kopin, 1976; Pedersen & Christensen, 1975; Sever, Osikawska, Birch, & Tunbridge, 1977; Weidmann et al., 1978; Ziegler, Coleman, & Kopin, 1977; Ziegler, Lake, & Kopin, 1976). Rising to an upright position, which normally results in a heightened blood level of norepinephrine, has a more pronounced effect on the norepinphrine level of older adults (Franco-Morselli et al., 1977; Saar & Gordon, 1979; Sever et al., 1977; Weidmann et al., 1978), as does isometric exercise (Sowers, Rubenstein, & Stern, 1973; Ziegler et al., 1976), exposure to cold stress (Palmer, Ziegler, & Lake, 1978), ingestion of large quantities of glucose (Young, Rowe, Pallotta, Sparrow, & Landsberg, 1980), and the stress of being placed in an experimental learning situation (Barnes, Raskind, Gumbrecht, & Halter, 1982; Eisdorfer, Nowlin, & Wilkin, 1970; Powell, Eisdorfer, & Bogdanoff, 1964).

This increase across age groups of adults in the blood level of norepinephrine is generally interpreted to be the result of greater SNS activity in older persons, which causes more of the neurotransmitter to "spill over" from the sympathetic neurons to their neighboring capillaries, and from there into the general circulation (Pfeifer et al., 1980). This interpretation is supported by the finding of corresponding increases across age groups of adults on simultaneous measures of sympathetic nervous activity and blood norepinephrine (Wallin et al., 1981).

However, when norepinephrine release into the bloodstream was directly measured in one cross-sectional study, no age differences were found in the norepinephrine spillover rate. Instead, it was found that the norepinephrine disposal rate was lower in older adults (Young et al., 1980). Rather than indicating a more active SNS in older adults, then, the rise in blood level of norepinephrine across adult age groups may reveal altered actions of norepinephrine on the body's cells.

Another line of investigation into age effects on SNS activity and the reasons for the increase in norepinephrine in the blood is oriented toward tissue responsiveness to sympathetic stimulation in older persons; in particular, the effects of SNS stimulation on heart muscle. A diminished sensitivity to norepinephrine in the cardiac muscles of older persons could partially account for the age-related cross-sectional decrease in the force of the left ventricle's contraction (Conway, Wheeler, & Sannerstedt, 1971; Kendall, Woods, Wilkins, & Worthington, 1982; Lakatta, 1979, 1980; Port et al., 1980; Schocken & Roth, 1977; Weisfeldt, 1980; Yin, Raizes, Guarnieri, Spurgeon, Lakatta, Fortuin, & Wasfeldt, 1978; Yin, Spurgeon, Raizes, Greene, Weisfeldt, & Shock, 1976). The norepinephrine increase would, according to this argument, be a compensatory mechanism to this reduction of cardiac effectiveness as a means of maintaining an adequate cardiac output (Vestal, Wood, & Shard, 1979).

The effects of age on the baroreceptor reflex lead to a second explanation for the increased blood level of norepinephrine. This reflex represents a regulatory mechanism that involves both the PNS and SNS. Baroreceptors are stretch receptors in the large arteries leaving the heart that are sensitive to sudden increases or decreases in the arterial pressure. When the walls of the arteries are stretched by an increase in the amount of blood flowing through them (which raises the blood pressure inside the artery), the baroreceptors send signals to the reticular formation area, which controls cardiovascular responses. Arterial pressure is reduced by a reflex in which the PNS is stimulated, thereby inhibiting cardiac activity, and reducing the pressure within the arteries. When the baroreceptors decrease their rate of response because the blood pressure is too low, the opposite reflex is stimulated by activation of the SNS, which increases cardiac output and constricts the blood vessels. One of the most important functions of the baroreceptor reflex is to maintain sufficient blood flow to the brain despite changes in posture from horizontal to vertical positions.

The baroreceptor reflex is progressively less effective across adulthood in reducing an overly high blood pressure (Gribbin, Pickering, Sleight, & Peto, 1971; Lindblad, 1977). The reason for this reduced effectiveness in blood pressure reduction is postulated to be a stiffening with age in the arteries, so that the baroreceptors are not stimulated as readily by the stretch of the artery when the blood flow (and hence pressure) increase. The dilation of the blood vessels that accounts in part for a lowering of blood pressure after the initiation of the baroreceptor reflex would, then, also be less effective. In the converse situation, when an increase in blood pressure is needed, the stiffening of the arteries would impede the process of constriction, which helps to raise the pressure within them. Greater SNS sensitivity, reflected in an increased blood level of norepinephrine, would serve to compensate for the critical situation in which SNS activity must be increased to ensure adequate blood supply to the brain when the person is in an upright position. The finding that norepinephrine increases more in the aged than in younger adults when changing from the recumbent to the upright position is consistent with this interpretation (Rowe & Troen, 1980).

If increased norepinephrine levels in the circulation is a compensatory device for decreased responsivity of heart muscle and the baroreceptor reflex, this would again suggest that SNS activity itself is not a locus of primary aging effects. Instead, the higher level of norepinephrine in the blood could be seen as an adaptive response that is directed toward maintaining the functioning of critical life support activities.

Parasympathetic Nervous System (PNS) Activity

Consistent with its role in energy production during periods of quiescence, the PNS is the division of the autonomic nervous system that controls digestive and restorative processes. The PNS transmits information by releasing acetylcholine to its effectors. Some of the functions served by the PNS include secretion by the salivary and other glands that produce digestive fluids, increased motility by the digetive muscles, and relaxation of the anal and urethral sphincters. Acetylcholine transmitted to the heart through the vagal nerve slows the rate of contraction.

There is considerably less information on PNS functioning across adulthood, compared to the extensive literature on the SNS. That which exists on the PNS is contradictory suggesting, on the one hand, relative stability (Finch, 1977) and on the other hand, age-related alterations in virtually every aspect of acetylcholine activity (Frolkis, 1982). Evidence in support of the stability interpretation is that the heart rate at rest, which is under PNS control, is steady across the adult years (see chapter 3). Also, many digestive activities under PNS control show minimum age effects (see chapter 5). However, the age differences in baroreceptor reflex sensitivity could be partially accounted for by alterations in PNS activity as well as by changes in sensitivity to SNS stimulation.

Adrenal Medulla

The inner regions of the two adrenal glands are comparable to the ganglia in the sympathetic system, where the neurons leaving the spinal cord synapse on neurons, which innervate the effectors. When stimulated by preganglionic sympathetic neurons, the cells in this organ, which are like modified postganglionic neurons in the rest of the autonomic nervous system, release norepinephrine and epinephrine. Unlike other autonomic neurons, though, the postganglionic neurons of the adrenal medulla secrete their neurotransmitters into the bloodstream, so that these transmitters act in an analogous fashion to hormones secreted by the endocrine glands. The effects of epinephrine are very similar to norepinephrine, and complement the mobilizing effects on the body of the norepinephrine secreted by stimulation of the SNS. However, epinephrine has a more powerful and prolonged effect on cardiac output and metabolic rate than does norepinephrine.

The activity of the adrenal medulla appears to be stable across adulthood, based on the findings that the amount of epinephrine in the blood in a resting or upright state, or under stress, shows no age diffrences (Barnes et al., 1982; Prinz, Vitiello, Smallwood, Schoene, & Halter, 1984; Weidmann et al., 1978). It is possible, though that a stable blood level could coexist with alterations in metabolism through increased secretion and decreased removal, or vice versa.

Regulatory Mechanisms of the ANS

In addition to the functions described so far of the autonomic nervous system, autonomic regulatory mechanisms of special importance to the individual's well-being have been studied in regard to their efficiency across adulthood: body temperature control and sleeping patterns. Age effects on the other regulatory mechanism, the baroreceptor reflex, have already been described.

Body Temperature Control

The regulatory mechanisms of the ANS maintain the core temperature of the unclothed body within the optimal temperature range for cellular metabolism (98.6 °F to 100.4 °F). This control is accomplished through integration by the hypothalamus of sensory information concerning the outside temperature (see chapter 11). Like a thermostat, the hypothalamus then initiates actions that raise or lower the body's core temperature. To increase body temperature, heat is produced by shivering (brought about by an increase in the contraction rate of the skeletal muscles), and conserved by raising the hairs in the skin and constricting the capillaries in the dermis to reduce loss of surface heat. Temperature reduction is accomplished by actions that maximize heat loss across the skin's surface, including dilitation of the capillaries and secretion by the sweat glands in the dermis. The individual can also take conscious steps to regulate body temperature, such as adding or removing clothes, adjusting the heat or air conditioning, seeking a change in environment, or immersing in a tub or pool of water. Obviously, in order to take optimal advantage of these behavioral temperature-controlling mechanisms, the individual's thermal sensory abilities must be operating effectively so that the outside temperature is accurately perceived.

Population health statistics on mortality rates during periods of intense heat and cold and experimental studies on thermal regulatory mechanisms converge almost entirely in their conclusions regarding the impaired adaptive responses of persons over 65 years of age to extremely hot and cold outside temperatures. The loss of adaptive mechanisms to temperature extremes actually appears fairly early in adulthood, but it does not seem to have serious consequences until later adult years.

Responses to Cold

A diminished response to cold in the elderly could reflect a reduction in sensitivity to cold, or an impaired set of mechanisms to produce and/or conserve heat. It appears that older persons are less aware that their core body temperature is low (Fox, MacGibbon, Davies, & Woodward, 1973), and are less able to raise their core temperature when the temperature at the periphery of the body is cold (Collins et al., 1977). There is some evidence that, in men, at least, the loss of ability to increase body heat production and decrease heat loss may be evidenced as early as the 40s (Wagner, Robinson, & Marino, 1974).

Responses to Heat

Compared to research on age differences in thermoregulatory responses to cold, there is considerably more information on the effects of aging on reactions to heat stress. The primary reason given for the reduced adaptability to heat is decreased secretion by the sweat glands in the skin (Ellis, Exton-Smith, Foster, & Weiner, 1976); Hellon & Lind, 1956; Hellon, Lind, & Weiner, 1956; Shoenfeld, Wassin, Shapiro, Ohri & Sohar, 1978). The decrease in the rate of sweating is thought to be related, in turn, to diminished autonomic input to the sweat glands as reflected by a higher threshold for sweating (Foster, Ellis, Dore, Exton-Smith, & Weiner, 1976). Paradoxically, the skin blood flow of men in their 40s seems to increase much more during heat exposure than is true for men in their 20s, but this increase apparently does not compensate for reduced sweat gland activity to lower body temperature effectively (Hellon et al., 1956; Lind, Humphreys, Collins, Foster, & Sweetland, 1970).

The evidence that older persons are more likely to feel warm in hot environments than young people because of lowered sweat gland activity is not easily reconciled with the frequently reported observation that the aged often report feeling "cold," even in warm weather. This response does not seem to be a result of normal aging, but instead may reflect impaired cardiovascular efficiency such that blood flow to the extremities is impaired.

Research on acclimatization of adults of different ages to prolonged exposure to heat provides important information on the adaptive capacity of the thermoregulatory systems of older persons over long periods of heat stress. In one study, four male physiologists in their 40s and 50s were found to adapt as well and as readily to the same conditions of working in the heat as those under which they were tested 21 years earlier (Robinson, Belding, Consolazio, Horvath, & Turrell, 1965). The effectiveness of their response was due to their ability to increase heat loss through the skin surface via higher skin blood flow. The ability to become adjusted to performing work in the heat over the summer months appeared to be well-maintained in a sample of healthy and active older men and women (Henschel, Cole, & Lyczkowskyj, 1968). Without a younger group with whom to compare their responses, this finding is hard to evaluate, particularly in view of the results of an age-controlled heat-acclimatization study in which men in their

50s and 60 did adjust to work in the heat over the summer months, but did so less well than did men in their 20s (Wagner, Robinson, Tzankoff, & Marino, 1972).

It is possible that an active life-style does serve to mitigate somewhat against age losses in heat adaptability, as is suggested by a cross-sectional study on women in which no age differences in sweat rate were found in relation to age (Drinkwater, Bedi, Loucko, Roche, & Horvath, 1982). On the other hand, this lack of an age effect in women may reflect the fact that young women have diminished sweat gland responsiveness compared to young men (Foster et al., 1976).

Sleep Patterns

Unlike other autonomic functions, the function sleep serves to the individual's survival is not known. It is thought, however, that sleep serves a restorative function for the organ systems of the body, and its regulation is a critical feature mediating the body's physiological adaptation. Moreover, the function of sleep in everyday life is clearly crucial to the individual's feeling of well-being.

During sleep, the state of arousal of the brain undergoes marked shifts, reflected in distinctive patterns of electrophysiological activity. These patterns, measured by electroencephalograms (EEGs), fall into two broad categories: slow wave sleep and paradoxical sleep. Slow wave sleep is made up of four stages, corresponding to lower and lower levels of reticular formation arousal. Stage 1 is the period of drowsiness, during which the EEG is rapid and desynchronized. During Stages 2 and 3, the EEGs become slower and more regular, corresponding to deeper and deeper sleep. Finally, Stage 4 is reached when the person is in a state of deep sleep. Stage 4 sleep is distinctively different from the preceding stages, with a pattern of large and peaked brain waves. No dreaming takes place during Stage 4 sleep, and most bodily functions assume low levels of activity.

Every 100 min or so, paradoxical sleep emerges. This type of sleep is called paradoxical because, although the EEG pattern resembles that of Stage 1, the person is harder to awaken than from any of the slow wave sleep stages. Dreaming takes place during paradoxical sleep, and when a person in this type of sleep is observed, movements of the eyeballs can be seen taking place underneath the closed eyelids. These "rapid eye movements" (REMs) uniquely characterize paradoxical sleep, which is also called REM sleep. Other features of paradoxical sleep are a decreased muscle tone, irregularities of heart rate and respiration, and in men, penile erections. The relative length of paradoxical sleep normally increases from the early part of the night toward the morning.

The results of research on age differences in sleep regulation provide a fairly consistent picture of the features that characterize the sleep of adults in middle and later adulthood. Across age groups of adults, there is an increase in the amount of time spent lying awake in bed at night (Coleman et al., 1981; Dement, Miles, & Bliwise, 1982; Feinberg, Koresko, & Heller, 1967; Hayashi & Endo, 1982; Prinz, Weitzman, Cunnhingham, & Karacan, 1983; Webb, 1982). While the total amount of time spent in bed increases cross-sectionally after the age of

60, the total amount of time spent sleeping actually remains fairly constant throughout adulthood (Feinberg et al., 1967; Williams, Karacan, & Hursch, 1974). Consequently, the proportion of time asleep to time in bed, an index of sleep efficiency, decreases cross-sectionally across adulthood, beginning after 30 years for men and 50 for women.

Persons over the age of 65 take longer to fall asleep when they first retire for the night (Feinberg et al., 1967), and an unbroken night's sleep is extremely rare in persons over the age of 50 years (Williams et al., 1974). An increase in periods of wakefulness appears to begin in the 30s and rise through the 40s. By the later adult years, almost 20% of the night may consist of periods of wakefulness, including the time spent falling asleep, awake at intervals during the night, and in lying awake before arising in the morning. In one study on persons from 73-92 years of age, an average of 21 awakenings were reported (Hayashi & Endo, 1982). Men have more frequent sleep disturbances between the ages of 40 and 70 than do women, possibly because of being more disrupted by REM sleep and its associated penile erections (Williams et al., 1974). By the later adult years, women catch up with men in the extent of sleep disturbances they experience. In old age, the primary immediate causes of sleep disruptions are sleep apnea, periodic leg movements, and heartburn (Coleman et al., 1981; Dement et al., 1982) and also frequent needs to urinate (see chapter 5).

There are some fairly distinct alterations in EEG sleep patterns with age, as revealed by cross-sectional studies (Coleman et al., 1981; Dement et al., 1982; Feinberg, 1974; Feinberg et al., 1967; Hayashi & Endo, 1982; Prinz, 1977; Webb, 1982; Williams et al., 1974). One prominent age difference, related perhaps to the phenomenon of increased wakefulness, is a rise across adulthood in Stage 1 sleep (drowsiness without actual sleep). Stages 2 and 3 appear to show some decreases after 60 years, but more striking is the large decrease of Stage 4 sleep in old age to the point at which it is not even detectable in many of the persons studied. The amount and percent of total REM sleep appears to remain fairly stable across adulthood until the 60s and 70s, and then it diminishes. The observable behaviors associated with REM sleep also become less evident. These age differences support the findings regarding wakefulness at night, signifying that the sleep of older adults is less restful, not only with more time awake, but with less time spent in deep, dreamless sleep or in heavy paradoxical sleep. Older adults also awaken more readily from all stages of sleep than do young persons.

The disturbances of sleep that increase in prevalence during adulthood may have their ultimate source in altered SNS activity. Increased norepinephrine in the blood in the aged is associated with more wakefulness and less REM and Stage 4 sleep (Prinz, Halter, Benedetti, & Raskind, 1979). Alternatively, it could also be argued that the norepinephrine level of older adults is heightened because of the increased disturbances that result from age-related alterations in other systems of the body (respiratory, muscular, digestive) which cause norepinephrine to rise because the individual is kept awake and stimulated. However, reports of these symptoms may be more frequent in the old who notice these disturbances more because of their heightened wakefulness caused by sympathetic arousal. In view of the striking age effects on sympathetic arousal, attribution of increased norepinephrine level as a primary cause of wakefulness in older adults seems to be a more compelling argument, a hypothesis receiving support from Prinz et al. (1984), who found no effect of wakenings on plasma norepinephrine levels in young and old men. Another possibility is that reduced deep sleep and REM sleep in the aged are due to lessened activity in the central nervous system structures responsible for controlling these functions (see Chapter 8).

Other questions regarding age effects on sleep relate to the significance of the finding that older adults spend more time in bed, but still sleep for about the same number of hours each night as do younger persons. Do older adults spend more time in bed because of a physiological need to accumulate "enough" sleep? The time spent in bed may be required in order to reach a nightly total of 7 h of actual sleep (the average number of hours asleep for adults of all ages). On the other hand, does increased time in bed reflect altered sleeping habits based on changed schedules in later life (Williams et al., 1974)? Older persons may be influenced in their sleep patterns by their relative freedom from the early morning demands of child-rearing and going to work and late-night social obligations (which can be shifted to daytime hours) that characterize the years of raising a family and working. The availability of more time to spend in bed may be taken advantage of by the older adult who had worked long hours for many years and enjoys the relaxation and freedom from pressure this time provides. Many younger persons certainly take advantage of vacations to "catch up" on their sleep. In the case of the retired person, the "vacation" continues indefinitely, and so more time may be spent in bed, although no more than 7 h of sleep a night are required from a physiological standpoint.

Although answers to these questions are not available, they raise the possibility of alternative explanations of sleep pattern changes in the aged which could challenge the conclusion that autonomic changes alone account for an age-related decrease in the amount of restful sleep at night. In addition, the questions raised about the causes of changed sleep patterns with age relate to the larger unanswered question of the role that sleep plays in the adult's physiological and psychological well-being and adaptation.

Psychological Consequences of Aging of the Autonomic Nervous System

As can be seen from the descriptions of age effects on regulatory mechanisms, the everyday life of the adult is influenced in some crucial ways by the adequacy of functioning of the ANS. In particular, the ability to adjust blood circulation, body temperature, and sleep habits to meet the particular needs of the situation are critical for engaging in pleasurable recreational activities, such as traveling, exercising, socializing, as well as the performance of work-related tasks requiring exposure to varied climates, demands for exertion, and shift schedules. Participation in some of these activities may be limited for reasons other than defective regulatory functioning—for instance, a skin disorder may keep someone out of the hot sun in the summer, cardiovascular disorders may place limits on exertion, and an individual may be a light sleeper for psychological reasons related to anxiety or tension.

Younger adults may react negatively to some of the demands of their schedules which create undue exertion, temperature discomfort, or loss of sleep. Although older persons may not have as great a capacity to make these adjustments, they may have fewer pressures to do so and therefore feel a certain sense of relief. However, to the extent that aging brings with it perceptible changes from otherwise stable adult levels, it may be speculated that the individual will begin to experience feelings of restriction that may require difficult psychological adaptations. In some cases, the precautions that must be taken may be essential to maintain life, such as seeking increased protection against the cold. Recognition of the need to monitor the effectiveness of the body's regulatory functions may lead to a reduced sense of independence, since the individual can no longer engage spontaneously in some activities, such as gardening in the sun, going for a walk in the winter, or even making rapid changes in posture. Awareness of these losses may be heightened by the fact that the regulatory systems over circulation and body temperature are designed to be ones that are very salient to the individual so that danger signs indicating some failure can be readily detected.

To the extent that they exist, fear and apprehension about the perceived efficiency of the autonomic regulatory systems might contribute to impairment of their functioning. Anxiety can increase the level of sympathetic arousal and in so doing, interfere with the functions it regulates. Many researchers have noted the deleterious effects of anxiety or at least unfamiliarity on indices of ANS functioning, such as blood norepinephrine, blood pressure, sleep patterns, and sweat gland activity. Consequently, undue concern about the body's effectiveness may impair its ultimate ability to adapt to variations in the internal and external environment. On the other hand, older individuals may take in their stride the need to attend to their body's signals and make the necessary adaptations in their behavior so that the changes in regulatory functioning do not have detrimental effects. The likelihood of this happening is increased by the fact that the changes in autonomic control systems take place over a period of many years, and the required adaptations can be quite gradually made.

Conclusions Regarding Age Effects on Control Mechanisms

In the endocrine and autonomic nervous system, age differences have been demonstrated in the various regulatory functions carried out by these integrated control mechanisms. In many cases, though, it is apparent that the locus of age effects is not in the systems themselves, but in the effectiveness of the body's response to the processes initiated by these systems in their efforts to control their various organs. In other cases, age effects on the control system seem to occur secondarily to age effects on the body's tissues, which necessitate compensation to maintain an intact response.

Nathan Shock, whose studies in the Baltimore City Hospitals in the 1950s and 1960s were among the first in each of the various areas of physiological functioning of the human body, has theorized (1977, 1983) that aging brings with it an adequate response to the need to preserve homeostasis in the internal environment under stable conditions. When the older individual is stressed, however, by increased intake of glucose, salts, water, or participation in exercise and exposure to extremes of temperature, the body's physiological adaptations are impaired by the aging process due to a limitation with age in the range in which the control mechanisms can work effectively.

However, it can be argued instead that stresses are an integral part of daily life, and the body's control systems must constantly adjust to them. Were this not the case, the range of situations in which older adults could live would be much more severely restricted than they are in the normal population. In the last few years, there has been a growing body of evidence demonstrating the ability of older persons to adjust to the requirements of physical exertion, and also on the changes in the normal aging process in the diminished sensitivity of the body's tissues to regulatory influences (such as insulin and norepinephrine). Based on this knowledge, it might be hypothesized that it is the body's control systems that are unimpaired with age, and that the primary changes are in the organs these systems regulate. The control mechanisms may serve as buffers or stabilizers against disruptions of efficient physiological functioning caused by deteriorations of the other systems.

Following from this interpretation, it may be predicted that the quality of adaptation to the physiological stresses of everyday life is preserved in the aged until the point at which disruptions in the life-support systems become so severe that the control mechanisms' strategies to preserve normal functioning are pushed beyond their endurance or range of effective influence. Age effects in the body's control systems are, from this perspective, secondary to the loss of adequate functioning by the systems that are being controlled.

CHAPTER 7

Reproductive System

In mature adults, the primary functions of the reproductive system are to make it possible for them to have their own children, and for them to derive pleasure from the sexual act. The feelings connected with sexuality, both in terms of reproduction and the expression of sexual drives, are important contributors to the adult's overall well-being and form a large component of intimate relationships with other adults. In addition, the adult's physical self-concept may be thought to be comprised, in part at least, by his or her self-evaluation of sexual attractiveness and responsiveness, and capacity to produce offspring.

Age Effects on the Female Reproductive System

Throughout the decade of the 40s a woman's capacity to bear children is gradually reduced, until by the age of 50-55, it ceases altogether. The climacteric is this transition period during which reproductive capacity diminishses. The monthly phases of ovulation and menstruation end by the time the climacteric is complete. The term "menopause" refers to the end of menstruation, that is, the woman's last menstrual period. Associated with the ending of the montly phases of the ovary and uterus is a diminution of the hormones produced by these organs. Decreased levels of these hormones affect the functioning of other organs in the external genital area and reproductive tract, as well as secondary sex characteristics. Other changes in sexual functioning take place as the result of aging processes that occur generally throughout the connective tissues of the body and the circulatory system.

Changes in Monthly Cycles and Reproductive Capacity

Starting with the menarche at puberty, women experience monthly cycles during which ova are produced by the ovarian follicles and an environment is created in the endometrial lining of the uterus in which a fertilized ovum can develop into an embryo. Regulation of the monthly cycle is accomplished by the pituitary hormones, FSH and LH, and the estrogen and progesterone produced by the ovaries. The monthly ovarian cycle begins with 10-20 of the 300,000 to 400,000 immature ova contained in the ovary at birth stimulated to ripen at the onset of the ovulatory cycle by FSH. During development of the follicles in the first 14 days of the cycle, estrogen begins to be secreted, under the influence of both FSH and LH. Ovulation occurs when the mature ovum ruptures the ovarian surface and begins its descent toward the uterus down one of the fallopian tubes which encircle the ovaries.

After ovulation, the remains of the follicle within the ovary become transformed into the corpus luteum. Stimulated by LH, it secretes estrogen and increasingly large amounts of progesterone for the next 13-14 days. The rising levels of these ovarian hormones in the blood provide negative feedback to the anterior pituitary to stop secreting FSH. Progesterone also stimulates changes in the uterus to prepare it to received a fertilized ovum. If fertilization does not occur, the corpus luteum becomes a corpus albicans, a mass of scar tissue, and no more hormones are produced. At the end of the monthly cycle, the low level of estrogen in the blood stimulates the hypothalamus to secrete FSH-RF, and the cycle begins once more.

The monthly cycle of the uterus involves changes in the endometrius in response to hormonal stimulation from the ovaries. As estrogen is secreted by the ovarian follicles, the lining of the uterus thickens with blood vessels. After ovulation, this thickening continues and, in addition, glycogen is stored within the endometrium. If fertilization has not occurred following ovulation, the endometrial wall breaks down in response to lack of hormonal stimulation from the deteriorating corpus luteum. Menstruation occurs as the vessels in the uterus pinch themselves off, the endometrial lining is shed, and blood flows from the broken tissues.

The menopause is the obvious signal that the changes associated with the climacteric in the ovary and the ovarian hormones have progressed to the point where their monthly cycles have ceased. The changes involved in the climacteric occur gradually over a period of at least 10 years preceding the menopause, however. One of the major changes is a shortening of the number of days in the monthly cycle, from about 30 days at age 30, to 25 by age 40, and 23 by the late 40s. This decrease in the length of the cycle is due to a shortening of the preovulatory phase during which the follicles are developing within the ovary (Sherman & Korenman, 1975). There are also more cycles of irregularly short or long duration in the years prior to the menopause (Treloar, Boynton, Benn, & Brown, 1967), and more cycles during which ovulation does not occur (Talbert, 1978). During this time, the reproductive organs become less and less capable of sustaining a pregnancy. Beginning at about 35 years and increasing rapidly after that, the ova released each month are more likely to be defective (Collman & Stoller, 1962; Zellweger & Simpson, 1973). If these ova are fertilized, they are more likely to produce infants with severe abnormalities (Kram & Schneider, 1978).

Associated with changes in the monthly cycle during the climacteric are alterations in the reproductive hormones. There is a decrease in the quantity of estradiol, the primary and most potent estrogen produced by the ovaries (Sherman & Korenman, 1975; Sherman, West, & Korenman, 1976). A reduction in the number of growing follicles in the ovaries, a process that progresses throughout the reproductive years (Block, 1952), appears to be what is primarily responsible for the diminished estrogen production. Those follicles that grow are less likely to produce estradiol in response to FSH stimulation. In addition, the corpus luteum becomes less stable (Talbert, 1978).

In response to diminishing negative feedback control from estradiol, FSH production by the anterior pituitary gland is increased, and so, to a lesser extent, is LH (Everitt, 1976a; Talbert, 1977, 1978; Wide & Hobson, 1983). The rise of FSH may be the cause of the more frequent and earlier growth of follicles in the ovaries of women approaching the menopause. In addition, although the prolactin producing cells of the anterior pituitary do not seem to atrophy in later adulthood (Kovacs, Ryan, Horvath, Penz, & Ezrin, 1977), there is a diminution in the amount of serum prolactin around the time of the menopause (Vekemans & Robyn, 1975), reflecting decreased need for mammary gland stimulation associated with the loss of reproductive capacity.

As the amount of estradiol produced by the ovaries diminishes (due to fewer growing follicles and corpora lutea) stimulation of the endometrium of the uterus eventually is reduced, and menstruation ceases (Schiff & Wilson, 1978), the event that constitutes the menopause. Following the menopause, estradiol production drops even further, and its cyclical variations end (Judd & Korenman, 1982). However, estrone, another estrogen, is found in the blood of post-menopausal women in greater amounts than in premenopausal women. Androstenedione, another hormone secreted by ovarian tissue (Judd, Judd, Lucas, & Yen, 1974) and the adrenal cortex is converted to estrone in cells throughout the body (Suiteri & MacDonald, 1973). The estrone produced in this manner does not, however, compensate for the quantity of estradiol that is lost, and as a result, other degenerative changes associated with the aging process occur, which reduce the size and vitality of the reproductive organs after the menopause.

Changes in Female Sexual Functioning

Appearance

Many changes that occur in the woman's body after the menopause are due to the diminution of estrogen and progesterone production by the ovarian follicles and corpora lutea. Other results of hormone depletion following the menopause combine with other effects of aging to produce gradual changes in the appearance of secondary sex characteristics.

The decreased elasticity of the skin that occurs throughout the body in later adulthood, creates a general sagging tone, which also appears in the woman's breasts. Drooping of the breasts is exacerbated by the replacement of mammary gland tissue with fat, which is not as firm. The alveoli become smaller, eventually disappearing late in adulthood. The nipples also decrease and do not become as firmly erect upon stimulation (Timiras, 1972). Any stretch marks remaining from previous pregnancies become darker, adding to the wrinkling of the skin to change the look and feel of the abdominal area. Other changes in a woman's sexual appearance are the result of generalized age changes throughout the body. The accumulation of subcutaneous fat in the torso, especially around the waist, creates uneven bulges which, in addition to the sagging and wrinkling of the skin on the neck, arms, and thighs, radically alters the shape of the woman's body. The woman's face also loses it resiliency and youthful appearance, due to wrinkling around the eyes and mouth, tooth loss and color changes, and drying and thinning of the skin. The hair on the head becomes grayer and thinner, and some hair may grow on the face (see chapter 2).

The appearance and functioning of the genital organs also becomes progressively altered after the menopause. On the mons veneris and around the vulva, the pubic hair thins and becomes coarser. There is a loss of subcutaneous fat and elastic tissue in the labia majora and minora, and they become thinner and wrinkled. The labia majora tend to shrink more than do the labia minora, so that the latter become more visually apparent. As in the rest of the body, there is atrophy of the dermal and epidermal layers of the skin in the vulva. In the vagina, the biochemical environment changes in response to estrogen loss, and infections are more likely to occur. The surface cells of the vaginal wall become thin, dry, pale, and smooth (Schiff & Wilson, 1978). The vagina also becomes narrower and shorter (Talbert, 1977), so that it is less capable of changing its size and shape to accommodate an inserted penis. As with the other changes in sexual appearance described so far, some of the changes in the genital area reflect alterations in surface texture and elasticity occurring in other parts of the body. Other changes reflect decreased production of estrogen, which is not compensated for by the small amount produced via the adrenal cortex.

Sexual Activity

Post menopausal changes in the genitals are of significance not only for their effects on the woman's sexual appearance but also for their effects on the woman's enjoyment of sexual intercourse. The older woman may experience some discomfort during intercourse, as the vagina and vulva are less well cushioned and adaptable in size to accommodate intromission by the penis. In addition, the rhythmic and smooth contractions of the uterus that characterize the younger woman's orgasm are irregular and possibly painful during orgasm in the older woman. A comprehensive description of the differences between pre- and postmenopausal women comes from Masters and Johnson (1966), in their classic research delineating the four phases of male and female sexual responses during sexual activity.

The description of female sexuality by Masters and Johnson (1966) was based on observations of almost 400 women ranging from 18 to 78 years old, 34 of whom were between 51 and 78 years of age. The respondents were given intensive interviews concerning their sexual behavior, their sexual responses were discussed, and they had medical examinations. Many of the respondents participated in the study over a period of several years. The unique contribution of this research to the understanding of the human sexual response was its basis in direct observation, in the laboratory, of men and women while they were having sexual activity. Both natural and mechanical sexual stimulation were used to make these observations. The female's sexual response during intercourse was observed and recorded by means of illuminated plastic penises, controlled by the respondent and electrically powered. The respondents practiced with this equipment, using fantasy and conditioning to achieve as much of a subjectively natural response as possible.

The results of comparisons of pre- and postmenopausal women in the four stages of sexual responsivity are shown in Table 7.1. In summarizing their research on the effects of aging on women's sexual responsivity, Masters and Johnson (1966) claimed that there is virtually unlimited potential for continued enjoyment of sexual relations well into old age, given that there is a willing and desirable male partner. The importance to women's sexual experiences of having such a partner in the form of a spouse in later life has been demonstrated in a systematic program of research on sexual activity in adults at Duke University. Sexual activity and interest were studied in a longitudinal sample of men and women from 60-94 years and a cross-sectional sample of men and women 46-71 years. In the longitudinal part of the research (Verwoerdt, Pfeiffer, & Wang, 1969), married women (compared to unmarried women) between the ages of 69 and 76 were found to have a consistently higher frequency of and greater interest

TABLE 7.1. Comparison of "young" (20- to 40-year-old) and "old" (50- to 78-year-old) women's sexual responsiveness reported by Masters and Johnson (1966).

Phase	Response or organ	Young women	Old women
Excitement	Time needed for lubrication of vagina after stimulation	15–30 s	1–5 min
Plateau	Vagina	Limitless capacity for expansion	Potential for expansion reduced but retained
	Uterus	Raised to allow greater expansion of vagina	Less elevation with less room provided for vagina to expand
	Minor labia	Redden as blood flow increases	No coloration change
	Major labia	Elevate and flatten against body	Hang in loose folds
	Clitoris	Elevates and flattens	No change
Orgasm	Vagina	8–12 contractions at 8 s intervals	4–5 contractions at 8 s intervals
	Uterus	3–5 contractions	1–2 contractions which might include painful spasms
Resolution	Timing of return to prearousal state	Rhythmic	Rapid

in engaging in sexual activity, and a lower proportion of continued sexual abstinence and disinterest over time. Some married women increased their sexual activity, over time, while none of the unmarried women became more active.

The cross-sectional results on a wider age sample supported the importance of marital status to women's sexual activity in middle and old age (Pfeiffer & Davis, 1974). Moreover, a large differential in a negative direction was found in the percent of sexually active women between the 46- to 50-year-olds and 51- to 55-year-old samples. In both the longitudinal and cross-sectional studies, married women who had stopped having sexual relations attributed this lack of activity to their husbands, and their husbands agreed with this attribution (Pfeiffer, Vervoerdt, & Davis, 1974; Pfeiffer, Vervoerdt, & Wang, 1969).

The reason given for the change in the women's sexuality apparent in the Duke studies in relation to their husband's lack of enthusiasm for sexual relations was the poorer health of the men, who were several years older than their wives. In general, the availability of male partners for women is limited by the fact that men have shorter life expectancies than women, so that there is a greater proportion of widows than widowers. In addition, there are fewer older unmarried men to be companions for widowed older women (Robinson, 1983).

The scarcity of available sexual outlets for older women is further complicated by an aversion to masturbation and homosexual relationships among current generations of older adults. In addition, it is unusual for a woman to become sexually involved with a younger man, so the age range of available partners is effectively restricted, increasingly so as the woman ages. A reluctance to become sexually intimate with a man outside the context of marriage may further inhibit some members of current generations of older women from seeking heterosexual outlets. A woman who maintains traditional benefits may also be reluctant to seek out a man's company if she is unattached, or to be the initiator of sexual activity with a man with whom she establishes a relationship (Croft, 1982).

Psychological Consequences of the Climacteric in Women

Because of the obvious and relatively dramatic way in which the climacteric is signaled to women by the menopause, it is virtually impossible for the woman in her 50s to be unaware of the fact that her reproductive capacity is gone. How this fact is interpreted will depend, it may be speculated, on the degree to which the woman's sense of bodily competence and self-concept is based on her ability to bear children. Her reaction to the menopause may also be influenced by the recognition that it brings of the aging of her body, and that she is moving closer toward her own death. Discomfort from symptoms associated with the climacteric, such as "hot flashes" (feelings of rapid increases in body temperature) and mood changes caused by hormonal imbalances, may make the transition a difficult one to which to adapt. On the other hand, these physical problems may be interpreted as no more bothersome or annoying than those which she experienced for all the years that she was having a monthly menstrual period. In fact, the postmenopausal woman may be relieved at knowing that her menstrual periods have come to an end, and in addition, that she no longer has to concern herself with contraceptive methods. By the time she is of menopausal age, it is doubtful that a woman would realistically desire to have more children; she may even be a grandmother. Therefore, the loss of reproductive capacity that is made clear with the menopause may not trouble her at all.

However, changes in sexual appearance that occur throughout the climacteric and into old age may have a more wide-ranging impact on a woman's self-concept. With the alterations in her face and body that accompany the aging process, a woman may believe that she is gradually losing her appeal to men, a belief likely to be exacerbated by Western society's tendency to place a great deal of value on a youthful appearance in women. Part of the feeling of loss of sexual appeal may be seen as due to the aging of her facial features and the general shape of her figure, both of which influence the image presented to the outside world of strangers and nonsexual friends, relatives, and acquaintances. Since these changes can be compensated for by cosmetics and clothing, particularly in their early stages, they would probably have only a minimal impact on a woman's self-image. In addition, a woman undergoing these sorts of changes can find ample opportunities, should she desire, to discuss and compare various solutions to the perceived need to look youthful. Friends and professional consultants, such as beauticians and salespersons, can advise her about the different styles and remedies she can use to disguise such age-related changes as graying of the hair, changes in skin color and texture, and the accumulation of fat around the hips and middle of the torso.

Other age-related changes that occur in a woman's sexual appearance, though, are ones that are not generally visible to anyone but herself and her sexual partner, such as sagging of her breasts and changes in the nipples, bulges around her waist and hips, thinning of pubic hair, and the changes in the appearance of her external genital area. Consultation with other women about these age effects is probably rarely if ever sought, because to do so would create embarrassment. Moreover, there is little that could be done anyway to disguise these changes, which are inevitably observed when the woman, unclothed, is involved in sexual relations. Short of avoiding this kind of exposure altogether, which is counterproductive and virtually impossible to arrange in the context of an intimate relationship anyway, the woman may be reassured by the fact that her partner is aging also. There may be a certain feeling of comfort and companionship between sexual partners in later life, particularly if their relationship is one that has endured for many years. Seeing each other changing may reinforce this sense of having been together for a long time and having had their relationship grow with them.

However, if age changes in the woman's vagina, vulva, and uterus occur such that intercourse becomes a painful experience, a woman may naturally become increasingly less willing to engage in sexual relations at all. Other than this perception of actual discomfort, if a woman feels that her sexual responsivity has become altered (through some of the changes described in Table 7.1), she may take this to be indicative of a loss of orgasmic capacity. A diminution of erotic desire may follow from the older woman's perceived waning sexual attractiveness, stemming from her own or her husband's overt or covert negative reactions to changes in her body. The acceptance of social stereotypes that older women do not have sexual desires (Weg, 1978) can also inhibit the woman's sexual expressiveness. Having to compete with younger women for a sexual partner following the loss of her mate may reinforce feelings of sexual unattractiveness (Butler, 1978).

The aging woman who avoids sexual relations for any of these reasons may posssibly come to view herself more and more as an asexual creature, who is not only incapable of reproduction, but also incapable of expressing her sexual feelings and needs. However, the older woman may find that her partner also is experiencing age-related changes in his timing and responsivity during sexual intercourse (see pages 119–121). These changes that occur in the aging male may result in the woman experiencing greater enjoyment from sexual activity than she was able to when she was involved with a younger man. While her responsivity slows, his may slow even more.

This complementary set of changes will allow the older adult couple to engage in a prolonged sexual encounter during which the woman may be more likely to experience an orgasm than when she was younger. At that time, intercourse may have been a more hurried affair, with the man anxious to reach orgasm and not allowing her sufficient time before then for her to do so herself. Another important factor that might result in greater freedom to enjoy sexual feelings is the release from concern over birth control and lack of interference from monthly menstrual periods that come with the menopause. At the same time, there are no longer interruptions from children and the heavy demands of work and family schedules, so that the older woman is freer to engage in and enjoy spontaneous sexual relations. Finally, since the clitoris has not been reported to undergo any involutional changes in structure or function, there is no physiological reason for a woman not to continue to derive pleasure from orgasmic release in old age.

Age Effects on the Male Reproductive System

The primary functions of the male reproductive organs are the production of sperm cells, the distribution of those sperm cells to the ova for fertilization, and the manufacturing of male sex hormones. The sperm cells mature at a continuous rate, so that fertilization of the ovum through insemination is always possible once a man has reached sexual maturity in adolescence. In addition to the lack of cyclical variation in male fertilization potential, the onset and loss of reproductive capacity lack the dramatic markers that characterize the female's reproductive period.

Changes in Reproductive Capacity

Just as women gradually lose the capacity to procreate throughout the middle years of adulthood, men also experience a climacteric. For men, the loss of reproductive capacity relates only to a reduction in the quantity of viable sperm. There are other changes in the male reproductive system, but these do not necessarily result in the inability to father children.

Sperm Production

The testes are the primary site of sperm production and male sex hormone production. Sperm production takes place within the seminiferous tubules, which are long tubes coiled within lobes inside the testes. Because of their number (about 1000) and length (1-3 ft), hundreds of millions of sperm can be produced. The sperm develop in concentric layers within the tubule, starting at the outermost section and progressing toward the middle, hollow portion through which they move on through a series of ducts out of the testes. Spermatogenesis is stimulated by FSH produced by the anterior pituitary in response to FSH-RF release by the hypothalamus.

Although the weight of the testis is not reduced by age (Harbitz, 1973), there is a reduction in the proportion of normal viable sperm by the seminiferous tubules, beginning in the 40s to 50s (Harman, 1978). The motility of the sperm in semen that has been thawed after freezing (considered to be a sensitive measure of sperm quality), after reaching a peak between 26 and 35 was reported to decrease to age 50, due to an increase in the number of sperm with coiled tails. However, there was no loss, up to this age, in the sperm count, semen volume, or percentage of normal cells (Schwartz et al., 1983). The reduction in sperm count later in adulthood appears to be related to structural changes in the tubules. The central core of the seminiferous tubules, through which sperm travel eventually to the vas deferens, has been observed to be narrower in the testes of older men, due to thickening of connective tissue around the inner circumference of the tubule. Some of the tubules become totally nonfunctional, degenerating to the point of collapse.

In response to lowered production of sperm in the testes, there is an increase in FSH produced by the anterior pituitary (Harman, 1978; Swerdloff & Heber, 1982), but the aging testis does not respond to this stimulation by producing more sperm. The climacteric for men, therefore, involves a reduction of reproductive capacity due to a decrease in the delivery of sperm from the testes. However, this is a reduction in, not a total loss, of capacity, so that paternity is possible at any age. An example of continued reproductive potential in older men frequently found in the literature is reference to a 94-year-old man who successfully fathered a child (Seymour, Duffy, & Koenen, 1935). There is very little information on humans about aging of the structures in the male reproductive system responsible for collection, storage, and delivery of the sperm from the testes to the penile urethra (the vasa efferentia, vas deferens, or ampulla). The penile urethra may be partially constricted by age-related increases in the size of the prostate gland (see next page). This constriction would interfere with urination, but it has not been reported to affect the passage of semen during ejaculation (Brandes & Garcia-Bunuel, 1978).

Production of Semen

The testes and the ducts in which sperm are delivered to the penile urethra are connected anatomically with three accessory structures that add important substances to the sperm, forming the semen. The seminal vesicles produce a thick fluid that contributes to the volume of the semen. This fluid is high in fructose, the sugar metabolized by sperm to produce the energy needed for its movement through the vagina toward the fallopian tubes. The prostate gland discharges a thin, milky, highly alkaline fluid that protects the sperm as it travels through the acid areas within the vagina. The prostate is continually producing this fluid, discharging part of it into the urine and storing the rest until ejaculation. The Cowper's glands constitute the third accessory male sex organs. During sexual stimulation, these glands secrete a clear, sticky, alkaline fluid that serves to neutralize the penile urethra from the acid environment created by the elimination of urine.

With aging, there are some structural changes in the seminal vesicles, which include development of amyloid deposits (Pitkanën, Westermack, Cornwell III, & Murdoch, 1983), a smoothening of the folds of its mucosal tissue, a thinning of the epithelial layer, and replacement of muscle fiber with connective tissue. Some changes of this nature may be observed as early as the age of 40. The fluid-retaining capacity of this gland appears to be smaller by a factor of 50% in men over 60 (Talbert, 1977). The functional significance of these changes in humans have not been reported.

Extensive data are available on the aging of the prostate gland, due to the frequency of medical problems associated with prostatic malfunction in older men (especially prostate cancer). In trying to understand the origins of these diseases, information that could lead to a better understanding of normal aging has become available. The glandular tissue in the prostate consists of ducts that terminate in sacs lined with secretory cells. The glandular tissue is surrounded by fibrous connective tissue and smooth muscle. The epithelial cells on the surface of the glandular tissue change in shape starting at about 40 years, and also start to become more irregular. There is an increasing incidence of atrophy of prostatic gland cells after 45 years, particularly in the rear section of the prostate (Talbert, 1977).

After 60 years, loss of secretory activity and deterioration of the glands has been observed to occur throughout the entire prostate. In men over 65 years, hard masses may appear in some of the glandular sacs. The presence of these masses has been attributed to stagnation of secretions that have not been eliminated through the ducts. At the same time, the connective tissue loses elasticity and contractility between the ages of 50 and 60 years as muscle fibers are increasingly replaced with collagen (Moore, 1952). The functional outcome of these structural changes in the prostate on prostatic secretion are reduced volume and pressure of semen expelled during ejaculation.

Another age-related change in the prostate is overgrowth or hypertrophy of glandular and connective tissue cells in the middle and side portions of the prostate that surround the prostatic urethra (the tubes through which the secretion is expelled). This condition is called benign prostatic hypertrophy, and it is found with increasing prevalence is progressively older men past the age of 50, rising to an estimated 50% of men 80 years and older. As noted earlier, the penile urethra may be constricted by this overgrowth of the surrounding prostatic tissue, and urinary retention may ensue. Urinary retention, in turn, may cause involuntary penile erections (Masters & Johnson, 1966). Even though benign prostatic hypertrophy is by definition not inherently a threat to health, it may cause discomfort and embarrassment. However, if urinary retention is chronic, more serious kidney problems may ensue.

It had been thought at one time that changes in testosterone production or metabolism (see the following) are causally related to benign prostatic hypertrophy. However, currently there is not agreement regarding whether any association exists between hormone levels and the condition of the prostate or if there is, how this relationship would be affected by aging (Brandes & Garcia-Bunuel, 1978; Harman, 1978).

Structural changes in the Cowper's glands with age have not been documented. The only information of the functional status of the Cowper's glands in older men comes from Masters and Johnson (1966), in their description of the effects of aging on the sexual response cycle. According to them, less or no fluid is secreted during the period of excitement from the Cowper's glands.

Testosterone Production

Of the male sex hormones, or androgens, the one primarily responsible for male sexual characteristics is testosterone. This hormone is produced by the Leydig cells, which lie between the seminiferous tubules within the testes. Testosterone production is anatomically and functionally separate from sperm production; that is, the secretion of testosterone is not dependent on the status of the sperm. The manufacturing of the sperm and male sex hormones both occur in response to stimulation by FSH and LH, the same anterior pituitary hormones that stimulate estrogen and progesterone production by the female's ovarian follicles and corpora lutea. The pituitary hormone FSH stimulates the development of sperm cells in the seminiferous tubules, and LH stimulates the production of testosterone by the Leydig cells. As in the female, these pituitary hormones are stimulated by releasing factors produced by the hypothalamus. Male pituitary hormones also operate in a feedback cycle with secretion of hormones by the Leydig cells of the testes. LH is produced in the anterior pituitary whenever testosterone levels in the blood are low. Testosterone has numerous effects on the development and maintenance of male sexual characteristics. Its influence is first apparent at puberty, when the anterior pituitary begins to release FSH and LH. Maintenance of adequate testosterone levels is needed for the continued physical expressions of masculinity. The size of the penis, scrotum, prostate, seminal vesicles, and epididymis depends on the production of testosterone. The accessory glands also require testosterone for their secretory activities. The surface corrugations and dark color of the scrotum require adequate testosterone levels in the blood. Many sex differences in bodily appearance are a function of the higher testosterone levels in men. These differences include the heavier facial and body hair of men, as well as pattern baldness in those men who have a genetic predisposition to develop it. Because it promotes muscle growth, testosterone is also responsible for the greater physical strength of men compared to women. Bone growth is also stimulated by testosterone. The deeper voice men have is due to the growth-inducing effect of testosterone on the larynx. While hormone levels in women have no clear-cut uniform relationship to variations in behavior or amount of desire to engage in sexual activity (Rossi & Rossi, 1980), it does appear that in men testosterone is linked to aggressiveness, to intensity of sexual desire (Davidson et al., 1983), and possibly hostility (Persky, Smith, & Basu, 1971).

Of great interest to researchers who study the aging male's reproductive system is the question of whether the reduction in virility associated with the effect of the climacteric on the seminiferous tubules extends to the production of testosterone by the Leydig cells of the testes. Several major sources state the conclusion that the male climacteric does involve a primary dysfunction of the Leydig cells, so that they are less capable of producing testosterone (Davidson et al., 1983; Harman, 1978; Swerdloff & Heber, 1982; Talbert, 1977; Vermeulen, 1976). In response to the lowered testosterone circulating in the blood more LH, secreted by the anterior pituitary, is present in the blood of these older men. However, either because not enough LH is secreted, or because the Leydig cells cannot respond to the increased LH, the level of testosterone in the blood (free or bound to plasma proteins) does not show a concomitant increase. This situation is analogous to that found in postmenopausal women, whose increased FSH in the blood is caused by lowered estrogen, but whose ovaries are incapable of producing estorgen in response.

In contrast to the universally documented effects of aging on the female's reproductive hormones, though, there is a growing body of evidence that forms a basis for doubting the actual effect of aging alone on testosterone production in men. Many of the frequently cited studies documenting age-related losses in testosterone involved comparing aged men living in old-age institutions or veterans hospitals, having surgery, or having chronic diseases, with healthy younger men living on their own in the community (e.g., Davidson et al., 1983; Hollander & Hollander, 1958; Pirke & Doerr, 1975; Stearns et al., 1974). When the sample of older men is specifically selected according to the criteria of living in their own homes and having no major health problems, comparisons with younger men reveal no age differences in testosterone levels (Harman & Tsitouras, 1980; Kent & Acone, 1966; Murono, Nankin, Lin, & Osterman, 1982; Sparrow, Bosse, & Rowe, 1980), although some degree of loss of Leydig cell reserve was observed in one study of healthy older men (Harman & Tsitouras, 1980).

In accounting for the disparity between results from healthy versus institutionalized older men, it might be reasoned that since the older men living in institutions almost certainly have little outlet for sexual expression, they experience atrophic changes within the Leydig cells, which decreases their testosterone levels. This possibility is supported by the finding of one study that sexual activity in men over 60 years was positively related to the amount of testosterone in the blood (Tsitouras, Martin, & Harman, 1982). Conversely, it is possible that older men living in institutions have lowered testosterone as a result of poor health. For instance, disorders of the cardiovascular system can reduce the blood supply to the testes, so that the Leydig cells do not receive adequate nourishment to perform their function (Harman, 1978; Swerdloff & Heber, 1982). Even those authors who describe age losses in testosterone levels acknowledge the wide individual variations that exist across samples of adult men, with overlap between young and old groups frequently occurring (Vermeulen, 1976), and few older men showing clinically subnormal levels (Davidson et al., 1983). It may be tentatively concluded, then, that lowered testosterone is probably not a function of the aging process in men, but is instead related to the occurrence of diseases (and lowered sexual activity) which become more prevalent in old age.

Changes in Male Sexual Functioning

Changes in Penile Erectility

Penile erection occurs as the result of stimulation by the parasympathetic nervous system during sexual arousal and during sleep, which causes increased blood flow through the arteries in the penis. Although no changes in the outward appearance of the penis in its nonerect state have been reported, there are numerous internal changes that could affect its erectility. There is a growth of connective tissue in the corpus spongiosum (the spongy tissue cavity on the underside of the penis), which may begin in the 30s or 40. The veins and arteries in the corpus spongiosum and corpora cavernosa (spongy cavities along the sides of the penis) have been reported to be rigidified in men by the age of 55-60 (Talbert, 1977). Consequently, the penis would be less easily made erect upon stimulation by the flow of blood through the arteries connecting it with the pelvis, and the constriction of blood through the veins.

Age differences in penile erectility (Solnick & Birren, 1977) and sensitivity (Edwards & Husted, 1976) appearing by the 40s may be functional consequences of these age effects on the penile erectile, vascular, and connective tissue. There is also a reduction in the number of episodes of penile erections during sleep after the age of 60 years (Karacan, Williams, Thornby, & Salis, 1975; see also chapter 6).

Sexual Activity

The sexual response cycle was studied in men over 50 by Masters and Johnson (1966), who had 39 men ranging from 50 to 89 years in their sample. In general, a slowing down was reported of the progression through the four sexual response phases. It was also found that older men perceived less of a demand to ejaculate during a single sexual encounter and across a series of sexual encounters. The results of their comparison between younger and older men in the sexual response are summarized in Table 7.2. Research on the effects of aging on male sexuality has yielded varying results about changes in older adult men in sexual interest and activity. The earliest cross-sectional drop was found in the Duke study to occur between the 46- to 50- and 51- to 55-year-old age groups (Pfeiffer et al., 1974). Davidson et al. (1983), in their cross-sectional study of hormone levels and sexual attitudes in Veterans Administration outpatient clinics, observed decreases after the age of 40 (the youngest group in the sample) in the frequency of sexual activity, but not drive or enjoyment. Only a small percent of the variance in sexual activity was accounted for by the decrease found in this study in testosterone levels, and none of the variance in sexual enjoyment. In another cross-sectional study, though, no age differences were reported until after the age of 65 in sexual interest and activity (Martin, 1977).

The longitudinal component of the Duke study revealed many variations in individual patterns of change in the sexual behavior of older men over time. The few unmarried men in the sample were more sexually active and interested than their married counterparts, and showed patterns of increases over the course of the study (Verwoerdt et al., 1969). Other variations were found to exist due to health and social class. For both sexes, but particularly for men, previous sexual behavior was found to be predictive of sexual activity in later life (Pfeiffer & Davis, 1974), suggesting that there is continuity throughout adulthood of patterns of sexual expression for men.

Psychological Consequences of the Climacteric in Men

The major component of the reproductive loss which constitutes the climacteric in men appears to be a function of structural changes in the sperm-producing tissue in the testes. Testosterone production, and its effects on the male sex organs and sexuality, would appear to be stable in adulthood in healthy men, particularly in those who remain sexually active. The reduction in sperm number in and of itself is not likely to have appreciable effects on the older man's sexual self-image. By the time a man is in his 60s and 70s, he probably is not desirous of fathering children. He may not even be aware of any loss in sperm number, since there is no way that he can know how much sperm he is producing without having a sperm count performed on his ejaculate, a test usually taken only if a man is concerned about not being able to father children. Changes in his fertility would not be evident to him as a result of sexual relations with his female partner if she herself is postreproductive. Should a man be interested in becoming a father late in life with a younger woman as his mate, it would still be possible for him to do so, although perhaps conception is not as readily achieved as in previous years.

Unlike the reduced sperm output in postclimacteric men, reduced semen production is a detectable feature of the man's sexual experience. If he holds the inaccurate but common belief that connects the volume of semen ejaculated with fertility, he may be made aware (but for the wrong reason) of his reduced potency. Even if he is not concerned about the volume of seminal fluid for this reason, he may feel a slightly diminished sense of enjoyment during the moment

TABLE 7.2. Comparison of "young" (20- to 40-year-old) and "old" (50- to 89-year-old) men's sexual responsiveness reported by Masters and Johnson (1966).

Phase	Response or organ	Young men	Old men
Excitement	Time needed for penile erection after stimulation	3–5 s	10–15 s to several min
	Degree of erection	Full	Reduced until close to ejaculation
	Elevation of testes	Occurs in late phase	Little or no elevation of testes
	Blood flow to testes and scrotum	Increase	No increase
	Cowper's gland	Secretes fluid	Less or no fluid secreted
Plateau	Increase of sexual tension	Rapid development of pressure for ejaculation	Can be prolonged, often indefinitely, before pressure for ejaculation felt
	Glans at coronal ridge	Increase in circumference just before ejaculation	Increase in circumference just before ejaculation
Orgasm First stage	Subjective experience of inevitable ejaculation	2–4 s	1–2 s, not at all, or for 5–7 s with irregular prostatic contractions
	Prostate	Begins contracting regularly at .8-s intervals	1–2 contractions, may be irregular
Second stage	Penile urethra	3–4 contractions at .8-s intervals, followed by contractions at longer intervals	1–2 contractions at .8-s intervals
	Expulsive force of seminal fluid measured in inches from meatus	12–24 in. distance	3–5 in. distance
	Volume of semen after 24–37 h. of continuence	3–5 ml	2–3 ml
Resolution	Return to full erection possible	Several minutes	Several hours
	Return to prearousal state	Two stages, lasting minutes to hours	One stage, frequently in several seconds

of ejaculation, due to the fact that less fluid is expelled. The force and length of prostatic muscular contractions prior to ejaculation is related to degree of pleasure felt during orgasm. Fewer and less intense contractions in the prostate of the post-climacteric man would, therefore, reduce the subjective quality of sexual enjoyment. Moreover, muscles in the prostate do not contract as regularly and smoothly during ejaculation in older men (Masters & Johnson, 1966). However, it should be pointed out that alterations in the glandular and connective cells of the prostate associated with the climacteric vary across men in their rate of progress and degree of severity. Moreover, some men may be more able to adapt to these physical changes, which in other men, would greatly restrict their involvement in sexual relations. Whether the man's sense of sexual competence is reduced by whatever changes occur in prostatic functioning also may depend on whether he perceives changes in sexuality as a signal of impending deterioration in other areas of functioning. For some men, feelings of health and vigor may be strongly linked to their perceived sexual potency.

Masters and Johnson described in some detail the impact of altered physiological responses on the older man's enjoyment of sex. According to them, the satisfaction an older man derives from sexual intercourse in the postclimacteric years depends in part on how fully he adheres to common myths about male sexuality. Belief in these myths can detract from the quality of sexual experiences because of the interference it creates in the ability to achieve an erection. Erectile failure in turn becomes an experience that further lowers the man's confidence in his sexual capacity and in turn creates more experiences of failure. The first such myth is the idea that men lose their sexual vigor after the age of 50. Related to this myth is the notion that it is inappropriate for older men to have an interest in sexual activity (Boyarsky, 1976; Butler, 1978; Weg, 1978); hence the derision implied in jokes about "lecherous" old men. Another myth is that sexual activity is dangerous because it increases the risk of cardiac failure. However, this rarely occurs during conjugal intercourse, and even men who have recuperated from myocardial infarction may engage in sexual activity without undue risk (Boyarsky, 1976).

The myth of sexual inadequacy in older men underlies concern about sluggishness of penile erection, so that any unaccountable delay is interpreted by the man as evidence that he is old and impotent. As can be seen from Table 7.2, such a delay is likely to occur during the excitement phase. Even if penile erection is slowed by only a couple of minutes, to a man who is accustomed to a more rapid response, the delay can cause panic the first few times it occurs. Masters and Johnson pointed out that if the man or his partner tries to force it, the erection will be less likely to occur than if the situation is approached with acceptance. If an erection begins but is less than complete, the man may feel as much concern as if he failed to achieve any erection at all. In this case, Masters and Johnson suggested that insertion of the semierect penis into the vagina could bring on a full erection. Again, though, if the man is overly concerned about his potency, his erection will be lost entirely before this reparative action can be taken.

The reduction of orgasm from two stages to one that the man over 50 begins to experience need not reduce the subjective quality of sexual relations. Masters and Johnson also noted that a reduced volume and pressure of ejaculation is not necessarily detrimental to pleasure. Similarly, the lessened demand to ejaculate during intercourse is not in and of itself a deterrent to sexual enjoyment. This lessened demand may even be a benefit, since it allows the older man to continue at plateau levels of sexual excitement virtually indefinitely, thereby allowing his female partner the time she may need to achieve her own optimal number of orgasms. The older man's erection, although initially slower, persists once it has been achieved. If he does not ejaculate, moreover, the man's time to achieve another erection after his one sexual encounter is greatly reduced compared to the time of a true refractory period (after ejaculation). None of these features of the older man's sexual experiences will be perceived favorably, however, if the man defines himself as sexually potent only as long as he maintains the sexual patterns of his youth. For this type of man, any change in sexual functioning will be seen as a signal of impotence.

Other changes in the older man's physique and secondary sex characteristics related to any diminution in testosterone level may also influence his feelings about his sexual competence. Belief in the widely held myths that men with hairy chests and large penises (Verinis & Roll, 1970) and muscular physiques (Montemayor, 1978) are more sexually potent and masculine may be damaging to a man whose reduced testosterone level has changed any of these features.

In addition to fears about their loss of sexual prowess, Masters and Johnson (1966, 1970) cited the negative effects on the man's sexuality after the age of 50 of career pressures, poor health, overindulgence in alcohol, and boredom with his marital partner. These are all factors that can, in Masters and Johnson's terms, make the older man less "interested and interesting" to his wife. A cyclical process may set in, as the man loses interest in sex, his wife in turn becomes less responsive, and the man consequently becomes even less likely to engage in intercourse. Cultural factors may play a contributing role for current generations of older persons who were raised with the belief that the man is responsible for initiating sexual activity, even between married partners. Lack of sexual stimulation of the husband by his wife's changed appearance as she herself ages may reflect his incorporation of socially based stereotypes about the lack of attractiveness and sexuality of the older woman. His lack of interest in his aging wife may have an inhibitory effect on her, so that she herself becomes less likely to try to stimulate him sexually. In addition, age differences between spouses may create real differentials in health status, with the man's ability to engage in sex reduced due to health problems even while his interest in sexual relations remains constant (Pfeiffer et al., 1974).

If the aging man enjoys sexual activity (and has no serious health problems), and values his emotional relationship with his partner, there may be little change in the frequency with which he engages in intercourse, even well into old age. Changes with age in his partner's appearance may be overlooked in favor of the psychological significance she has to him as the woman with whom he has shared his life and raised his family. His long-term partner may become more sexually responsive after the menopause due to freedom from the need for contraception, or a sense of liberation from child-care obligations. In either case, a relaxed attitude toward the changes in his sexual functioning may be easier to maintain if he regularly participates in sexual relations and can consistently demonstrate his potency as a sexual partner.

In the face of loss of a long-term partner, an older man may seek new companionship if he had grown accustomed to frequent sexual activity. A man may seek a younger companion to try to renew his sexual vigor, requiring for stimulation the visual and tactile cues of the sexual partners who remind him of those he encountered in his youth. Any new relationships, with younger or same aged partners, may lead to an increase in sexual activity due to the romantic feelings that stimulate sexual arousal.

From this discussion of the effects of age on male sexuality, it may be concluded that changes in the structure and functioning of the man's reproductive organs associated with the climacteric are important but not exclusive influences. The physical changes take place within the context of a man's previous sexual experiences, his health, and his beliefs in myths about male sexuality and cultural stereotypes about female attractiveness. For both the man and the woman, a continuation of or a rise in sexual activity in old age can constitute an important contributor to feelings of competence and a positive sexual self-image that can in itself help to offset changes with age that potentially could detract from the quality of their sexual experiences. Participation in sexual activity could also serve to reinforce feelings of mutuality, sharing, and joint concerns over each other's well-being that have developed in the context of an enduring intimate relationship (Weg, 1983).

CHAPTER 8

Central Nervous System

Complex actions, thoughts, and emotions are made possible by the actions of the central nervous system, which coordinates all the incoming sensory information from inside and outside the body, and prepares and instructs the effectors throughout the body to make a coordinated response. The effect of aging on this system apparently underlies many of the age differences observed in cognitive functioning, including memory, learning, problem solving, information processing, and intelligence.

It is beyond the scope of this book to cover these areas from a behavioral perspective, which is adequately treated in many reviews of the cognitive psychology of aging (e.g., Botwinick, 1978; Kausler, 1982; Salthouse, 1982). This chapter will deal with the neurological basis for these age effects on intellectual behavior: the changes at the cellular and structural levels that could serve as important influences on the ability to learn new information and remember that which has been learned, and to apply both types of information to solving novel problems. In subsequent chapters, the physiological basis for age effects on perception will be examined, again focusing on the neurological substrate of behavioral and cognitive changes, with only minimal attention to these processes except insofar as they have implications for centrally based neural mechanisms.

Issues in Conducting Research
on Aging of the Brain

Research on aging of the central nervous system has been directed at determining the effects of the aging process on the individual components of nervous tissue as well as the impact of aging on the major neural structures. Underlying this research is the attempt to identify the basic cause of aging in the nervous system. There are many practical problems, though, in this research, which often require compromises between the type of data needed by investigators and that which is available.

The major methodological problem is that it is impossible to study the brain in a living, healthy person whose nervous system is functioning normally. The majority of researchers have had to rely upon postmortem samples of brain tissue in order to perform analyses of, for instance, the effect of aging on the number of neurons in a given area of the brain. Although samples of tissue obtained from brain biopsies have occasionally been used in studying the aging process, they are not particularly representative of persons with normal, healthy brains. Moreover, this procedure is obviously so invasive that biopsies cannot be routinely conducted for research purposes. As a consequence, the number of available subjects is, realistically, limited to those who give permission for an autopsy to be performed on them after they die (or whose families do so after death). In addition to the limitation this requirement places upon the number of available subjects, the use of brain tissue obtained at autopsy creates other technical problems. Individuals do not die if they are "normal" and even if death is due to a nonneurological trauma or disease, the brain is often altered by the general upheaval to the body's systems, or the deterioration of the organ system whose failure precipitated death (Buell, S. personal communication, 1983). Moreover, some aspects of brain functioning are very unstable after death. The autopsy must be performed within a few hours in order to preserve nervous tissue, and even this interval may be too lengthy for some functions.

Because of the difficulty of doing research on living, healthy adults, researchers interested in brain structure and function must often study laboratory animals who can be sacrificed shortly after they have been observed. This method is especially useful for studying the effects of particular kinds of experience on the brain. Some simple organisms have nervous systems that are of interest because they illustrate some fairly universal feature of neural functioning. These organisms are ideal, for instance, for studying the neurotransmitters, some of which are degraded so quickly after the organism dies that the brain must be dissected almost immediately after death in order to ascertain the level of the chemical substance that was present. However, the question inevitably arises of how much generality to humans is permissible from research on nonhuman species (John & Schwartz, 1978). Added to this issue is the question of how old is "old" for various species in relation to the age of humans.

Some of these problems apply to the other areas of physiological functioning and so are not restricted to studying the aging nervous system. However, research on the nervous system is unique in that the brain tissue is extremely sensitive to experience. True longitudinal studies would be particular desirable, especially if these could be combined with behavioral observations. By virtue of the dependence of research on humans on autopsied brains, though, it is only possible to conduct cross-sectional studies. Usually these studies are conducted by researchers who do not have access to information on what their subjects were like when they were alive. Comparative research with nonhuman subjects allows for brain-behavior correlations to be made, but their validity for humans is always a matter of careful and final verification.

The Fallout Versus Plasticity Models of the Aging Nervous System

Describing the effects of age on the central nervous system is an awesome task, since the effects of experience and intrinsic physical changes add many layers of complexity to a system that is in itself enormously complex. Yet, as in all aspects of physiological functioning in adulthood, there is no point of stasis at which it may be said that aging is not occurring. In the case of the nervous system, the effects of aging have qualities found in few other bodily tissues. When a neuron dies, it cannot be replaced by another, since no new neurons are generally formed once the nervous system has been established in the perinatal stage of development. As a result, once a neuron dies, all the synaptic connections formed by that neuron are lost, possibly along with the behaviors based on those synapses, depending on their redundancy with other synapses in the central nervous system.

Neuronal Fallout Model

The loss of neurons due to death without replacement forms the basis of the "neuronal fallout" (Hanley, 1974) model. A major question of interest that has guided researchers whose work is based on this model is how the rate and extent of neuronal fallout varies across different brain regions. Following from this question is the functional question of how much neuronal fallout has to occur and where it occurs before its effects appear in behavior. There is a psychological version of the neuronal fallout model, on the basis of which researchers have sought to demonstrate the extent and time of onset of loss of a variety of memory and cognitive abilities.

Those who conduct research based on the assumptions of the neuronal fallout model recognize that since there is a great deal of redundancy in the cerebral cortex, with the same information being represented in multiple sites, loss of synapses in only one or two areas would not result in loss of that specific information, since it is retained intact elsewhere. Nevertheless, it is assumed that progressive neuron loss would ultimately reduce the reserve capacity represented by the alternate sites of information storage, and it is beyond this threshold that neuron fallout would have deleterious functional consequences (Henderson, Tomlinson, & Gibson, 1980; Ordy, 1975; Scheibel, 1979). The effect of neuronal fallout would be expected to be more severe when there is less reserve capacity to begin with, in parts of the nervous system where there is little redundancy (Finch, 1982).

A variant of the neuronal fallout model includes discussion of what happens as the result of the loss of "pacemaker" neurons (Brizzee, 1975), which control many other neurons. This loss can be especially damaging if it causes a cascade of degeneration, whereby neurons die when the neurons which are innervated by them die (Greenough & Green, 1981). In such an intricate network as the nervous system, the loss of a few critical neurons, or many redundant pathways, may thereby change the pattern of communication throughout the system so that the harmful effects spread far beyond the damaged areas (Ordy, 1981), as when the foveal cones are lost in the retina (Ordy, Brizzee, & Johnson, 1982).

Plasticity Model

Although its rate and extent may vary, the loss of neurons described by the neuronal fallout model suggests that decline of sensory, motor, and integrative functions served by the central nervous system is an inevitable consequence of the aging process. An alternate interpretation is suggested by the plasticity model. Plasticity is the growth of new structures and functions in response to unusual and unexpected challenges in certain kinds of neurons. Of course, much of the human nervous system is composed of "nonplastic," genetically programmed circuits, such as those reponsible for spinal reflexes. The connections involving neurons with highly specialized functions are also fixed because these neurons do not form ready substitutes for neurons serving other functions. However, there are millions of neurons that do have the "plastic" quality of being able to grow new axons or longer dendrites when other neurons are damaged through trauma or degeneration.

According to the plasticity model, the neurons that do not die take over the functions lost by degenerated synapses, establishing new synaptic connections through elaboration of their dendritic trees (Curcio, Buell, & Coleman, 1982). Plasticity, it is argued, may occur not only as compensation for neuronal fallout, but also due to stimulation from the environment to process new information (Greenough & Green, 1981). There is even the suggestion that as a reuslt of the proliferation of dendrites, there is a net gain in the density of synapses with increasing age in adulthood (Brody, 1982). The neurologically based plasticity model corresponds to psychological models of intellectual development in old age, according to which intellectual decline is not inevitable, and new learning may occur as the result of training or experience (e.g., Baltes & Schaie, 1976).

The proposition that plasticity may compensate for or even overtake the rate of neuronal fallout is counter to decades of literature produced by researchers who almost exclusively sought to, and usually did, demonstrate declines in neuron numbers throughout the central nervous system (Hanley, 1974). In this research, demonstration of plasticity would have been precluded by virtue of the methods used, in which cell bodies but not dendrites were the focus of measurement. While these studies contain important information about cell loss during aging, their a priori lack of applicability to a plasticity model should be taken into account in interpreting them.

Another complication in comparing the neuronal fallout and plasticity models of neural aging is derived from the increased prevalence of certain kinds of neurological and cardiovascular disorders among persons who survive into their advanced years. Studies purportedly dealing with normal aging may in fact represent documentation of the damaging effects on brain tissue of these diseases.

One such disease is senile dementia, a neurological disorder that causes neurons to undergo structural deterioration and death (Tomlinson & Henderson, 1976). It has been suggested not only that neuronal fallout is exaggerated in the presence of senile dementia (e.g., Ball, 1977), but also that the degree of plasticity in surviving neurons is vastly reduced (Buell & Coleman, 1979).

Failure to separate demented from normal, healthy elderly people is often found in the research on the effects of aging on the central nervous system. One example is a report on the "progressive" changes in the "aging" nervous system in which six out of the nine cases examined at autopsy were diagnosed as having had senile dementia (Scheibel, Lindsay, Tomiyasu, & Scheibel, 1976). Possibly erroneous interpretations from research such as this overestimate the degree of neuronal fallout that occurs as the result of the normal aging process.

Hypothesized Reasons for Neuron Loss

The study of Buell and Coleman (1979) mentioned in the previous section is one that has had a far-reaching impact on the understanding of aging in the nervous system. This study was the first, and so far the only one, in which plasticity in the nervous system of aged humans was demonstrated. Other research on the plasticity model has been conducted on rats exposed to differing amounts of environmental complexity (e.g., Green, Greenough, & Schlumpf, 1983). The other important features of Buell and Coleman's work was that it pointed to the coexistence of neurons that were dying along with those showing plasticity. The dying neurons were present in the brains of all adults studied, whose ages ranged from 44 to 92 years. While affirming the possibility of plasticity in the aged human nervous system, this study confirms the need, implied by the neuronal fallout model, to understand why neurons die.

Lipofuscin Hypothesis

One explanation of neuron death is based on the frequently cited observation in research on aging that neurons accumulate lipofuscin, a yellowish pigment, within their cell bodies, and this accumulation over the course of adulthood is what causes them to malfunction and die. It has been suggested that lipofuscin impairs the normal activity of neurons by interfering with protein metabolism (Mann & Yates, 1974). The presence of lipofuscin has been associated with learning deficiencies in rats and nonhuman primates (Ordy, 1981). However, this suggestion that lipofuscin is harmful is not universally accepted, and it has been counterargued that lipofuscin has benign (Bondareff, 1981) and possibly even beneficial effects (Davies & Fotheringham, 1981).

Evidence that strongly mitigates against the damaging effects of lipofuscin has come from observations of the inferior olive, a structure in the brain stem involved in motor coordination. Neurons in the inferior olive contain large quantities of lipofuscin, which begins to accumulate very early in life. If lipofuscin causes neurons to die, the inferior olive cells are ones that should be particularly vulnerable to loss over time. The crucial evidence on cell numbers in the inferior olive over the adult years does not support this hypothesis, however, since no decrease is observed in either total cell counts (Moatamed, 1966; Monagle & Brody, 1974) or number of neurons per unit volume (Sandoz & Meier-Ruge, 1977). Moreover, the neurons in the inferior olive seem to retain their ability to respond effectively to stimulation, as demonstrated in a study on live rats (Rogers, Silver, Shoemaker, & Bloom, 1980).

Even if lipofuscin accumulation was circumstantially linked to neuron loss, though, it would still not be possible to implicate this process as a primary cause of neuron death. It may be that lipofuscin collects in the cell body of the neuron as a by-product of other intracellular degenerative processes that themselves are what cause neurons to die.

Circulatory Deficiency Hypothesis

A second hypothesized cause of neuron loss is deprivation of cerebral blood supply through circulatory defects associated with aging of the cardiovascular system. This explanation places the primary locus of neuronal aging outside the nervous system, and onto factors that affect the adequacy of the heart and arteries to distribute blood to the neurons in the brain. A major concern of research in this area is to delineate the distinction between whatever normal age changes occur in the circulatory system from diseases such as atherosclerosis and coronary heart disease, which reduce the efficiency of blood delivery to the neurons in the central nevous system. However, because of the high incidence of cardiovascular disorders in the age, this distinction becomes very difficult to make in a practical sense.

The presence of circulatory defects that can affect the nervous system is inferred from a reduction in cerebral blood flow, the rate of blood circulation through the brain. If the cerebral blood flow is reduced, neurons would die because their requirements for oxygen and glucose were not being met. Several studies using radioactive tracer techniques have in fact established that the cerebral blood flow of healthy older persons is reduced compared to that of their younger counterparts (Frackowiak, Jones, Lenzi, & Heather, 1980; Frackowiak, Lenzi, Jones, & Heather, 1980; Lavy, Melamed, Bentin, Cooper, & Rinot, 1978; Meyer et al., 1976; Möller & Wolschendorf, 1978; Thomas et al., 1979; Wang & Busse, 1975). This evidence would support the interpretation that a significant factor in neuron death is a reduction of blood supply to nervous tissue resulting from the aging of the cardiovascular system. However, this interpretation is challenged by recognition of the fact that the blood supply to the brain, as to other organs, is regulated by the cellular demands of neurons for oxygen and glucose.

Thus, an alternative interpretation of the cerebral blood flow data is that a decrease in this variable reflects an adjustment made by the microcirculation of the brain to the diminished metabolic needs of a neuron population reduced in quantity by primary aging of the nervous system. This is the interpretation made on the basis of a study in which brain volume (as determined by computed tomography) was correlated with cerebral blood flow. Both measures were found to decrease across adult age groups, but they were not significantly correlated in samples younger than 70 years of age. It was concluded that prior to this age, neuron degeneration had not proceeded to a significant enough degree to be reflected in lower cerebral blood flow. In the advanced age groups, there was sufficient neural loss to be reflected in diminished circulatory demands (Yamaguchi et al., 1983).

More evidence to support primary neural aging as the cause of decreased cerebral blood flow is cited in studies on the rate of oxygen uptake by nervous tissue, a measure of how much oxygen is being used by the neurons. The decrease with age observed in this variable signifies that there is less metabolizing tissue in the brains of older persons (Dekoninck, Jacquy, Jouqet, & Noel, 1976; Frackowiak, Jones, Lenzi, & Heather, 1980; Frackowiak et al., Heather, 1980). A similar decrease across age group of adults has been observed for a measure of neuronal glucose uptake (Dekoninck et al., 1976). Still, it could be argued that neither the correlational findings on brain volume nor the data on oxygen or glucose uptake provides definitive rejection of the cardiovascular hypothesis, since a causal sequence has not yet been established. What remains to be shown is that there is a temporal progression of neuron loss followed by a reduction of cerebral blood flow.

Although the mechanism through which it would have its effect, circulatory or neural, has not been established, there is evidence that the aged can enhance the blood supply to the brain and hence psychomotor performance through exercise activity (Clarkson, 1978; Spirduso, 1980). A mechanism through which this enhancement could occur might be a growth of the capillary network supplying cortical neurons (Hunziker, Abdel'Al, & Schulz, 1979). The finding that psychomotor activity, which is regarded as a universally declining function in the aged, can be improved by exercise is a behavioral demonstration of plasticity in the nervous system, and logical extension of the conclusions regarding exercise training as having the potential to compensate for age-related losses in the cardiovascular, muscular, and respiratory systems (see chapters 2, 3, and 4).

Metabolic Changes and the Cascade Hypothesis

Other hypotheses concerning the source of neuron loss are less thoroughly investigated. One is that there are metabolic disturbances such as decreased protein kinase activity, which would indicate diminished nerve cell responses to neurotransmitter or neurohormonal activity, and a reduction in carbonic anhydrase activity, which would signify impaired glycolytic turnover (Meier-Ruge, Hunziker, Iwangoff, Reichlmeier, & Schultz, 1980). Another hypothesis is that protein synthesis within the nucleus of the neuron is increasingly subject to error over time (Lynch & Gerling, 1981). Since there is no turnover of neurons, these errors are more likely to accumulate over the lifespan, as contrasted with cells such as the epithelial cells of the intestinal villi, which are replaced every few days. Moreover, since neurons do not replicate themselves, these errors would not be erased when new neurons take over for old, error-prone ones, as occurs in cells that replace themselves by division.

The "cascade effect" hypothesis, referred to earlier in terms of effects of neuron loss, is that a deafferation process in the aging brain causes a progressive sequence in which localized brain damage spreads through loss of synaptic connections to have widespread effects on many neurons throughout the brain (Greenough & Green, 1981). Minor insults to afferent and efferent pathways of the brain, which can accumulate over time, could thereby come to have progressively damaging effects on the structural integrity of the nervous system as a whole.

Evaluation of Neuron Loss Hypotheses

At present, there is no sound basis for ruling out any of the possible causes of neuron loss which have been proposed in the literature on aging of the central nervous system. Each one describes an age-related process, or at least cross-sectional differences, and while no one process can yet be said to be the ultimate cause of aging, each seems to be a potentially important one that could contribute to aging of the neuron.

The Aging Neuron

The neuron is the basic unit of structure and function within the nervous system that receives and transmits information, and so most research is based on the assumption that any effect of aging on the nervous system would have to be manifest at this level. Not all neurons degenerate and die as a function of time or aging. Those that do will be referred to here as "aging neurons." Aging neurons are ones that have undergone various debilitative changes in their structure which could eventually lead to their death. Some of the abnormalities in the aging neurons are the same as those that characterize neurons diseased by senile dementia, but the abnormalities are observed with less frequency in the brains of persons who do not contract that illness. As in many other areas of physiological aging, it is difficult to place into completely separate categories changes caused by diseases that are more prevalent in the aged and changes due only or primarily to aging.

The Cell Body

The cell body is the site where processes that support cellular activity take place. The size and shape of the cell body of cortical neurons does not seem to be altered, at least up until the mid-70s (Schulz & Hunziker, 1980). However, various abnormalities within the cell body have been observed with increasing frequency in the brains of older persons. The aging neuron has been seen to contain distortions in the surface of the membrane surrounding the nucleus, deterioration of the Golgi complex, and a reduction in the amount of substance contained in Nissl bodies (Brizzee, Klara, & Johnson, 1975; Brody & Vijayashankar, 1977). The accumulation of lipofuscin pigment in the cell body constitutes another feature of the aging neuron, described earlier as a hypothesis for neuronal fallout.

Other changes in the cell body have been studied in research comparing senile dementia and normal aging (Tomlinson, Blessed, & Roth, 1968; Tomlinson & Henderson, 1976). In a process called granulovascuolar degeneration, neurons in the hippocampus develop spaces, or vacuoles, in their intracellular fluid outside the nucleus. These vacuoles contain small granules of unknown matter. Also in the hippocampus, neurofibrillary tangles develop, in which small fibrils in the cell body and dendrite increase tremendously in number, pushing the nucleus to one side. At the same time, the neurofilaments twist together in pairs. It is thought possible that neurofibrillary tangles obstruct the flow of cellular fluids within the cell body of the aging neuron.

Dendrites

Dendrites are the extensions of the cell bodies that receive information from other neurons. They often form elaborate and intricate patterns, called aborizations, and may be covered with dendritic spines. The more extensive this dendritic network, the larger the receptive surface area of the neuron.

A dramatic pattern of loss of dendritic elaboration was reported to occur in a variety of neurons as a function of age in a series of investigations by Scheibel and associates (Scheibel, 1979; Scheibel, 1982; Scheibel, Lindsay, Tomiyasu, & Scheibel, 1975, 1976; Scheibel & Scheibel, 1975; Scheibel, Tomiyasu, & Scheibel, 1977). This pattern was described in qualitative fashion, based on the appearance of dendrites stained by the rapid Golgi method from the brains of older persons, but not on actual quantitative measurements. These observations led Scheibel and associates to conclude that dendrites are lost progressively from the aging neuron, beginning in the outermost sections of the dendritic tree and eventually spreading throughout the entire tree. In the final stage of deterioration, the neuron is reduced to a stump with no dendrites at all. An example of a set of neurons in varying degrees of degeneration is given in Figure 8.1.

In interpreting the observations from the Scheibel et al. studies about the aging neuron, Buell and Coleman's findings should be recalled. The description by Scheibel and associates is of the aging, that is, dying neuron, but not necessarily all or even the majority of neurons in the aging nervous system. Buell and Coleman (1979) had reported the existence of dying neurons in all tissue samples they examined. The net increase in dendritic extent they found in normal aged brains was detected only when dendrite lengths were measured quantitatively on a random sample of neurons. Since the conclusions from the Scheibel research were based on qualitative observations, undue emphasis may have been given to the prevalence of dying compared to normal neurons; the former being particularly salient. Data from Buell and Coleman (1981), shown graphically in

FIGURE 8.1. Summary of progressive changes in dentate granules during senescence. *Note.* From "Progressive Dendritic Changes in the Aging Human Limbic System" by M.E. Scheibel, R.D. Lindsay, U. Tomiyasu, and A.B. Scheibel, 1976. *Experimental Neurology, 53*, p. 426. Copyright 1976 by Academic Press. Reprinted by permission.

Figure 8.2, present a striking contrast to the picture of dendritic degeneration given by Scheibel et al. (1976).

Another qualification that must be made in drawing conclusions on normal aging from the Scheibel et al. studies is that the samples used included demented patients whose data were not separated from the data of persons who were normal. Since Buell and Coleman (1979) observed more dying neurons in the brains of demented persons, this distinction is extremely important. Finally, a technical problem with Scheibel's research is that the rapid Golgi procedure used for staining dendrites may not be suitable for use on autopsied tissue with prolonged postmortem delay before fixation (Buell, 1982). Since Scheibel and his co-workers did not used young controls to compare with the older persons, they could not have separated the effects of the procedure from the effects of aging.

The Environment of the Aging Neuron

Nervous tissue is composed of a complex network of neurons, glial cells, and blood vessels. Communication among neurons and support of their metabolic activities are highly dependent upon the adequacy of the extracellular environment provided by the glia and cerebral circulation. The degeneration of aging neurons creates an abnormal environment for the neurons which survive and may impair their functioning. Other changes in the glia arise spontaneously to alter further the surroundings of the neuronal population of the brain.

Senile Plaques

Parts of damaged and dying neurons are found to collect around a central core of hard abnormal tissues in what are called senile plaques. These abnormalities have been observed in samples of tissues taken from the hippocampus and cerebral cortex with increasing frequency in old age (Tomlinson et al., 1968). Because they are composed of parts of degenerated neurons, it seems logical to conclude that senile plaques are a product rather than a cause of neural aging. However, to the extent that senile plaques accumulate, their presence could impede the normal functioning of intact neurons.

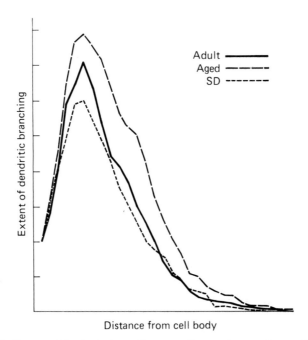

FIGURE 8.2. Extent of dendritic branching according to distance from cell body. *Note.* From "Dentdritic Growth in the Aged Human Brain and Failure of Growth in Senile Dementia" by S.J. Buell and P.D. Coleman, 1979, *Science*, *206*, p. 855. Copyright 1979 by AAAS. Reprinted by permission.

Glia

It is thought that glial cells serve as structural support and protection for neurons and, in addition, actively interact in the metabolic and electrochemical process of neurons. Glia outnumber neurons in the brain by a ratio of 10 to 1, although this proportion varies by brain region. By virtue of their large number and their role in supporting neurons, aging of these cells would have significant implications for the integrity of nervous tissue. Two of the most numerous glial cells are astrocytes and oligodendrocytes. Astrocytes are star-shaped cells that make contact between neurons and capillaries through their "sucker feet." These cells expand in size and number to form a scar after neurons are damaged or lost through degeneration. The myelin sheath that encloses the large axons of some neurons is formed by spiral loops of oligodendrocytes. This type of glial cell also reacts to injury by swelling, and in addition, has the potential to engulf neurons that are in the process of degenerating.

Commensurate with the degree of neural death as a function of age, there should be an increase in the population of astrocytes (Brizzee, 1975; Frolkis & Bezrukov, 1979). Increases in the proportion of glia have in fact been found in some parts of the brain affected by neuronal fallout such as the visual cortex

(Devaney & Johnson, 1980) and parts of the brain stem (Blinkov & Glezer, 1968). However, at the same time, there is also a decrease in the proportion of oligodendrocytes, in parallel with the destruction of the myelin sheath accompanying neural degeneration (Brizzee, 1975). Perhaps because of these opposing tendencies of different kinds of glia to increase and decrease with age, there have not been consistent reports of total glial population changes (Blinkov & Glezer, 1968; Henderson et al., 1980; Moatamed, 1966).

Quite apart from the issue of numbers of glia is the question of how adequately they perform their functions. Various structural abnormalities have been described within several area of tissue taken from the brains of older persons. These glial abnormalities include swelling and deterioration of the sucker feet of astrocytes, and of fibrils within the cell (Ravens & Calvo, 1966). The degeneration of these gial cells may interfere with the process of supplying neurons with adequate nourishment from their surrounding capillaries.

Changes With Age in Communication Among Neurons

Synapses

Behavior is the outcome of communication among neurons, communication that takes place at their points of contact, called synapses. The synapse is where the terminal branches of one neuron's axon come into contact with part of another neuron (another axon, a dendrite, or a cell body). Electrical signals are transmitted chemically across the synapse over the synaptic cleft, the gap at the point of junction between neurons. Through this transmission, neurons form interconnected pathways along which information travels from one part of the nervous system to another. Each neuron has multiple synapses, sometimes totaling many thousands. The patterns formed by multiple synaptic connections have important functional consequences: Through divergent patterns, one neuron can have widespread effects throughout the nervous system, and through convergent patterns, a great deal of information about activity in other regions of the system can be integrated.

Each synapse has an effect that is either excitatory or inhibitory. A synapse with an excitatory effect renders the postsynaptic neuron more likely to send an electrochemical impulse down its axon which may lead to excitation of neurons that it innervates. Inhibitory synapses reduce the postsynaptic neuron's impulse-generating potential. When neurons with inhibitory and neurons with excitatory effects converge onto one postsynaptic neuron, the generation of an impulse by that neuron will depend on whether the algebraic sum of all input received reaches a level necessary to initiate the electrochemical impulse.

The existence of these two kinds of synapses is a fundamental requirement for the coordination of activity in the central nervous system. For instance, focusing of attention on one sensory input is made possible by inhibition of input from other sensory pathways. Another example is the simultaneous excitation and inhibition of skeletal muscles that have opposing effects on limb movement. In order for the elbow to be flexed by the excitation of the biceps muscle, the triceps muscle, which would otherwise extend the elbow, must be inhibited at the same time. An extreme imbalance between inhibition and excitation can have disastrous consequences. For example, organisms injected with certain nerve poisons that block inhibitory synapses die within a matter of minutes from severe convulsions caused by the unimpeded activity of excitatory synapses.

New learning invariably involves the formation of new synapses, but it is not possible to pinpoint the neurological cause of a particular behavior to a given synapse. The synapses that are formed in response to environmental stimulation and that may underlie learning are formed diffusely throughout the brain. It is the case that certain brain structures do have fairly definite functions. Even in these rather specialized structures, though, information is represented diffusely across many synapses, and there is a great deal of redundancy or multiple representation of information throughout the central nervous system. It is this redundancy that can offset the effects of loss of particular synapses involving aging neurons. At the same time, the dendrites of surviving neurons may be showing plasticity by growing. Whether age is associated, on the one hand, with a decrease or, on the other hand, no net change in synapse number will depend on whether the neurons showing plasticity are able to compensate for those which are dying.

Counts of synapses in the human brain in relation to age have been performed only in areas of the frontal cortex, where neither an increase nor a decrease with age is observed well into old age (Cragg, 1975; Huttenlocher, 1979). This finding of stability is compatible with the suggestion that neurons showing plasticity are able to compensate for neuronal fallout in the aging nervous system (Buell & Coleman, 1979). What is not addressed by counts of synapse number is the possible change in organization within the nervous system caused by loss of certain synapses, such as excitatory ones that balance inhibitory ones, or convergent versus divergent ones. Changes in these patterns may have broader implications than the simple loss of quantity of synaptic connections on the integrity of information transmission.

Neurotransmitters

Communication of information at the synapse occurs (in most cases) through the transfer of neurotransmitters across the synaptic cleft. Neurotransmitters are released by the presynaptic axon from synaptic vesicles into the synaptic cleft. When a neurotransmitter travels across the synaptic cleft and reaches the postsynaptic site, it renders the postsynaptic neuron more or less likely to trigger an impulse, depending on whether the synapse is excitatory or inhibitory. The nature of the receptor surface on the postsynaptic neuron is what determines the synaptic effect of the neurotransmitter.

The neurotransmitters identified in the central nervous system include: acetylcholine (ACh), the monoamines (serotonin and the catecholamines—dopamine, epinephrine, and norepinephrine), neutral amino acids (gamma amino butyric acid [GABA] and glycine), acidic amino acids (glutamic acid) and peptides (substance P). The neurons that release these substances when stimulated are referred to in terms of their respective transmitter substances (e.g., cholinergic neurons release ACh). Pathways of neurons are identified according to the neurotransmitters released along afferent (ascending) and efferent (descending) chains of synaptic connections.

Assessing the levels of neurotransitters available for synaptic communication rather than counting synapses is potentially a more informative approach to studying the organization of the aging nervous system. Changes in the amount of neurotransmitters would signify alterations in the efficiency with which information travels through specific pathways and hence alterations of the excitatory and inhibitory regulation necessary for synchronizing complex functions. Several neurotransmitter pathways, when studied in this manner, do seem to undergo alterations as judged by reductions in the amount of particular neurotransmitters in aging brains.

A set of investigations by Drachman and his co-workers have provided fairly strong circumstantial evidence for an age-linked loss of activity of cholinergic neurons (Drachman, 1977; Drachman & Leavitt, 1974; Drachman, Noffsinger, Sahakian, Kurdziel, & Fleming, 1980). The designs used in this research include the following conditions. Young adult subjects are given scopolamine, a drug that is known to block acetylcholine transmission for a short time. Their performance on a memory test is then compared to the scores of normal healthy elderly subjects and undrugged young adults.

The consistent finding of the Drachman research is that the young adults given scopolamine perform at levels exactly comparable to the performance levels of the elderly. Moreover, both of these groups receive lower scores than the non-drugged young adults. By inference, the elderly would appear to have lower memory scores due to a reduction of cholinergic neurons with age, a decrease in the extent or number of synaptic connections, or a functional impairment of cholinergic synaptic transmission, particularly in the hippocampus, where cholinergic pathways seem to underly memory acquisition. If cholinergic transmission is, for any of these reasons, deficient in the elderly, then administration of an acetylcholine agonist should reverse some of the memory defects in the older subjects. However, physostigmine, which is such an agent, was not found to improve significantly the memory or cognitive scores of a small sample of older persons in a preliminary study reported by Drachman and Sahakian (1980).

Some support of the "cholinergic" hypothesis of memory loss in the aged is provided by anatomical evidence that there is a loss of hippocampal cells with age (Ball, 1977; Mouritzen Dam, 1979) as well as lowered activity of cholinergic neurons and receptor binding sites (Nordberg, Adolfsson, Marcusson, & Winblad, 1982; Perry, Perry, Gibson, Blessed, & Tomlinson, 1977; White et al., 1977). A study on the functioning of cholinergic neurons in living rats provided further support favoring the cholinergic system in memory loss in the aged: Acetylcholine produced less of an evoked response in hippocampal cells of older rats than younger ones (Lippa et al., 1980).

Evidence also exists for a reduction of synaptic activity mediated by the monoamine group of neurotransmitters in some brain regions. One of these regions is a pathway in the basal ganglia, a set of nuclei located deep within the brain that is responsible for the programming of motor movements initiated by plans for action in the cerebral cortex. This pathway connects a nucleus in the midbrain, the substantia nigra, with the striatum of the basal ganglia. The striatum is the structure that receives input indirectly from sensory receptors and it is on this input that programs for movement are based.

A reduced number of binding sites (Severson, Marcusson, Winblad, & Finch, 1982) and lowered activity of the dopaminergic neurons in the substantia nigra-striatal pathway has been observed in samples of tissue obtained over the adult age span (Carlsson et al., 1980; McGeer & McGeer, 1975, 1976, 1980). When present in an extreme form, a reduction of dopaminergic activity is the cause of Parkinson's syndrome. Persons who suffer from this disorder have rigid facial expressions, difficulty communicating with gestures, shuffling of their gait while walking, and trembling of their hands while at rest. Although this is a specific disease and not a normal aging process, reduced dopaminergic activity in the substantia nigra-striatal pathway may have similar, but more moderate, effects on motor control in the normal aged.

Age-related decreases in the activity of other monoamines have been described by Frolkis and Bezrukov (1979). Norepinephrine reduction in motor areas of the brain stem may have detrimental effects on the functions controlled by these areas: postural balance and muscle tone. A diminished capacity to regulate behavioral arousal with age has been associated with cross-sectional alterations of catecholamine metabolism in the reticular formation. Sleep functions mediated by norepinephrine (in the locus coeruleus) and serotonin (in the raphe nucleus) appear to be altered by changes in these neurotransmitters. A decrease with age of norepinephrine would have the effect of reducing deep slow wave and REM sleep, whereas an increase of serotonin would increase the amount of light sleep. Other changes in the reticular activating formation may result from con-comitant changes of monoamine oxidase and acetylcholine activity. Catecholamine alterations with age in the hypothalamus and other areas of the limbic system may have some effects on emotional reactivity.

An unresolved issue in the literature is reconciling the findings of studies indi-cating increased norepinephrine levels in the blood as the result of changes in the autonomic nervous system (see chapter 6) with the observations based on studies of the central nervous system of diminished norepinephrine activity. One hypothesis that would be consistent with the literature discussed in chapter 6 regarding diminishing sensitivity of various tissues to norepinephrine is that the locus coeruleus is for some reason (e.g., decreased tissue binding of norepinephrine) less responsive to norepinephrine and shows decreased activity of norepinephrine in its cells as a result. The heightened norepinephrine level in the blood may include, as was discussed in chapter 6, attempts by the sympathetic nervous system to compensate for reduced tissue responsivity.

The impression created by the array of findings on age differences in neuro-transmitter activity is one of reduced synaptic connectivity in the nervous systems of older persons. However, counteracting this image are data from these and other studies in which no effects of age were observed. In the case of acetylcholine, which is generally recognized to decrease within the hippocampal pathways, there are contradictory data regarding its level of activity in the cerebral cortex of the aged (McGeer & McGeer, 1980; Nordberg et al., 1982; Perry et al., 1977; Sims, Bowen, & Davison, 1982; White et al., 1977). Even in the hippocampus, though, decreased cholinergic synthesis has not been consistently shown (Carlsson et al., 1980). The level of acetylcholine activity has not been found to be altered in later adulthood within the basal ganglia (Carlsson et al., 1980; McGeer & McGeer, 1975; Perry et al., 1977). Dopamine reduction in the substantia nigra-striatal pathway appears to be very specific to this transmitter in this area of the brain. Reduced GABA activity has been found in the thalamus, and this age-related decrease could account for some of the sensory losses in the aged, since the thalamus is a site of synaptic relay of afferent information. Again, this transmitter reduction is specific to this particular area. Other neurotransmitters have not been shown to diminish appreciably with age in the thalamus, and GABA levels are maintained in older adults in a number of other brain areas (McGeer & McGeer, 1975; Perry et al., 1977).

The most striking age reductions in synaptic activity as indexed by transmitter levels appear, then, for acetylcholine in the hippocampus, dopamine in the substantia nigra-striatal pathway, norepinephrine and serotonin in some parts of the brain stem, and GABA in the thalamus. These reductions with age, however, take place in the context of stability across adulthood in other neurotransmitters within the same brain region and the same neurotransmitters in different pathways within the central nervous system. It does not, then, appear that aging brings with it universal synaptic loss and hence disorganization at a general level.

However, an important implication of the findings established at present is that aging may be associated with some disruptive changes in the patterns of neural communication because different neurotransmitters appear to age at different rates (Frolkis & Bezrukov, 1979; Ordy, 1981). Inhibitory and excitatory synapses need to operate in an integrated fashion in order to produce their intended effects on behavior. When the synaptic activity is disparately altered, an imbalance is created, and can result in a lack of coordination in the functions subserved by interacting neurotransmitter systems. For instance, if the transmitter system serving an inhibitory function is impaired with age, the excitatory system it normally holds in check will stimulate an excess of neural activity. For this reason, it is important to consider not only isolated pathways, but also the mechanisms of communication between these systems in the structures of the brain and how they might be affected by the aging process.

Age Effects on Major Neural Structures

Organization of the Central Nervous System

Neurons are organized in the central nervous system into structure consisting of pathways or tracts made up of axons and clusters of nerve cell bodies that form nuclei. Ascending (efferent) tracts transmit information from the sensory structures to the higher brain centers, and descending (efferent) tracts carry integrated instructions for action to the effectors (skeletal muscles). The major structural units of the central nervous system are organized such that "higher" functions are served by structures within the cerebral cortex. These higher functions are ones involving judgment, control, and flexibility, because they are based on the integration of multiple sources of input and multiple means of control. The lower functions are served by structures situated underneath and deep within the cerebral cortex. These lower functions operate much of the time in the service of the cerebral cortex, preparing sensory information to be interpreted and carrying out the orders for action initiated by the cortex.

Some of the lower functions, though, are fairly self-contained and are capable of operating automatically, without much communication to or from the cerebral cortex. Sensory input-motor output information loops make it possible for many activities to be carried out in a routine fashion, adjusting responses to the situation as it is perceived on a moment-by-moment basis, but without deliberate intent. Many activities carried out at the subcortical level may be done without conscious awareness by the performer of these actions. However, there are pathways connecting the subcortical and cortical areas that make it possible for the cerebral cortex to enter into an automatic process and override it, based on the appraisal of a situation or opportunity for a response made possible only by the more complex and integrated cortical structures.

General Age-Related Morphological Changes

The integrity of the nervous tissue of the brain is reflected in a very general sense in its weight, volume, and appearance. The fully developed brain in the 20- to 25-year-old adult human male weighs about 1300-1400 g and 1200-1350 g in the adult female. Numerous cross-sectional studies of brains weighed at autopsy have yielded evidence that the brain loses, on the average, from 5% to 10% of its weight in the age span from 20 to 90 years (Dakaban & Sadowsky, 1978). Although it is a universal feature of aging (Ordy, 1975), brain weight loss is moderated by health (Tomlinson et al., 1968), and it varies according to brain area (Frolkis & Bezrukov, 1979).

Brain volume at age 20 averages about 1300 cm³ in men and 1000 cm³ in women (Blinkov & Glezer, 1968). A decrease of about 200 cm³ is reported cross sectionally from 20 to 80 years (Ordy, 1975), with a pronounced drop beginning after 50 years (Yamaura, Ito, Kubota, & Matsuzawa, 1980). As the volume of nervous tissue shrinks, the amount of space widens in the ventricles (cavities in the brain filled with fluid). This shrinking of brain volume reported to occur in old age is reflected in the brain's external appearance as well as in the larger ventricular volume. The gyri (swellings) of brain tissue become smaller, while the sulci (valleys between gyri) widen. As with reductions in brain weight in old age, volume loss is a function of health status (Tomlinson et al., 1968). In both cases, differences among generations in nutritional patterns that influence height and weight may produce differences in brain status that appear to be a function of age alone.

Regional Variations in Neuron Counts

Variations with age in total brain weight, volume, and appearance mask the differences in the rates of aging of the neurons within individual brain structures. The neuronal fallout due to the presence of aging neurons which is responsible for loss of brain weight and volume in adulthood occurs at different rates throughout the brain, and in some cases it is not apparent at all. Moreover, different estimates of loss are obtained depending on whether the measure of loss is simply the total neuron count or whether it is in terms of packing density, the neuron count per unit of volume. Various discrepancies in the literature appear to have arisen because different researchers have employed different indices of neuronal fallout.

Brain Stem

The brain stem is the area of transition between the brain structures and the spinal cord. The three structures within the brain stem are the medulla oblongata, pons, and midbrain. Running vertically through these structures is the diffuse collection of ascending and descending pathways called the reticular formation. As a result of the activity in this structure, attention can be directed toward specific incoming information and to particular patterns of response. This structure also regulates the stages of brain activity during sleep. Other pathways in the brain stem contain specific nuclei (such as the inferior olive) which serve as points of signal amplification and integration among interacting tracts.

In general, throughout the brain stem, there seems to be little or no structural basis for impairment of motor or arousal functions mediated by these nuclei and connecting pathways. Of the brain stem nuclei, the locus coeruleus is the only one in which neuronal fallout has been reliably observed. However, loss is not apparent in this structure until the mid-60s, at which point it is reported to amount to about 40% (Vijayashankar & Brody, 1979). Loss of cells in the locus coeruleus may have some bearing on disturbances of sleep patterns in old age, perhaps as a cause of decreased norepinephrine level and subsequent reductions in the amount of deep and REM sleep. As discussed earlier, the relationship between the diminished locus coeruleus norepinephrine activity and heightened serum norepinephrine levels has not been addressed in the literature. The evidence of decreased cell numbers in this nucleus supports the speculation that sympathetic activity increases to compensate for losses in the locus coeruleus.

Cerebellum

One of the major functions of the cerebellum is the control of finely tuned voluntary movements initiated by the cerebral cortex. The cerebellum programs these movements and regulates them while they are taking place, and also adjusts posture to change required to maintain equilibrium during movement. These functions are made possible by afferent tracts providing an extensive amount of sensory input, and pathways linking the cerebellum to the brain stem and cerebral cortex which prepare and initiate the motor instructions. A variety of specialized neurons in the cerebellum permit this performance of complex integrative functions.

One of the specialized neurons in the cerebellum, the Purkinje cell, has been found to decrease by 25% cross-sectionally over the human life span. This decline is particularly noticeable after the age of 60 years (Hall, Miller, & Corsellis, 1975). The elaborate dendritic arborizations of the Purkinje cell permit it to receive the great quantity of sensory information needed to evaluate the body's movement status. The Purkinje cells then integrate this information and transmit it to nuclei deep within the cerebellum, where it is then fed into the motor centers of the brain stem. Loss of Purkinje cells with age may have disproportionately large effects on motor functioning because there are relatively few of these highly specialized cells, and because each one integrates so much wide-ranging sensory information.

Some of the motor defects associated with clinical loss of cerebellar functioning have similarities to certain disorders seen in the elderly. Cerebellar loss is marked by difficulty in coordinating a series of fine movements, especially during rapid succession, impairment in balance and timing of muscle action during movement, and loss of muscle tone leading to premature muscle fatigue upon exertion (Schmidt, 1978). Possibly these functional losses, when seen in the aged, have their functional origin in Purkinje cell number decline.

Basal Ganglia

The nuclei that make up the area known as the basal ganglia serve a complementary function to the cerebellum (see the description of the substantia nigra-striatal pathway). Apart from decreased volume, age does not seem to have a consistent effect on the number of neurons in the nuclei of the basal ganglia (Bottcher, 1975; Bugiani, Salvariani, Perdelli, Mancardi, & Leonardi, 1978). The control of gross movements initiated by the cerebral cortex and programmed by the basal ganglia would not, then, appear to change with age in later adulthood. Some functions controlled by the substantia nigra-striatal pathway may be impaired in old age due to neurotransmitter activity decreases, however. In addition, the implementation of movement programmed by the basal ganglia may be faulty due to neurotransmitter disturbances in the brain stem. Cortical cell loss in the motor area may also impair the quality of the instructions received by the basal ganglia. There is stability in the structural integrity, however, of the basal ganglia themselves throughout adulthood.

Diencephalon

Thalamus

This portion of the diencephalon is a relay point for sensory stimulation coming from the eyes, ears, skin, and joints on its way to different projection regions of the cerebral cortex. Specific nuclei are contained within the thalamus that serve as projection areas for the sensory structures responsible for vision, hearing, touch, and movement. These sensory modality-specific nuclei are organized somatotopically, that is, so that each point of the receptor surface projects to a different point in the thalamic nucleus. Other nuclei in the thalamus receive stimulation from multiple sensory sources. These nonspecific nuclei serve as a functional continuation of the reticular formation, serving to control the individual's degree of conscious awareness by regulating the overall arousal level within the cerebral cortex. Still other nuclei in the thalamus are motor structures serving, in part, to link the cerebellum and the cerebral cortex. There are no reports in humans of cell loss in the specific or nonspecific sensory thalamic nuclei or motor structures.

Hypothalamus

As described in chapter 6, the hypothalamus controls the activities of the autonomic nervous system and is the connecting structure between the neural and endocrine systems. It is also involved in the limbic system, and so serves to integrate learning and autonomic responsiveness. There are no reports of cell number decreases in humans in this structure as a function of aging.

Limbic System

Most of the central nervous system functions described so far are in the realm of automatic or stimulus-controlled behavior rather than that which is motivated or stimulated by emotions. The limbic system is the set of structures that provides the neurological basis for the interaction and communication between the "rational" and "irrational" sides of human behavior.

The limbic system consists of a set of circular networks of pathways and nuclei with tracts leading to and from the brain stem, diencephalon, and cerebral cortex. At the center is the hypothalamus, which functions within the limbic system to link learning with emotional states of pleasure or pain. The hypothalamus contains reward "centers" which animals such as rats will stimulate for long periods of time by learning to press electrode-connected levers, and comparable "centers" that cause pain to be felt when they are stimulated. There are no available studies on how these parts of the limbic system might be affected by age in humans.

The registration of information from short-term into long-term memory has been linked to cholinergic pathways in the hippocampus, as described earlier. The amygdala and septal area are limbic system structure connected to the hippocampus which are involved in the control of emotional reactivity. Other tracts connect limbic system structures with the reticular formation, and circle among the limbic system, brain stem, and cerebral cortex. It is through these pathways that the limbic system serves as the basis for integrating memory, learning, and motor behavior with the emotional state of pleasure, pain, and interest.

Neuronal fallout has been demonstrated in different regions of the hippocampus, as was mentioned earlier with regard to the age effects in the cholinergic pathway in this structure. The extent of neuron loss appears to be about 30%, beginning after the age of 30 (Ball, 1977; Mouritzen Dam, 1979). The other limbic system structures have not been studied with respect to age in humans. However, it is possible that there are some alterations in the relationship between learning activity and emotional arousal in older persons. There is evidence that older adults react with greater autonomic arousal to learning situations (as was discussed in chapter 6). The structural basis for this reaction may reside in the limbic system, with some type of cyclical process operating whereby poorer cognitive performance and greater anxiety feed upon and reinforce each other.

Cerebral Cortex

The cerebral cortex is a mantle of nervous tissue organized in layers and entirely covering the cerebral hemispheres. It has a volume of about 2000 to 2500 cm³ and a thickness of from 1.5 to 4.5 mm. There are an estimated 14-15 billion neurons in the cerebral cortex, which form perhaps as many as 1 billion synapses/mm³. At this rate, there could be as many as 50 trillion synapses in the total cortex.

The functions of the cerebral cortex are usually described with reference to its topography, that is, the frontal, parietal, occipital, and temporal lobes, each of which contains specific primary and secondary sensory areas, and the motor and premotor areas. It should be recognized, though, that there is a great deal of linkage among the areas by cortical pathways running transversely through its layers and also multiple representations of the same information across different brain regions (redundancy). This is because most of the cerebral cortex is made up of association cortex, which does not have a specific, point-by-point topographic organization but instead forms the basis for interrelationships among sensory areas, between sensory and motor areas, and connections with the limbic system. These interrelationships serve to give cortical functioning a higher degree of synthesis than exists at any other level of the nervous system. This integrative quality is probably what accounts for the interrelation of different sensory aspects of experience, abstract reasoning, long-term memory, future planning, and the use of symbols; in short, the capacity for thought.

The amount of neuronal fallout varies considerably among different areas of the cerebral cortex. Decreases of about 50% over the adult age range have been shown for cell density in the primary visual area located in the occipital cortex (Devaney & Johnson, 1980), and in the primary somatosensory area located in the parietal lobe (Henderson et al., 1980). Although he demonstrated a lower rate of neuronal fallout in his research, Brody (1955) observed that most of the loss that did occur in the cortex took place before the age of 40. In the case of the somatosensory area, there was actually a slight cross-sectional increase between the mid-40s and the 80s (see chapter 12 for further discussion of this). While the primary auditory area (in the temporal lobe) has not been studied in relation to cell counts by age in adulthood, investigations of the secondary auditory area have demonstrated a consistent decrease in neuron numbers on the order of from about one-third to one-half from the 20s through the 80s (Brody, 1955; Henderson et al., 1980). Losses in the sensory areas of the cortex add to changes in the receptor structures with age for these sensory systems (described in chapters 9 through 12) to influence the quality of input used in the association areas for the higher-order processing of information concerning environmental stimulation.

In the motor areas of the frontal cortex, estimates of neuron loss range from 20% to as much as 50% (Brody, 1955 Henderson et al., 1980). In the motor area are located the Betz cells, large pyramid-shaped neurons that are relatively few in number (about 34,000 in each hemisphere). These cells were studied by Scheibel and his associates, and found to undergo progressive loss of dendrites with age. Given that the dearborization of dendritic trees may not be an inevitable consequence of normal aging, it is still plausible to argue that synaptic losses in these key cells may play a role in causing some of the cortically based motor disturbances in the age. Some of these disturbances, such as diminished motor readiness (Mankovsky, Mints, & Lisenyuk, 1982) and loss of fine control of timing of responses (Rabbitt, 1980) would seem to be particularly closely related to the functions of the Betz cells, as well as the stiffening, slowness, joint pain, and leg cramps which often afflict older people (Scheibel, 1979, 1982).

The cortical association areas seem to be less vulnerable to neuronal fallout than are the primary sensory or motor areas. In the prefrontal region, which is the site of associative memories for combinations of different sensory features of an experience, three areas studied have shown no adult differences (Higatsberger, Budka, & Bernheimer, 1982; Huttenlocher, 1979). Two other areas had more pronounced age-associated cross-sectional decreases of about 50% over the adult years: the frontal pole (Brody, 1970), and the area just below it (Henderson et al., 1980). Small cross-sectional decreases, on the order of about 20%, have been noted for three of the association areas located in the temporal lobe (Brody, 1955; Henderson et al., 1980; Higatsberger et al., 1982; Shefer, 1973).

It would appear that the higher cortical functions served by neurons in the association areas of the cerebral cortex can be retained by plasticity, since the neuronal fallout that has been demonstrated is not so high as to render recovery of function by other neurons a practical impossibility. The overlap or redundancy of neuronal function in these areas adds further support to the likely of there being plasticity that overcomes the deleterious effects of neuronal fallout. The results of neurotransmitter studies also provide reinforcement for the notion of functional plasticity, given that most neurotransmitters in the cortex appear to remain stable across adulthood in their activity. This constancy, it should be recognized, may reflect increased activity by fewer functioning neurons as well as constant output from a constant number of active, healthy cells. Even so, the neurological capacity for the functions served by higher cortical structures of judgment, foresight, reasoning, and "wisdom" does not seem to be threatened by neuronal fallout in old age.

Psychological Consequences of Neural Aging

A primary concern that adults have about the aging process is that as they grow older they will become "senile." The term "senile" conjurs up visions of mental incompetence, emotional instability, a return to childlike dependency on others for the monitoring of bodily functions, and physical immobility. Adults over 60 or 65 years of age commonly take as evidence of impending senility a brief memory lapse, such as forgetting the name of an acquaintance, misplacing an object, or losing track of one's train of thought in a conversation. Although these behaviors frequently occur earlier in life, they are regarded with considerably less alarm and probably never as a sign of senility. As mentioned in earlier chapters, some other physiological changes associated with aging may similarly be taken as evidence of senility by the older person, including changes in motor coordination, kidney and bladder functioning, and certain gastrointestinal symptoms (see chapters 2 and 5).

The disease called "senile dementia" (of the Alzheimer's type) is not a general condition associated with the aging process; it affects 5%-7% of the over-65 population. Those older persons who develop this disease exhibit progressive impairments of the cognitive functions of memory, comprehension, and judgment, and deterioration of voluntary control over their elimination patterns and motor coordination. It is the fear of developing these terrifying conditions that makes even a minor mental or physical disturbance seem so ominous to the aged adult. The term "senile" has no real medical or psychological definition and its use leads only to confusion regarding a disease that is not typical of normal aging. The erroneous linkage among "aging," "senility," and "senile dementia" is commonly made by older persons because this is a myth about aging perpetuated in the lay community.

Other psychological consequences of aging of the central nervous system relate more directly to its effects on various sensory, motor, and cognitive capacities and behaviors. It was shown that aging appears to affect certain specific functions served by the central nervous system, including perception, short-term memory, sleep patterns, fine motor coordination, and large muscle control. The effects of aging on sensation will be covered in subsequent chapters. In general, it may be said here that, apart from their meaning as signs of "senility," age-related alterations in these functions can affect the adult's adaptation to the environment in serious ways.

In general, sensory changes associated with aging have the effect of reducing the quality of input received about the state of the outside world. Central nervous system changes, in particular, affect the degree of refinement of sensory information that has been processed by the receptor cells in the sensory structures of the eyes, ears, nose, tongue, and skin.

Secondly, short-term memory losses based on age changes in the cholinergic system in the hippocampus make it more difficult for the older adult to carry out routine daily functions that demand remembering things over periods ranging from a few seconds to several minutes. The learning of new skills and facts would also be less efficient, to the extent that learning depends on retention.

Third, loss of motor coordination and control due to aging of the central nervous system augments age effects on the muscle cells themselves, and lowers the older adult's ability to plan, as well as to implement, desired actions. Finally, sleep pattern disturbances, as noted in chapter 6, can hamper the adult's freedom to carry out normal patterns of daily routines, requiring increased time in bed to compensate for greater numbers of periods of wakefulness at night.

However, this picture of losses is consistent only with a neuronal fallout model, and does not take into account what is known about the compensatory processes of redundancy and plasticity. The impact of these processes is most likely to be found in the association areas of the cerebral cortex where higher-order or abstract thinking processes are mediated. The effect of aging may be less noticeable, then, to the older individual in those functions such as problem solving, reasoning, and judgment. To a certain degree, such abstract thinking abilities may be affected by perceptual and memory losses, since higher-order decisions depend on a minimum quality of information on which to base judgments. However, the ability to make decisions given sufficient information should not be altered by aging of the central nervous system. These abilities may even improve, as the individual stores more experiences into long-term memory association areas that augment the body of knowledge to be drawn from in future decision-making situations.

The distinction between two types of abilities, one based on sensory input and learning new information, and the other involving experience and judgment is in some ways parallel to the psychological description of age differences in intellectual functioning based on the two-factor model of intelligence described by Horn and Donaldson (1980). According to this model, and the research on which it is based, the two types of human intellectual ability show divergent age patterns. Fluid intelligence, which is the ability to form novel associations (as measured by tests of spatial relations and induction), shows a steady decrease throughout adulthood, after a peak in the early 20s. This may be the ability that reflects neuronal fallout. The other type of ability is crystallized intelligence, which is the knowledge of specific information (vocabulary and "Trivial Pursuit" types of facts) and the ability to use judgment (in social situations, for instance). It seems reasonable to postulate that it is crystallized intelligence that represents plasticity in the nervous system; the growth of new synaptic connections as the result of experience, and the increase in "wisdom" based on decision-making skills that have been practiced over a lifetime. Although there have been challenges to this theory, particularly the notion that fluid intelligence cannot be "learned" (Plemons, Willis, & Baltes, 1978; Willis, Blieszner, & Baltes, 1981), it has withstood subsequent analyses based on some previous criticisms of it (Horn & Cattell, 1982), and there seems to be growing acceptance of it as a heuristic device for organizing the life-span literature on intellectual development (e.g., Botwinick, 1977).

It should be reemphasized that there is a great deal of variation among adults in the ways in which even normal aging (apart from disease) affects central nervous system functions. Some of these functions in old age are, in addition, affected by the degree to which they were developed and utilized all through the adult years, especially the functions involving higher-order thinking and reasoning. Exposure to sources of perceptual, cognitive, and motor stimulation in adulthood and old age through mental and physical exercise can also influence the degree to which abilities that depend on these functions are preserved. The amount of demand for their use, and importance placed by the individual on abilities affected by aging of the central nervous system, will contribute further to the effects on the individual's sense of competence and self-concept of their change or stability.

The variation among individuals in the rate, locus, and importance of changes in the central nervous system means also that their psychological impact will have wide individual variations. A reasonable generalization to make is that, because of the critical nature of the central nervous system for virtually all behavior, changes with age in the functions it serves have potentially broad-ranging effects on adaptation and feelings of competence in many aspects of daily life. What determines these effects may be the extent to which the aged brain is able to overcome neuronal fallout by taking advantage of the multiple sources of plasticity.

CHAPTER 9

Visual System

Awareness of the appearance of people and objects in the environment is made possible by the visual sensory system. Eyesight is a critical feature of adaptation to the environment, serving as the basis for registering key information about how people and things are arranged in space and time. The eye makes vision possible by transforming light energy into neural impulses that travel through sensory pathways to the central nervous system. When these impulses reach the cerebral cortex, they are further processed, serving as the basis for refined and integrated perceptual judgments.

Age Effects on the Eye

Refractive Properties of the Eye

The optical structures in the eye are transparent structures that transmit and focus the light that stimulates the neural cells on the retina. As the light rays reflected from distant people or objects travel through these media, their rate of speed becomes slower than when they were traveling through the air because the structures of the eye are denser than air. The bending of these parallel light rays, called refraction, causes the light rays eventually to converge at the focal point when they pass through the eye. Under normal circumstances, the focal point is on the retina, and so the light rays reflected off the object stimulate neural signals that accurately describe the stimulus's position in the visual field. The light reflecting off stimuli at closer ranges, in order to be focused on the retina, must be refracted more because it does not enter the eye in parallel rays. The increase in refraction needed to focus on closer objects is made possible by a change in the curvature of the lens of the eye, so that it is more convex and the light rays are bent at a steeper angle. Accommodation is this increase in curvature of the lens to focus light from objects near the eye onto the retina.

The refractive power of a lens is measured in diopters, the reciprocal of focal length (the distance from the lens to the focal point). A convex lens with a focal length of .10 m has a diopter value of $+10$. A spherical lens, such as that in the eye, is completely circular, with the same type of refractive surface on each of its two halves. The total refractive power of the eye when focused on objects far away is 67 diopters. Of this, the cornea contributes 52 and the lens 15 diopters.

The lens can be made to be more convex, though, through the process of accommodation, which can increase its refractive power by as much as 14 diopters. Focusing on near objects is also assisted by convergence of the two eyes, so that the pupils move closer together by the action of the ocular muscles, which control eye movements.

Age Effects on the Structure of the Eye

The optical structures of the eye transmit and focus the light reflected off outside stimuli onto the retina. Many of the age effects on these optical structures and on the retina itself serve as at least partial explanations for age differences observed in studies of basic visual functions.

Cornea and Sclera

The outside surface of the eyeball is formed by three concentric layers: the cornea and sclera, the uveal tract, and the retina. The outermost layer forms a protective coating around the other layers and the eye's soft interior. In the front part of the eye, this coating is the cornea, transparent to allow light to enter. The "white" of the eye is the sclera. The outward appearance of the eye changes in old age, due to alterations in the visible portions of the cornea and part of the sclera. The cornea loses some of its luster (Scheie & Albert, 1977; Weale, 1963). Arcus senilis, described in chapter 2, is a white ring that can form around the outer perimetry of the cornea. A yellowing of the sclera and development of translucent spots allowing the blue and brown of the underlying pigment to show through can also affect the outward appearance of the older person's eye (Edelhauser et al., 1979). Apart from their effects on appearance, there is probably little impact of these changes in the cornea and sclera on visual functioning. Another change that has little effect on vision is in the cornea's sensitivity to pain. The cornea has many free nerve endings within it, so that it is very sensitive to pain. With increasing age in adulthood, less pain is experienced from pressure applied to the cornea's surface (Fozard, Wolf, Bell, McFarland, & Podolsky, 1977).

There are several more significant changes in the cornea that do affect its optical properties. Increased translucency with age of the cornea has the effect of increasing the refractive power of the eye. Other changes in corneal tissue lead to the greater scattering of light rays within the eye of the older adult, which can have a blurring effect on vision. Another age effect on the cornea that influences the eye's optical properties is in its curvature. The cornea in young adults tends to be more highly curved and so has more refractive power in the horizontal than the vertical plane. The opposite pattern is observed in the corneas of adults over 60, whose corneas are flatter on the vertical than horizontal plane. Most eyes are astigmatic to a certain extent, meaning that their refractive surface is not completely spherical. Age effects on the cornea have the consequence of changing the plane of astigmatism from the vertical to the horizontal. If this change is severe enough, it will lead to the need for a different type of corrective lens for the older adult than he or she may have been used to during young adulthood (Weale, 1963).

Uveal Tract

The middle layer of the eye, called the uveal tract, serves both nutritive and optical functions. This layer includes the choroid, ciliary structures, and iris.

Choroid

The choroid is brown in color, and contains blood vessels that carry nutrients and waste products to and from all parts of the eye except the lens, near the cornea, and inner layer of the retina. With increasing age in adulthood, the choroid seems to undergo some changes whose main impact is to make its surface uneven. The inner membrane adjacent to the retina thickens, becoming less elastic and more easily torn. These changes may begin relatively early in adulthood, starting in the 30s (Weale, 1963). The back of the eye when viewed through an ophthalmoscope may appear patchy and irregular in the older person's eye, and small yellow dots may appear where thickening of the choroid has occurred (Scheie & Albert, 1977). These changes may interfere with the quality of the visual image reaching the retina, and have detrimental effects on the circulation of blood in the eye (Kuwabara, 1977).

Iris

The iris is the colored or pigmented portion of the eye. As it dilates and contracts, it increases and decreases the size of the pupil, and in so doing, regulates the amount of light reaching the retina. Control of the movements of the iris is exerted in part by the autonomic nervous system. Some discoloration of the iris may occur in old age due to atrophy of tissue in this structure, contributing further to the age effects on the eye's appearance of changes in the cornea and sclera. A much more significant age effect on the iris, in terms of functional implications, is atrophy of the iris dilator (Carter, 1982b). The effect of this atrophy is to reduce the size of the pupil, a condition called senile miosis. The size of the pupil is largest in adolescence, decreasing cross-sectionally until age 60, after which it remains stable (Loewenfeld, 1979). The iris also becomes more rigid in old age, so that pupil size is less readily altered according to lighting conditions (Weale, 1963).

Ciliary Body

The ciliary body is a mass of muscles, blood vessels, and connective tissue. Processes extending out of the ciliary body connect to the lens via ligaments called the zonule fibers, which form a ring around the lens. The contraction and relaxation of the ciliary body, under control of the ciliary muscle, alter the shape of the lens by varying the pressure on the zonule fibers, thereby controlling the focusing of light rays through the lens. There are age effects on the ciliary body, both in its function as the producer of the aqueous and in its muscular structure capacity.

Less aqueous is secreted by the ciliary body (Marmor, 1977), so that the nourishment and cleansing of the lens and cornea are reduced.

Ciliary Muscle

The ciliary muscle is the mechanism that permits accommodation to take place. When this muscle contracts, the ciliary body moves forward, reducing the tension on the zonule fibers, which are attached all around the perimeter of the lens. These fibers are straight and stiff, and when the tension on them is relaxed, they slacken their pull on the lens capsule. The lens capsule is elastic, having an intrinsic tendency to assure a spherical shape. When the zonule fibers reduce their pull on the lens capsule, the lens becomes more convex. When an object to be focused on is far in the distance, the ciliary muscle relaxes, the ciliary body moves back, and the zonule fibers increase their pull on the lens capsule. The lens capsule, in turn, becomes flattened, or unaccommodated.

The ciliary muscle undergoes a series of alterations in adulthood. After the age of 30, the mass of the muscle in the front of the eye increases, reaching a peak at 45 years. The rear part of the ciliary muscle atrophies, though, with connective tissue replacing muscle fibers. After 45 years, the net result of age effects on the front and rear part of the muscle is loss of muscle mass along with an overall increase in connective tissue. A uniform decrease in size of the ciliary musculature occurs after 50 years, when no new muscle fibers are formed.

The initial increase in mass in the front of the muscle is thought to be due to the greater work energy needed by this part of the muscle to alter the shape of the lens, which becomes decreasingly pliable in adulthood (see pages 157-158), during accommodation. When the lens reaches the point where its shape cannot be transformed at all, the ciliary muscle then atrophies from disuse (Weale, 1963). According to this interpretation, age effects on the ciliary muscle are secondary to the increasing stiffness with age of the lens.

The Retina

Visual processing begins in the retina, where receptor cells, stimulated by light energy, trigger impulses in other retinal cells. As a result of retinal processing, the information leaving the retina is already partially organized before it reaches the central nervous system structures where higher level analyses take place.

The visual receptor cells are called rods and cones (the names correspond roughly to their shapes). Both types of receptors manufacture pigments, which are broken down by light energy and, in the process, generate a nerve impulse. Because of differences in the pigments made by the rods and cones, vision in dim light (scotopic vision) is due to the activity of the rods, while in bright light, the cones are primarily responsible for transmitting visual information (photopic vision).

Rods are distributed most densely on the periphery of the retina, dropping off in frequency toward the macula, the central portion of the retina. In the very center of the fovea (the foveola), there are no rods at all. The cones located here have the property of being in one-to-one correspondence with higher-level neurons, rather than having their inputs combined with those of other receptor cells, as occurs elsewhere in the retina. Another special feature of the fovea is that the blood vessels and other layers of neurons that lie in front of receptors on the rest of the retina are pushed aside, allowing light to reach the cones there directly. As a result, the best vision in a well-lighted setting (where there is photopic vision) is at the fovea. Eye movements are normally directed at keeping the fovea lined up with the objects to be viewed, so that they can be discriminated most acutely. Vision in the periphery of the retina, served by the rods, is never as distinct as that on the fovea, even in dim light (scotopic vision), because of the many rods converging on single neurons leading to the higher visual pathways.

There are many more retinal receptors (between 6 and 7 million cones and 110 to 125 million rods) than there are ganglion cells, whose axons form the 1 million or so fibers in the optic tract. In order for all the information processed by the millions of rods and cones to be used in the cortex, it is necessary for that information to be put into condensed form as it passes from the receptors to the ganglion cells. This condensation is made possible by the convergence which is found at successive layers of retinal neurons where neurons from one layer synapse onto neurons at the next. At each set of synapses, the neurons process information from larger and more encompassing receptive fields. A receptive field of a visual receptor is the area on which light stimulation will cause it to trigger an impulse. Within the retina, the greater size and larger information-handling capacity of higher-order neurons makes it possible for the relatively small number of ganglion cells to retain and even improve the quality of information received by the rods and cones.

The receptive fields of neurons at increasingly higher levels of organization in the visual system become progressively larger and more complex, from the retina through to the cortex. The first level of convergence on the retina is at the point where the rods and cones synapse onto the bipolar cells. Although some bipolar cells receive stimulation from only one cone (in the foveola), most are activated by either several rods or several cones. Bipolar cells also receive stimulation from the horizontal cells, whose effect is to alter the nature of the input that reaches the bipolar cells from the rods and cones. Because of horizontal cell activity, there is a concentric receptive field for the bipolar cell with a center that responds one way (e.g., inhibition) to light stimulation and a surrounding ring that responds in the opposite way (e.g., excitation). The center of the bipolar cell's receptive field corresponds to its dendritic tree, which receives stimulation directly from the rods and cones that synapse onto it. The ring of the bipolar cell's receptive field reflects the inhibitory influence of horizontal cells on neighboring receptors.

The response at the center of the bipolar cell's receptive field may be either excitatory, or "on" (increasing its frequency of firing in response to light stimulation), or it may be inhibitory or "off" (decreasing its firing rate when light hits it). "On" center receptive fields of bipolar cells have "off" center surrounds, and "off" centers have "on" surrounds. The effect of this pattern of stimulation is to augment the bipolar cell's response to the contrast provided by edges of light against dark. A field of light without any edges produces a smaller total response in the bipolar cell than does an edge, because the "on" and "off" parts of the receptive field have a cancelling effect on each other. Due to the nature of the receptive fields of bipolar cells, the information from rods and cones about lines and borders is presented to the ganglion cells in greatly enhanced and condensed form.

The receptive field of the ganglion cell follows the same principle of organization as that of the bipolar cell, that is, antagonistic responses to light of center and surround ("on" and "off"). There are a variety of ganglion cells, whose receptive fields incorporate more information than the bipolars about the stimulation reaching the receptors. The information coded by the ganglion cells includes the amount of brightness, the presence of edges, orientation and movement of lines, the duration of a light stimulus, and color. The receptive fields of the ganglion cells can incorporate this quantity and variety of information because for many of them, information is fed in from many bipolars. The ganglion cells also receive stimulation from amacrine cells, adding further to the diversity and extent of information the ganglion cells can process. The amacrine cells cover wide areas of the retina, providing synaptic input to ganglion cells regarding the spatial and temporal qualities of retinal stimulation.

While both rods and cones contribute to the responses of ganglion cells, the predominant influences are the foveal cones. There is little convergence from foveal cones at either the bipolar or ganglion cell layers, a factor that accounts, in part, for the greater clarity of vision in the fovea.

There are contradictory data on the effect of age on the numbers of photoreceptors. Decreases across adulthood are reported for the number of rods (Kuwabara, 1977; Ordy et al., 1982; Weale, 1978), foveal cones (Ordy et al., 1982), and cones outside the fovea (Weale, 1963). Loss of horizontal, amacrine (Kuwabara, 1977), bipolar, and ganglion cells (Ordy et al., 1982) is also claimed. A contrasting view is that there is no significant photoreceptor cell loss with age, since the rod and cone tips are constantly replacing themselves in a process not affected deleteriously by aging (Young, 1976). To the extent that it occurs, the effect of retinal cell loss on vision is also disputed. While any reduction of photoreceptors is seen by some authors as causing poorer visual acuity (Ordy et al., 1982; Weale, 1982), it is also claimed that many photoreceptors can be lost without having a detrimental effect on vision (Marmor, 1982).

Removal of the material being shed from the rods and cones when they replace themselves is accomplished by the outermost layer of the retina (pigment epithelium). Although the rate at which this process takes place is not affected by age, the efficiency of debris removal from the photoreceptors by the pigment epithelium is reduced. Debris and lipofuscin (a residual from the cell renewal process) therefore accumulate in the pigment epithelium and also in the choroid. The collection of these waste products may contribute to reduced retinal cell functioning and possibly degeneration (Marmor, 1980).

Vitreous

The vitreous is the transparent, gelatinous mass that makes up the inner substance of the eye from the lens and zonule fibers back to the rear part of the eye. It is made up of a collagenous fiber network through which other long protein molecules are wound, giving the vitreous its viscous quality. The aqueous is a clear fluid that circulates through the front part of the eye, around the lens to the cornea. It is formed from blood plasma by cells in the ciliary body. The aqueous serves many functions, the primary one being to nourish and carry away waste products to and from the lens and rear of the cornea, which do not have blood vessels.

The vitreous undergoes a process of liquefaction, beginning at the age of 40, in which due to collapse of the collagenous fiber network liquid replaces the gelatinous substance (Balazs & Denlinger, 1982). As a result of this process, parts of the vitreous shrink away from the surface of the retina and may detach from it. The tension on the parts of the retina to which the vitreous remains attached may then increase, leading to visual disturbances, such as the appearance of "floaters" and light flashes (Scheie & Albert, 1977). Apart from vitreous detachment, other age-related changes in the vitreous can interfere with vision. Opacities may be produced by disruptions in the fibrous network of the vitreous, reducing the amount of light reaching the retina and producing the appearance of floating bodies. Light may be scattered more diffusely through the vitreous before it reaches the retina, making the eye of the older person less able to detect dim light (Weale, 1963).

Lens

The lens is made up of transparent fibers arranged in concentric layers. Lens fibers are continuously being formed around the outermost layer, beneath the lens capsule. The new lens fibers replace the ones formed previously and, in the process, compress the older fibers toward the center where they eventually form part of the nucleus. Since existing lens cells are not shed as new ones are produced, the size of the lens is continuously increasing.

Having profound effects on many aspects of visual functioning is the aging of this structure. The effect of aging on the lens may be primarily attributed to the continued laying on of new fibers without the concomitant shedding of old lens fibers. The growth of new lens fibers slows in later adulthood, and the old lens fibers shrink (Warwick, 1976). Nevertheless, the net impact of the continued overlaying of new fibers onto the old is an increase in the front-to-back diameter of the lens which is almost linear from the 20s to age 90 (Paterson, 1979). The mass of the lens triples from its original value by the age of 70, outstripping the rate of an increase in volume, so that the lens becomes more dense.

The greater density of the lens is particularly evident in the nucleus, where old lens fibers become compressed when new ones are formed in the outer portion of the lens (Weale, 1963). Because the lens nucleus in the older adult is more dense, it is more resistant to pressure from mechanical forces (Cotlier, 1981). In addition, alterations in the structural proteins that make up the lens fibers are believed to lead to hardening of the lens substance (Paterson, 1979). The lens capsule becomes less elastic, so that its shape is more resistant to change (Fisher, 1969).

One major impact of these changes in the structure of the lens nucleus and capsule is reduced accommodative capacity. The change in the refractive power of the lens in shifting focus from far to near objects becomes reduced over the life span from a peak of 14 diopters at 8 years, 11 diopters at 20 years, 9 diopters at 30 years, and less than 2 diopters at age 50 (Moses, 1981). By the age of 60, the lens is completely incapable of accommodating to focus on objects at close distance. Presbyopia is the name given to the condition in which the eye has lost its accommodative capacity.

Correction of presbyopia requires a spherical convex spectacle lens to make up for the loss of refractive power of the lens in focusing on near objects. Corrective lenses may be needed only for reading or, if the individual is also myopic, they may take the form of bifocals, with a convex lens on the bottom (for reading) and a concave lens above (for distance viewing). The age at which presbyopia requires correction depends on the extent of discomfort the individual experiences when performing close work, as well as the amount of dependence on near vision involved in the person's vocational and leisure pursuits. Also, the age at which presbyopia begins varies according to the environmental temperature, occurring earlier on the average in people living in warm climates (Weale, 1981).

A second effect of the age changes in the lens structure is a reduction in the transmission of light through the lens as it becomes denser and less transparent (Spector, 1982). It is estimated that after 65 years, about 60% of the population experiences reduced lens transparency (Cotlier, 1981). This age effect compounds senile miosis in reducing the amount of light reaching the retina. In addition to cutting down light transmission, age-related changes in the density of the lens have the effect of scattering light rays before they reach the retina. As a result, when light rays reflect off objects, they strike diffusely over the retina, producing a blurred retinal image (Carter, 1982b). The reduction in light transmission through the lens is especially pronounced for the short-wavelength light in the blue-violet range of the spectrum which is absorbed as it passes through the lens.

A greater absorption of blue and violet light in the lens of the older adult is attributed to an age-related increase in yellow pigment. By the age of 35, the lens has a yellowish tinge, particularly evident in the center, and this color progressively deepens in adulthood, often becoming amber in old age (Warwick, 1976). The effect of lens yellowing is to impair the older adult's ability to discriminate colors in the green-blue-violet end of the spectrum. Moreover, since shorter wavelength light is what maximally stimulates the rods, these receptors will be particularly affected by the lowered transmission of blue and violet light through the lens, impairing the quality of night (scotopic) vision (Weale, 1963).

Age Effects on Basic Visual Functions

The optical properties of the eye and the neural components of the visual system determine a number of critical visual functions. The functions to be described here are called "basic" because they are relatively straightforward applications of the structural properties of the eye and the peripheral levels of the visual nervous system (the primary cortex and below). Many of these functions are affected in fairly direct ways by the aging of optical structures.

There are a number of well-documented effects of aging on visual functioning, based on clinical and experimental evidence, some of it extending over several decades of cross-sectional analyses, and a smaller number of longitudinal studies. Many of the age effects to be described can be predicted on the basis of the structural changes already reported. Some age effects on basic visual functions have more than one structural cause. The basis for others is a matter of speculation. In some cases, nonperceptual factors may be the sole or a contributing cause of age differences in functioning. It is also important to realize that while a distinction is often made between normal aging processes and disease in these studies, confusion still exists due to the presence of individuals with uncorrected visual disorders in the samples of older adults who have provided the data.

Refractive Power

The aging of the lens and the cornea causes shifts in the eye's refractive power during adulthood. From about age 30 to the mid-60s, the eye gradually becomes more hypermetropic (farsighted). This shift is due to the increae in size of the lens as new lens fibers are added, causing the shape of the lens to flatten. Consequently, there is less refraction of light rays as they pass through the lens. After the mid-60s, there is a reversal of the trend toward hypermetropia, and the eye becomes more myopic (nearsighted). The compacting of lens fibers in the lens nucleus that accompanies the growth of new lens fibers in the outer portion of the lens throughout adulthood eventually increases the refractive power of the eye, bringing light rays into focus in a more forward direction.

These refractive changes are not related to the age-related changes in accommodative capacity that lead to presbyopia. Both hypermetropia and myopia are defined in terms of parallel light rays reaching the eye from distant objects, while presbyopia is defined in terms of the eye's ability to change its focus from far to near objects. The two types of change interact, though, as the eye becomes more hypermetropic in adulthood due to increasing size of the lens. As a result of presbyopia, the usual compensation made by the hypermetropic eye through accommodation for viewing close objects becomes decreasingly effective.

Acuity

The ability to detect details on objects at varying distances is referred to as visual acuity. Seeing the details on an object is a function of the size of the visual image, that is, the amount of light reflected off the object onto the retina. The smaller the visual image, the harder it is to discriminate its details. The size of the visual image is measured in terms of the size of the visual angle it subtends when the light rays from it converge at the lens and land on the retina. The size of the visual angle depends on the size of the object and its distance from the eye.

Snellen Chart Measurements

The most common way to measure visual acuity is by use of the Snellen chart. The individual whose acuity is being tested with this method is seated 20 ft away from the chart and told to read the letters, starting with the one at the top. At this distance, accommodation is not brought into play because the light is reflected into the eye off the letters in parallel rays. Each row of the chart has letters that would subtend an angle of 5° if they were seen from the distance shown on the side of the chart corresponding to that row. For instance, letters in the 50-ft row would have a 5° angle if they were seen at 50 ft. At 20 ft, though these letters in the 50-ft row subtend a somewhat larger visual angle.

Visual acuity as measured by the Snellen chart is expressed in terms of a ratio. The numerator is the distance of the person from the chart (usually 20 ft). The distance corresponding to the smallest row of letters that the person can read is the denominator. A person with 20/20 acuity is considered to have normal vision, in terms of being able to see letters with a visual angle of 5° at 20 ft. The denominator for the myopic person would be greater than 20, indicating that the letters that subtend a 5° angle at 20 ft are too small to be discriminated. Only bigger letters (which subtend more than a 5° visual angle at 20 ft) can be clearly discriminated on the Snellen chart by the myopic person. The hypermetropic person can see letters on the Snellen chart that subtend less than an angle of 5° at 20 ft, but if the hypermetropia is extreme, he or she cannot see the bigger letters that subtend more than a 5° angle at 20 ft.

As a measure of visual acuity, the Snellen chart has the advantages of being convenient to administer in a variety of settings, and also, its result are readily interpretable. There are some disadvantages to the Snellen chart, though. One disadvantage is that examiners might vary in when they consider a row to be "passed" since there is no set criterion. Also, the scores from the Snellen chart are not on a scale suitable for statistical analyses; 20/40 vision is not the "average" between 20/20 and 20/60 vision. A very serious disadvantage of the Snellen chart in terms of its practical applications is that it yields a measure of acuity only under the limited set of viewing conditions in which it is obtained and not for the range of visual circumstances in which acuity is required in everyday life.

Age-related changes in the optical structures of the eye and the visual nervous system together account for much of the loss of visual acuity in later adulthood that is observed via Snellen chart measurement. Scatter in the cornea and lens serves to deflect light rays entering the eye (Weale, 1963), spreading stimulation over wider receptive fields of the bipolar and ganglion cells and thereby reducing the contrast effects necessary for detecting details. Senile miosis, the decrease in pupil size, reduces the amount of light reaching the retina, so that stimulation of the retinal cells is less likely to occur (Pitts, 1982b). Whatever loss there is in later adulthood of retinal cells (Weale, 1982) and neurons in the visual cortex (Devaney & Johnson, 1980), add to structural changes in the retina to reduce visual acuity because of a lessening in the capacity to encode visual information.

The pattern of age differences in the ability to detect details when viewing objects at a distance is one of an increase in acuity to the 20s-30s which holds steady until 40-50 years, and then progressively declines (Pitts, 1982b) until, by age 85, there is an 80% loss of what the acuity was in the 40s (Weale, 1975). The marked decline between ages 70 and 80 was confirmed in a longitudinal study (Anderson & Palmore, 1974). The loss of acuity is especially severe at low levels of illumination, such as driving on a dark road at night (Richards, 1977), and also for moving objects (Panek, Berrett, Sterns, & Alexander, 1977; Reading, 1968).

Contrast Sensitivity Function

A better measure of visual acuity that overcomes the deficiencies of the Snellen chart is the contrast sensitivity procedure. Using this procedure, a measure of acuity is obtained that represents the individual's sensitivity for a range of patterns of light and dark gradations that resemble a variety of patterns seen in the real world, from faces to fabric textures. The stimuli used in the contrast sensitivity procedure are grids of black and white lines that vary in what is called spatial frequency, which may be thought of as the number of lines in the grid. Grids with low spatial frequencies have few black and white lines, while those with high spatial frequencies have more lines, which are also narrower. Usually, the grids that are used do not have sharply distinct black and white lines, but have instead sinusoidal gratings, where the edges between black and white are shaded and blurry.

The measure derived from acuity tests using sinusoidal grids of different spatial frequencies is the contrast sensitivity function, a plot of the relationship, for a given observer, between the spatial frequency of a grid and the amount of contrast between the black and white lines needed to detect that there is a grid pattern present. Contrast sensitivity is the reciprocal of the contrast needed to see the grid, that is, the more sensitive the observer's eye, the less contrast there must be. High contrast sensitivity is analogous to good visual acuity. In general, a fair degree of contrast is needed to detect gratings of very low frequency, so that sensitivity is low at low spatial frequencies. Intermediate frequency gratings require the least contrast between the black and white lines to be seen; in other words, sensitivity is highest in this range. Gratings with high frequencies yield lower sensitivities than do gratings with intermediate frequencies. A typical contrast sensitivity function of a person with good acuity is illustrated in Figure 9.1. Individual differences in contrast sensitivity are expressed in terms of this kind of plot, which makes it possible to represent the person's acuity as it varies across the range of spatial frequencies, and hence a variety of visual situations.

There is currently a lack of agreement concerning the effects of age on the

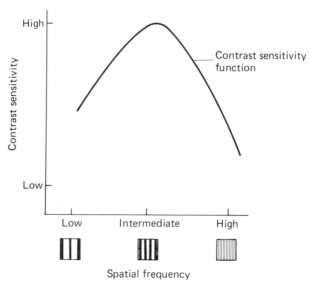

FIGURE 9.1. Contrast sensitivity function showing relationship between contrast sensitivity and spatial frequency of grid patterns. *Note*. From "The spatial vision of older humans" by R. Sekular and C. Owsley, 1982. In R. Sekular, D. Kline, and K. Dismukes (Eds.), *Aging and human visual function* (p. 190). New York: Alan R. Liss. Copyright 1982 by Allan R. Liss.

contrast sensitivity function, that is, at which spatial frequencies older adults are unable to detect contrast. Fairly consistent adult differences have been reported in contrast sensitivity to gratings of low and intermediate spatial frequencies (McGrath & Morrison, 1980; Sekuler, Hutman, & Owsley, 1980; Sekuler & Owsley, 1982) and also faces, which have low spatial frequencies (Owsley, Sekuler, & Boldt, 1981). In contrast, other reports document age differences in the intermediate to high spatial frequency ranges (Arundale, 1978; Derefeldt, Lennerstrand, & Lindh, 1979; Kline, Schieber, Abusamra, & Coyne, 1983; Owsley, Sekuler, & Siemsen, 1983). In yet a third group of studies, age effects were either negligible compared to the effects of disease (Arden & Jacobsen, 1978) or nonexistent (Arden, 1978; Dressler & Rassow, 1981). Some of the discrepancies in findings may be due to differing age ranges of the samples (with "old" being defined as 60, 70, 80, or 90), to the presence of older respondents with visual disorders, to uncorrected refractive errors among the older adults (Pitts, 1982b), and to different ways of showing the gratings (using slides, computer terminal monitors, or pictures). At present, it is not possible to locate precisely the range of spatial frequencies in which diminished contrast sensitivity in the aged is most severely reduced.

Sensitivity to Levels of Illumination

The lighting conditions in which people and objects are seen vary from being as dark as one's bedroom in the middle of the night to as bright as a beach on a sunny summer day. Adaptation is the process of adjusting changes such as these in the lighting conditions of one's environment. Both light and dark adaptation are based on the underlying principle that light striking the retinal surface causes degeneration of the pigments in the rods and cones so that they are incapable of initiating the neural impulses responsible for visual sensations.

Dark Adaptation

Dark adaptation refers to the process whereby the visual pigment that has been destroyed by light exposure is restored. When moving from a bright to a dim environment, good vision depends upon producing enough pigment to make up for that which had been destroyed by the light. It has been found that sensitivity to dim light decreases across age groups of adults from adults 20 years and up, both in terms of the time it takes to achieve restoration of rod pigment, and the quality of adaptation that is achieved after prolonged exposure to dim lighting. Decreased dark adaptation is particularly evident after 60 years (McFarland, Domey, Warren, & Ward, 1960). Rather than being due to aging of the process of pigment production, it appears that the decreased sensitivity to dim light in the dark is a function of factors already described in connection with other visual functions: reduced light transmission through the lens and senile miosis (Pitts, 1982a; Weale, 1975).

Sensitivity to Light

Light adaptation is the process of achieving a balance between pigment destruction and regeneration when illumination levels are high. Light and dark adaptation differ in their time courses because of variations between the rod and cone pigments. In cases where the rod pigments degenerate as the result of prolonged light exposure, it requires a relatively long time (at least 40 min) for them to be regenerated after the illumination dims. The cone pigments regenerate more quickly than do rod pigments in the dark, but the cones do not become as sensitive to low levels of light as do the rods. When going from dark to light, the cone pigments quickly reach a steady state between regeneration and destruction by light, and reach a greater sensitivity to light than does the rod pigment, which does not regenerate in bright light.

Senile miosis and the yellowing and increased density of the lens reduce the amount of light that enters the eye so that older adults are less sensitive in detecting low levels of illumination (Carter, 1982b; Weale, 1975). Aging also affects sensitivity to glare. There are various forms of glare that share in common the characteristic that they introduce extraneous light into the eye, reducing the clarity of the visual image. Veiling glare is due to the scattering of light over the retinal surface, resulting in reduced contrast of the retinal image of the object being focused. The age changes in the lens that produce greater light scatter create heightened vulnerability to veiling glare, augmenting its effects by further diffusing light across the retina (Carter, 1982b).

Disruptions in vision are also caused by exposure to a bright light source such as a flashbulb, or being approached by the headlights of an oncoming car on a dark road. This effect, called scotomatic glare, is due to loss of retinal sensitivity caused by overstimulation of the photosensitive pigments in the rods and cones. The reduced brightness of the retinal image resulting from senile miosis (Carter, 1982a) and greater density of the lens make the adult increasingly susceptible to scotomatic glare, particularly after the age of 40 years (Wolf, 1960; Wolf & Gardiner, 1965).

Color Vision

The ability to detect different colors is due to the fact that there are three types of cones, each of which has a pigment sensitive to one of three colors: red, blue, or green. When light reflected off a colored object passes through the lens onto the retina, each of these cones is stimulated to a different degree, depending on the amount of each color in the light entering the eye. The coding of color information in the visual system is accomplished by ganglion cells whose receptive fields have centers and surrounds that are antagonistic with respect to colors. The output of the ganglion cells reflects the mixture of colors that has stimulated the receptive fields on the retina.

The yellowing of the lens in adulthood reduces the adult's ability to discriminate the colors at the green-blue-violet end of the spectrum, as noted earlier. The thickening of the lens with age further impairs color vision at low levels of illumination by decreasing the amount of light reaching the retina and in the process reducing the participation of the color-sensitive cones in the photoreceptor response to light (Carter, 1982b). The effect of lens changes on color vision is heightened by senile miosis, which restricts light entry through the eye to the central region of the lens, its thickest and yellowest portion.

Temporal Summation and Resolution

The retinal receptors have the capacity to collect light stimulation over time, a property referred to as temporal summation. A weak light stimulating a set of retinal receptors may trigger a neural response from all when the amount of stimulation has accumulated beyond the threshold value for initiating the degeneration of neural pigment. At high levels of illumination, temporal summation is much less important a factor in influencing visual sensations, since each light stimulus will be strong enough to elicit a separate response. The effect of temporal summation is to reduce the observer's ability to judge the temporal duration of light flashes at low luminance levels, since the flashes are cumulated over time.

Temporal resolution is a complementary visual function to temporal summation. Flashes of light from a flickering light source cease to be sensed as separate stimuli when their frequency and brightness increase because more temporal summation occurs. The point at which an intermittent light stimulus appears to the observer as continuous is called critical flicker fusion. A person with high critical flicker fusion is able to detect separate flashes of light at high flicker frequencies that would lead to fusion in a person with lower temporal resolution ability.

With increasing age in adulthood there is a decrease in the point at which a flickering stimulus fuses into a continuous light (Shickman, 1981). Since the point of flicker fusion is reduced under lower levels of illumination, lens and pupillary age changes that reduce the amount of light entering the eye contribute to the age difference in flicker fusion. Moreover, it appears that age effects in the visual pathways in the central nervous system (Kline & Schieber, 1982; Weale, 1963) and in factors affecting the observer's tendency to report fusion (Fozard et al., 1977) can contribute to the loss of ability to differentiate light flashes.

Depth Perception

Judging distances and knowing the relationships among objects in space is a basic requirement for being able to maneuver about successfully in the environment. Depth perception is the visual function responsible for seeing spatial relationships. There are at least four known factors that contribute to the ability to see objects in depth. All or some of these come into play for perceiving depth in any given situation. For distances of less than about 300 ft, spatial relationships are judged on the basis of disparities between the retinal images that reach the two eyes. The eyes are separated from each other by a distance of about 2 in., and so within a certain viewing range, each eye receives slightly different input from the same scene. This disparity between the retinal images is referred to as stereopsis. When the information from the two retinal images is combined (possibly in binocularly stimulated cells in the visual cortex), the effect produced is to provide a three-dimensional representation of the scene.

The distance of individual objects from the viewer is determined in part by the angle of convergence between the foveas of the two eyes. In order to place the image of an object into maximum focus, muscles surrounding the eyeball move it so that the image is directed onto the fovea. The angle of convergence between the eyes becomes greater the closer an object is to the observer. The size of this angle is what serves as a cue for the object's distance.

A third factor contributing to depth perception is based on the visual experiences of the observer with objects of different sizes. Knowing the relative size of an object is a cue to its distance. If an object known to be very large (such as a tractor trailer) forms a small image on the retina, the observer will infer that the large truck must be far away. As the size of the image produced by the truck grows larger, the observer will know that it is moving closer.

Movement parallax is the fourth type of information used in depth perception. When the observer's head moves from side to side or when the observer's entire body is in motion, objects of different distances appear to move at different rates across the retina. Those that are far away appear to remain stationary, while nearer objects move rapidly from one side of the retina to the other.

Stereopsis is the only contributing factor to depth perception studied with regard to age. This visual function appears to be maintained through the mid-40s

(Hofstetter & Bertsch, 1976) but shows cross-sectional age differences subsequent to that period (Bell, Wolf, & Bernholtz, 1972).

Practical Applications

It is apparent that many of the age effects on basic visual functions may be accounted for by reduced pupil size and increased lens thickness, opacity, and yellowness. These age changes have the effect of reducing available light to the retina, which in turn impairs the quality of the visual image under a variety of conditions. One obvious remedy to compensate for decreased retinal illumination is to turn up the light levels in the older adult's environment. What is not obvious, though, is the proper amount and type of illumination that will have the desired compenatory effect without creating glare. Appropriate increases in ambient lighting vary by the adult's age and by the degree of contrast in the visual array. At 40, a person may require no light increase for medium- and high-contrast displays, but twice the light for low contrast. These figures rise somewhat for the 60-year-old, and by 80, increases of about 1.5 and 5, respectively, are needed to offset age effects on retinal illumination (Richards, 1977). To compensate for reduced dark adaptation in the old, higher levels of illumination are particularly important in parts of houses and other buildings that tend to be dimly lit, such as halls, staircases, entrances, and landings (Weale, 1963).

To reduce glare as much as possible, it is suggested that the light used to provide more illumination for the older adult be yellowish, since bluish light is scattered more by the lens (which absorbs yellow light). For the same reason, activities requiring visual accuracy should not be performed in outdoor light (which has a high blue content). However, if color judgments are required, the older person will be at a disadvantage in yellowish light, because colors will appear more yellowish, and greens, blues and violets will be increasingly difficult to discriminate (Weale, 1963).

Age effects on depth perception can impair the older adult's mobility by leading to misjudgments of distance and height of obstacles and barriers. These misjudgments may occur in familiar settings, but are probably most likely to occur in new situations, where the individual is not familiar with the layout of an interior or exterior setting. Errors in perceiving verticality and horizontality were found in one study to relate to frequency of accidental falls, even when age was controlled (Tobis, Nayak, & Hoehler, 1981). Since the elderly adult is more vulnerable to fractures from falling (see chapter 2), this age effect on visual perception can have serious consequences.

Another set of practical implications concerns the driving ability of the adult as affected by the aging of the visual system. Reduced retinal illumination with age has the overall effect of placing the older adult driver in a position of having to make decisions in critical situations with incomplete and possibly inaccurate stimulus information (Panek et al., 1977). This problem is exacerbated when the outside light is minimal, during the evening and nighttime hours (Richards, 1977). During that time, reading road signs and avoiding pedestrians and obstacles may be less efficiently done. The older adult's greater vulnerability to scotomatic glare reduces the ability to recover from the headlights of oncoming cars on dark roads at night, and to drive during the daytime hours on streets that have alternating shade and bright sunlight. Poorer dark adaptation in the older adult adds to the problem that even younger adults have of changing from cone to rod vision when the light suddenly fades (Fozard et al., 1977), as in a tunnel, under bridges, and in going from a well-lighted highway to a small darker street. The lower acuity for moving objects demonstrated in older adults has deleterious effects on the ability to detect and avoid people and other cars approaching the driver at intersections, on or along the road, and coming out of driveways. A loss of sensitivity to movement in the periphery of the visual field means that the older adult will be less prepared to react to oncoming people and cars that appear to emerge suddenly onto the scene because their approach was not observed by the driver.

The loss of accommodation ability in adulthood has the effect of making it more difficult for the adult to read the speedometer and other important dashboard indices after focusing on objects in the distance on the road (Panek et al., 1977). Age-related changes in color vision may impair visual acuity for green, blue, and violet objects if the person's car has blue-gray or violet-gray tinted windows (as almost all cars do).

While it would seem logical that these age effects would result in higher accident rates for older drivers, this is generally not the case (Panek et al., 1977). It may be that older adults spontaneously reduce their driving overall, or at least under conditions known to them to create visual problems (e.g., night driving). Alternatively, older persons may continue to drive, even with some or all of these reductions in visual ability, but compensate for visual losses either by their many years of driving experience or by being more cautious than they were when they were younger.

Psychological Consequences of Aging of the Basic Visual Functions

The changes brought about by aging in the appearance of the eye and in its function may alter the individual's self-image as well as adaptation to the physical environment. Since the eyes have such a critical role in determining facial appearance, it would seem that the alterations with age in the eyes could have effects on self-image that are disproportionately large in comparison to the actual magnitude of the changes. Some of these changes—such as wrinkling, puffiness, and discoloration in the skin around the eyes, arcus senilus, and bifocals—were discussed in chapter 2. It was also observed in the present chapter that the cornea becomes cloudier in adulthood, the iris may show patches of depigmentation, brown patches, and the sclera becomes yellowed. These changes together have the effect of making the eyes look less distinctive in color and brightness. The adult experiencing some or all of the age effects on the eye's appearance may feel less attractive and be regarded as such by others. There is little, if anything, that can be done by the individual to ameliorate these age effects.

Age changes in the eye's function may have a variety of psychological effects. The yellowing of the lens, which alters color perception, may make the older adult less capable of deciding on proper color combinations of clothing and interior design, causing others to question his or her aesthetic sensibilities. In addition, appreciation of works of art, movies, scenery, and home decor would be impaired. Another aspect of visual functioning important for everyday life is the recognition of familiar faces. If it is true that older adults are less capable of discriminating facial features, their social worlds may become restricted to the extent that they fail to be able to learn the faces of new acquaintances or recognize old ones. Loss of distance acuity in general has the effect of making the visual environment smaller and less accessible.

To the extent that disordered depth perception reduces the person's ability to move freely without running into furniture, tripping on curbs, or falling down stairs, the individual may feel insecure and constrained within the environment. Inability to use the visual sense to compensate for decreased somesthetic and vestibular functioning (see chapter 11) combined with knowledge about the heightened vulnerability of older persons to bone fracture may lead to considerable fear about moving around in unfamiliar settings. Consequently, the older person may suffer from reduced physical mobility, becoming effectively isolated from the outside environment. As a result, the adult who becomes restricted in this way may suffer from a lowered sense of bodily competence.

For adults of any age, reduced range of accessibility of visual information can result from either uncorrected refractive errors, or astigmatism, or both. The eye strain and blurriness that arise from these sources create discomfort. Even if the visual disorders are corrected with glasses or contact lenses, residual symptoms may remain in special circumstances such as after overworking, or while performing visual tasks that place unusual strain on the eye such as trying to focus on the small print of a crossword puzzle. For persons wearing spectacles, there are difficulties in adjusting to their removal and replacement and usually poor peripheral vision. These discomforts may be perceived as a nuisance but have little effect on the adult's self-image in terms of perception of bodily competence unless they interfere with the performance of critical adaptive functions. When a noticeable pattern begins to emerge as the person gets older, the discomfort may be added to by concerns over the body's aging. Presbyopia, although reached after a gradual process taking place over a number of years, is often perceived by the individual as taking place relatively suddenly (Carter, 1982a). The immediacy of this apparent change and its impact on the adult's visual abilities, as well as the association many persons have between presyopia and the infirmities of age may make it more likely that the adult will interpret this change in a negative way.

Many age changes in visual functioning can be compensated for by corrective lenses, increases in ambient lighting, and efforts to reduce glare and heighten contrasts. With these measures, the older person's adaptation to the visual environment can often be maintained at a reasonably high level. Nevertheless, age changes in vision narrow substantially the person's range of movement in the sense of increasing the amount of dependence on compensatory devices. When the situation does not permit compensation (such as driving at night), the older adult's adaptation to the environment is unquestionably reduced. Awareness of greater dependence on visual aids and ideal viewing conditions may have the effect of leading the older individual to feel a loss of competence that can detract from self-image. This loss may be particularly pronounced in people who have always had good vision. The effects of age will probably be emotionally less disruptive for the many adults who have had some form of visual problem requiring correction for a number of years.

Age Differences in the Neural Basis for Visual Perception

Visual perception is not a simple matter of reproducing point-for-point the image of the outside world onto the visual receptors and higher-order neural cells. Instead, at each level of organization in the visual system, specialized cells respond to specific individual features of the visual image, such as edges of light and dark. A total picture of the visual scene is perceived when these features are integrated by cells in the secondary areas of the visual cortex which are capable of responding to complex combinations of the specific features already processed into lines, curves, and angle. These combinations then form the basis for perceiving form and meaning in the visual image.

Higher Visual Pathways and Projection Areas

From the retina, visual information passes through the optic nerve, where the ganglion cells' axons synapse on neurons in the pair of lateral geniculate nuclei in the thalamus. Prior to reaching the thalamus, most of the optic nerve fibers cross over at the optic chiasma. The receptive fields of lateral geniculate cells have similar shapes and characteristics as those of the ganglion cells, but in the thalamus there is greater sensitivity to contrast than in the ganglion cells.

Fibers from the thalamus pass to the occipital lobe of the cerebral cortex, synapsing on neurons in the primary visual area. The cells in the primary visual cortex are of two types: simple and complex. Unlike the concentric receptive fields of the retina and thalamus, the antagonistic portions of the simple cortical neuron receptive fields are shaped like long, narrow rectangles. The orientation of the line striking the cortical receptive field determines the intensity with which the simple cortical cell responds. The maximum response occurs when the line's width and orientation matches up exactly with the center of the rectangular receptive field. All simple cells respond to a line stimulus moving in the same orientation as the receptive field, but some simple cortical cells respond to the rate and others to the direction of movement.

Complex cortical cells receive input from many simple cells having the same sensitivity to orientation. A moving stimulus in the right orientation stimulates the complex cortical cell, but the exact location in the field of the moving line is not important. The receptive fields of complex cells do not have the on-off antagonistic organization of simple cortical, thalamic, and retinal cells. Complex cortical cells are also more sensitive to the rate and direction of movement than are simple cortical cells. Another difference between simple and cortical cells is that while simple cells are influenced primarily by one eye or the other, complex cells receive convergent stimulation from the two eyes.

The retina is represented in a point-by-point fashion on the cortex. Cells representing each region of the retina in the cortex are vertically organized in columns. All cells within each column have receptive fields with the same axis of orientation. Cells from the primary visual cortex project to the secondary association areas, which are laid out like the retina and organized in columnar fashion according to the angle of orientation of the receptive field. The topography of the retina is also reproduced in these areas. In addition to complex cells, the secondary association areas also contain hypercomplex cells, which respond to combinations of the stimulus features that trigger the complex cells, that is, curving borders, the angles between two lines, and lines of different lengths. The level of visual processing in the secondary visual areas by complex and hypercomplex cells is, then, at its most sophisticated level, allowing for the detection of complex forms and movement on all parts of the retina. Transmission of this information to other areas of the brain, in the temporal lobes, makes possible the attachment of meaning to these patterns of stimulation.

Age Effects on Higher Neural Functions

Age-related changes in the pupil and lens underlie many of the observed age differences in the basic visual functions. Less is known about how age affects the visual pathways in the brain, because the functions that depend on processing in the visual cortex are less easily measured than functions such as, for example, acuity and light sensitivity. This is partly because the kinds of functions that depend on cortical operations involve what are called "cognitive" factors. In searching a visual display, to take one case, the individual's expectations about what to look for, reluctance or unwillingness to report having seen something, and ability to make a decision, are all influences that do not have a basis in strictly visual processes. Some of these influences may have a basis in the adult's motives and needs, so that the observer avoids seeing things that would be disturbing or upsetting (Erdelyi, 1974). Other cognitive factors relate to the way that attention is allocated to incoming visual stimulation (Hoyer & Plude, 1980, 1982). The study of cognitive and other factors influencing vision falls outside of the scope of this chapter. However, there are some visual functions thought to relate more directly to activity in the higher cortical levels of the nervous system. Changes in these functions may have some relation to a decreasing number of neurons in the primary visual cortex (Devaney & Johnson, 1980).

Backward Masking

The visual functions studied by "masking" techniques appear to have a basis in processes taking place at both the retinal and higher cortical levels. Recently, researchers on aging of visual functions have used these techniques extensively to determine the relative impact of aging on processing taking place at peripheral levels from the retina up to the primary visual cortex and at central levels, from the primary to visual association areas. Several theories of why masking has the effects it does can be applied to studies on aging to yield an understanding of the possible reasons for age effects, at both peripheral and central levels.

The masking procedure involves, very simply, showing the observer one stimulus called the "target" (usually a flash of light or a letter or letters) and trying to obscure the observer's perception of that stimulus by another stimulus, called the "mask" (another flash of light, a circle, a random array of lines, or a pattern), shown in the same visual field. The critical variables that are measured using this procedure primarily concern what it takes for the observer to "escape" from masking, that is, to be able to see clearly the target, uninfluenced by the mask. Most of the research on aging has used the backward masking technique, in which the target stimulus is shown first, followed by the mask. The observer is then asked to report on what the target stimulus was. (In the forward masking technique, the mask is shown first.)

Based on a long series of studies on young adults, Turvey (1973) set forth a number of criteria for the stimuli and relationships among time-dependent variables for discriminating between peripheral and central visual functions in the context of masking. Peripheral processes involve extracting the intensities of the stimulus pattern, and their efficiency is determined by the target energy. The target represented as a result of peripheral processing in the primary visual cortex according to its contours and orientation (Breitmeyer & Ganz, 1976) is then processed centrally, that is, analyzed and identified at higher cortical levels. According to Turvey (1973), central processing is not affected by the target energy, but it is a function of the time permitted for processing the target before the mask is presented, that is, the interstimulus interval. To escape from masking, the target must be presented long enough to reach the higher cortical levels where, once it has been analyzed, it is not degraded by the mask.

Using these (and others) of Turvey's criteria, investigators of age effects on backward masking have tried to determine whether peripheral or central processes are more affected by aging. It has been found that to escape masking, older adults need a longer target duration (Walsh, 1976), greater target energy (Till & Franklin, 1981), or a longer interstimulus interval (Cramer, Kietzman, & Laer, 1982; Kline & Birren, 1975; Kline & Szafran, 1975; Till & Franklin, 1981; Walsh, Williams & Hertzog, 1979). These results suggest that both levels of processes seem to be less efficient in older adults, beginning at least in the 50s and 60s. Age differences in peripheral processes are probably due, in part, to lower perceived brightness caused by decreased stimulation of retinal cells resulting from less light entering the eye (Walsh, 1982). Age differences in the time-dependent central processes may be a function of the longer time required by older adults to integrate the visual information in the target (Coyne, 1981; Eriksen, Hamlin, & Breitmeyer, 1970).

Another explanation of backward masking effects is based on the possibility that, just as there are neurons in the visual system that are sensitive to intensity, orientation, and movement, there are also pathways of neurons from the retinal to cortical levels sensitive to the temporal qualities of stimulation (Breitmeyer & Ganz, 1976). According to this view, "transient" cells respond to a change in light intensity, firing only when there is a decrease or increase. In contrast, "sustained" cells respond continuously as long as there is stimulation. In addition, the sustained cells are sensitive to the spatial frequency of the stimulation, with those responding to stimuli with higher spatial frequency responding later and persisting longer than those sustained cells responding to lower spatial frequencies. Transient cell activity has the effect of inhibiting sustained cell activity. As is illustrated in Figure 9.2, inhibition between these two channels (transient and sustained) can explain backward masking effects. The sustained activity that builds over the course of the time the target is shown is inhibited when the transient channel is stimulated by the mask. The later the mask is presented, the more that fine details of the target are lost. There are no research applications of this model to aging in the context of backward masking studies. It is suggested, though, that aging differentially impairs transient channel activity, so that the sustained channel becomes dominant (Kline et al., 1982). The older adult would be more susceptible to backward masking effects because transient channel activity resulting from presentation of the mask would not interrupt effectively the sustained pattern of the target (Kline & Scheiber, 1981).

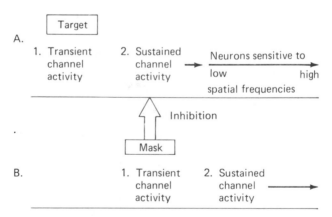

FIGURE 9.2. Masking effects as explained by two-channel model. When the target is presented activity is initiated in the transient channel followed by activity in the sustained channel, which persists over time. The first neurons to respond in the sustained channel are those sensitive to low spatial frequencies, followed by neurons sensitive to high spatial frequencies. Consequently, detailes of the target become resolved with more clarity over the coruse of the target duration. Transient channel activity stimulated by the mask inhibits sustained activity of the neurons responding to the target, so that the visual representation of the target is degraded. If the mask is presented relatively soon after the target, the gross outlines of the target will be lost, but if the mask is presented later only the details will be obscured.

Stimulus Persistence

The continued representation of a stimulus in the visual system after the stimulus is no longer present is called stimulus persistence (Haber & Standing, 1969). Stimulus persistence is demonstrated in cases where the observer reports "seeing" a stimulus after it is no longer being shown. Stimuli presented in sequence would appear to blur together. According to the "stimulus persistence" hypothesis of aging, visual processing in older adults is slower because the neural trace of a stimulus takes more time to clear through the system. Consequently, with increasing age, adults become less efficient at rapidly processing new information due, in effect, to a backlog of traces from prior stimuli preventing the processing of new ones.

The stimulus persistence hypothesis was used by Botwinick (1978) to account for a number of findings of visual perception studies and aging (as well as studies on other senses). At that time, the support in favor of the stimulus persistence hypothesis was regarded by Botwinick as equivocal. Since then, there is more research in favor of this hypothesis (reviewed in Kline & Scheiber, 1982).

Although stimulus persistence would seem to provide a plausible explanation for age effects on critical flicker fusion, it is generally agreed that fusion is based on retinal factors. The stimulus persistence hypothesis applies instead to central processes (Kline et al., 1982) and could, perhaps, be related to the possible disruption with age in transient/sustained channel activity. The diminished activity of the transient channel would mean that the activity of the sustained channel would continue over a longer time without being inhibited by subsequent stimulation (Kline & Scheiber, 1982).

Psychological Consequences of Age Effects on Higher Visual Functions

When discussing the relevance to the individual's adaptation in adulthood of higher-order neural processes in vision, it is when cognitive factors are introduced that the most applications are found. This is because cognitive factors often reflect and are influenced by the individual's life experiences (Hoyer & Plude, 1982; Whitbourne & Weinstock, 1979). These cognitive factors probably are a function of neural events taking place in the secondary visual areas, in intersensory association areas, and in connections with the limbic system. Other perceptual functions (hearing, touch, and so on) also involve cognitive factors, but these factors are most thoroughly investigated in the visual area.

The cortical functions discussed here may have the most direct relevance to the individual's sense of time. It is speculated that the amount of time required to integrate visual stimulation into a coherent image represents the "psychological moment." The increased central visual processing time presumed to occur in the aged may, according to this view, have the effect of lengthening a unit of perceived time (Eriksen et al., 1970). A slowing of the individual's "psychological clock" would result in "real" clock time seeming to go by quickly. For instance, when two units of psychological time have gone by, actually five units of real time have elapsed. The clock, therefore, would seem to speed ahead of the individual

(Fraisse, 1963). The effect, over the long term, to the older individual would be one of calendar time rushing by, causing the individual to feel that time left to live is "running out." Given the lack of agreement in studies of time estimation in the aged (e.g., Feifel, 1957; McGrath & O'Hanlon, 1968; Salthouse, Wright, & Ellis, 1979; Surwillo, 1964), this interpretation of the results on visual perception are probably somewhat premature. However, since many of these studies on time perception have serious and perhaps unavoidable flaws in the way they were conducted (Hendricks & Hendricks, 1976), the research on the visual psychological moment may ultimately prove to be more useful for theorizing about aging and the perception of time.

The types of studies on which the research on central visual processing and age is based do not, on the surface, appear to have a great deal of practical relevance, except as a way of learning about how aging affects cortical functions. However, the transient/sustained two-channel model used to understand some of these studies does have ramifications for the individual's adaptation (Breitmeyer, 1980). Transient stimulation is critical in order for the individual to be alerted to rapid changes in the environment caused by approaching people and objects or as a result of the individual's own bodily movements. If there are age decrements in the ability to interrupt sustained processing when new stimulation is introduced, then the older person will be less sensitive to environmental changes. Conversely, the accurate perception of details is made possible by activity of the sustained channel. Age effects in transient and sustained channels at the central level could augment the effects of aging on the structures of the eye to impair functions such as static and dynamic visual acuity, which are critical to maintaining the individual's mobility and sense of independence. However, offsetting some of these age-related changes may be higher cognitive functions, which can compensate for impaired vision through their reliance on a lifetime of processing visual information.

CHAPTER 10

Auditory System

Hearing is a function that governs the individual's ability to exchange communication with others and to appreciate the variety of natural and artifically created sounds that form an essential part of the matrix of everyday life. The ears contain the sensory organs responsible for converting the energy of waves of sound pressure into neural impulses that are ultimately interpreted in the brain as words, musical notes, bird calls, and the grinding of machinery. Like the visual system, the mechanisms through which the auditory system accomplishes its analyses of sounds are studied in terms of its peripheral and central features.

Age Effects on the Ear

The structures in the outer and middle ear that conduct sound waves and the inner ear mechanism that transforms this stimulation to neural impulses are affected to differing degrees by the aging process. In general, aging has less of an effect on the sound conduction pathways than on the sensory and neural components of the ear. This generalization applies to the structures themselves and also to the auditory functions served by these structures.

Receptive and Sensory Structures of the Ear

The ear contains a three-tiered set of structures that serves to transform sound into a form that is usable by the nervous system. The structures within the ear accomplish their functions of converting sound into neural signals by conducting waves of sound pressure through an elaborate series of channels and moving parts in the outer and middle ear until they reach the cochlea, the sensory organ in the innermost part of the ear.

The Outer Ear

The pinna is the visible flexible part of the ear which serves to direct captured sound waves. This structure plays an important role in localizing the source of sound, and so helps the individual in making directional judgments. The concha is the outside opening of the external auditory canal, in which sound waves of moderately high frequency are enhanced by resonance. Cells in the outer ear canal secrete cerumen (earwax).

The effects of age on the pinna are essentially cosmetic ones, affecting how the ears look and feel. With increasing age in adulthood, the pinna loses flexibility, becomes longer and wider, can become freckled, and develop longer, thicker, and stiffer hairs. With age, the outer ear canal becomes wider, the skin lining of its walls thin, it loses elasticity, and may become brittle and dry, so that it is more vulnerable to cracking, crusting, and bleeding.

Hearing is not affected by these particular changes in the outer ear, since they do not have an impact on the conduction of sound waves from the outside air to the tympanic membrane. Some of the effects of age on the outer ear structures, such as the growth of longer and thicker hair and thinning and dryness of the outer ear canal can have the secondary effects of contributing to the accumulation of an excessive amount of cerumen (Anderson & Meyerhoff, 1982; Corso, 1981; Schow, Christensen, Hutchinson, & Nerbonne, 1978). Due to a diminution of sweat gland activity, which normally keeps the outer ear moist, the accumulated cerumen may also be drier and less easily removed in the course of normal activities. The build-up of cerumen in the outer ear can have a drastic effect on hearing loss, especially for tones at low frequencies. It is estimated that about one-third of the cases of hearing loss in aged persons are the result of cerumen accumulation (Fisch, 1978).

Middle Ear

In contrast to the relatively simple sound wave conduction that occurs in the outer ear, the structures in the middle ear transfer sound energy through a delicately regulated series of conversions. These conversions are needed because sound waves travel more readily through the air in the outer ear canal than through the fluid-filled inner ear. These sound wave vibrations are the stimulus for neural impulses in the inner ear. The vibrations would be lost (reflected back toward the outer ear) unless the middle ear structures equalized the differing impedances (resistance to movement from sound pressure waves) of the outer and inner ears. The middle ear includes the tympanic membrane, Eustachian tube, and the ossicles as well as membrane-covered openings to the inner ear called the oval and round windows.

Sound waves passing through the outer ear reach the thin, elastic, and slightly tensed tympanic membrane, which is stretched across the end of the outer ear canal. The tympanic membrane vibrates in a swinging motion, causing the ossicles to be set into motion. The hammer strikes the anvil, causing the stirrups in turn to displace some of the fluid in the cochlea and start a wave traveling down through the cochlea. The middle ear enhances the sound vibrations in the outer ear to overcome the high impedance of the fluid in the cochlea by concentrating the energy reaching the tympanic membrane onto the much smaller area of the oval window as the stirrup hits it. When the oval window moves inward, a "relief valve" in the cochlea moves in the opposite direction so that the fluid has a place to which to flow.

The membrane over the round window serves as this relief valve. Since the force on the oval window membrane is stronger than that on the round window membrane, the fluid in the cochlea is forced to move in a forward direction when the oval window membrane is pushed in by the stirrups. The round window membrane bulges outward from the pressure of the cochlear fluid on it. Negative pressure on the tympanic membrane causes the process to be reversed.

There is a category of hearing loss, called conduction deafness, that has its source in sound transmission problems caused by structural defects in the middle ear. This form of deafness does not appear to be one of the types of hearing loss more common in the aged, though, despite the fact that the age effects would seem to have the same consequences as the structural defects in conduction deafness. Instead, it appears that the age changes in the middle ear structures have some, but not a significant effect on the hearing problems of the aged (Schow et al., 1978; Yarington, 1976).

These changes occur in both the tympanic membrane and the ossicles. The tympanic membrane, to be maximally responsive to sound pressure waves, must be firm yet elastic. With increasing age in adulthood, this membrane becomes thinner and less rigid. Its loss of resiliency is compounded by degeneration and atrophy of the muscles and ligaments that support it and enable it to move (Schow et al., 1978). Another change in the middle ear attributed to the aging process is calcification of the ossicular chain. The joints between the ossicles (hammer, anvil, and stirrups) stiffen, so that they transfer sound vibrations less efficiently from the tympanic membrane to the oval window of the cochlea. Here again, though, this apparent age effect does not translate into a functional effect, since aged people whose ossicles were found to be calcified did not show differences between audiograms taken by air and bone conduction methods (Etholm & Belal, 1974). Moreover, there is no relationship between age and the air-bone conduction gap (Fisch, 1978), a measure of middle-ear dysfunction as signified by different auditory thresholds for pure tones transmitted through the conductive pathways to the ear and vibrations delivered directly to the bone by a vibrator placed on the skull.

Contraction of the middle ear muscles attached to the hammer and stirrups causes the acoustic reflex, in which the ossicles stiffen, preventing sound energy from reaching the inner ear. The acoustic reflex is stimulated by a very loud noise, serving the function of protecting the sensitive inner ear from damage. Another benefit of the acoustic reflex is that it minimizes auditory distractions caused by the movement of the body, and by the sound created by one's own voice during speaking.

Weakness of the muscles and ligaments in the middle ear (Schow et al., 1978), in addition to the hardening of the ossicular chain, could have the effect of diminishing the acoustic reflex. The results of research on age differences in sensitivity to the acoustic reflex have been contradictory (Marshall, 1981), but appear to have been at least partially clarified in research by Silverman et al. (1983). It was found that the acoustic reflex in response to pure tone activators was not correlated with age for tones of varying frequencies (0.5 to 2 kHz), in agreement with other reports in the literature (Gelfand & Piper, 1981; Silman, 1979; Thompson, Sills, Recke, & Bui, 1980). However, the finding by Silverman et al. (1983) of no age differences in the acoustic reflex in response to a higher frequency tone (4 kHz) was not consistent with other findings (Wilson, 1981). In contrast to these findings of relative stability in adulthood, there was an early and significant reduction in the acoustic reflex response to broadband noise, beginning in the 40s. This finding was interpreted as suggesting that responses to more complex stimuli, as in the case of speech perception (see the following) may provide a more sensitive indicator of the "diffuse" effects of aging than responses to pure tone signals.

The Eustachian tube connects the middle ear cavity and the nasopharynx. Although not directly involved in hearing, the Eustachian tube serves the important function of adjusting the pressure in the ear to the outside air pressure. During swallowing, the Eustachian tube opens into the nasopharynx, equalizing the pressure between the two passageways. This is why during an ascent, up a mountain or in an airplane, swallowing helps to "open up" the ears. The Eustachian tube also serves to drain off excess mucus that collects in the ear canal due to inflammation. There appears to be little effect of age on the Eustachian tube. This structure can impair sound transmission if it becomes filled with fluid in the ears of anyone, regardless of age, with upper respiratory disturbances. The only possible age effect is the development of negative pressure in the middle ear cavity due to muscle weakening (Schow et al., 1978).

Inner Ear

The vestibular system (vestibule and semicircular canals) are located in the inner ear. This system contains cells that are sensitive to change in gravity and accelerative forces (see chapter 11).

The sensory organ of the auditory system is the cochlea, located in the inner ear. It is a tightly coiled spiral encased in a bony core with three fluid-filled cavities. The base of the cochlea is the portion nearest the middle ear, and the apex is the most central portion of the spiral. The cochlea's width decreases from the base to the apex. The elasticity of the scala media also varies across the length of the cochlea, with the base being much stiffer than the apex. Together, differences down the length of the cochlea in width and elasticity result in the cochlear partition varying in its sensitivity to high- and low-pitched sounds.

The varying sensitivity of the cochlea to different sounds is based on the nature of the pressure waves generated by sounds of high and low pitch. Sound waves have two principal features: frequency and amplitude. It is the frequency that determines the perception of pitch (high or low) and the amplitude that forms the basis for the perception of intensity (loud or soft). Sounds that are perceived by the listener as high in pitch (e.g., the high "C" sung by a soprano) generate pressure waves that have a high frequency (numerous cycles per second). In contrast, low-pitched sounds (e.g., the rumbling of thunder) generate low-frequency pressure waves. The wave that travels down the cochlear fluid in response to movement of the oval and round windows changes in amplitude from base to apex due to variations along the length of the cochlea in width and elasticity. Waves of different frequencies reach their highest amplitudes at varying points on the cochlea. After reaching their highest amplitudes, they quickly diminish. High-frequency waves reach their maximum amplitudes at the base, whereas waves of low frequency travel all down the length of the cochlea (through the spiral), reaching maximum amplitude at the apex. Consequently, the base of the cochlea is most sensitive to high-frequency sounds and low-frequency sounds are received with the greatest sensitivity at the apex.

The sensory cells that respond to sound pressure waves are located on the organ of Corti, situated on the basilar membrane. The tectorial membrane is like a flap and is made up of a jellylike, fibrous material. It is firmly attached to the inner edge of the bony cochlear wall and loosely connected by small fibers to the organ of Corti. The hair cells and their supporting cells are attached to the basilar membrane. When a wave of sound pressure travels through the cochlear duct, the basilar membrane moves up and down, causing relative movement between the tectorial membrane and the organ of Corti.

The sensory cells are called "hair cells" because they have cilia at their tops, which, when stimulated, begin to generate neural impulses to the auditory nerve. There are two types of hair cells: inner and outer. The inner hair cells are arranged in a single row near the inner portion of the cochlea. Each inner hair cell is completely surrounded by supporting cells. The inner hair cells make up most (95%) of the input to the afferent nerve fiber that lead to the brain, with each inner hair cell being innervated by about 20 afferent nerve fibers. There are longer and more numerous hairs on the outer hair cells, most of which are embedded in the tectorial membrane. The outer hair cells are arranged in three to four rows around the outer portion of the cochlea, supported only at their tops and bottoms by surrounding cells. The outer hair cells feed into the remaining 5% of the auditory nerve fibers, with each one being innervated by about six fibers from the auditory nerve. Since the outer hair cells are directly connected to the tectorial membrane, they receive mechanical stimulation from the relative movement of this surface caused by displacement of the basilar membrane when sound waves travel down the cochlea. The inner hair cells have a lower threshold of stimulation, and so can generate an impulse from the less intense stimulation created by movement of the thick cochlear fluid in response to movement of the basilar membrane.

The blood supply to the cochlea originates in the cochlear artery, which divides into two portions: the stria vascularis and the spiral vessels. The stria vascularis runs along the outer wall of the scala media. It is an extensive network of capillaries, and also is the site where the fluid in the scala media is secreted. Supplying most of the blood supply to the organ of Corti are the spiral vessels, which travel longitudinally beneath the basilar membrane.

Auditory sensitivity is defined in terms of how soft a tone of a given frequency can be and still be heard. Sensitivity varies with frequency, because for the normal listener, high frequency tones can be heard at very soft levels, while low frequencies must be louder in order to be detected. The audiogram is the format used to describe a person's auditory sensitivity. The individual being tested is presented with pure tones at differing frequencies, and for each tone, the threshold is determined: the point of loudness when the tone can be heard on 50% of the trials. A normal listener can detect sounds with frequencies ranging from 20 to 20,000 Hz.

The intensity dimension used in testing auditory sensitivity is represented in decibels (dB). The decibel is a ratio (on logarithmic units) expressing the relationship between sound pressure (which is a function of intensity) to a standard, "reference," pressure, which is the loudness at which most normal listeners can hear the tone. Decibels represented on this scale are referred to as "dB HL" (decibels relative to hearing level). Lower numbers on this scale represent better hearing.

There is a gradual loss of hearing that begins relatively early in adulthood (in the 30s) and continues progressively at least until the 80s. The noteworthy feature of this kind of hearing loss in addition to its relationship to age is that in most cases, the sensitivity to tones of high frequency is impaired earlier and more severely than loss of sensitivity to low-frequency tones. The term used to describe this kind of hearing loss is presbycusis. The anatomical source of presbycusis is the inner ear. Presbycusis is not, however, due to one single age-related change in the cochlea. There are several kinds of presbycusis, each believed to have a different structural origin within the inner ear, and also to differ in their functional effects as represented on audiograms (Anderson & Meyerhoff, 1982; Corso, 1981; Fisch, 1978; Gacek, 1975; Marshall, 1981; Pickett, Bergman, & Levitt, 1979; Schow et al., 1978; Schuknecht, 1964).

Forms of Presbycusis

Sensory Presbycusis. Sensory presbycusis is due to degeneration of the sensory cells in the organ of Corti. The loss of receptor cells associated with sensory presbycusis is more pronounced in the basal portion of the organ of Corti, especially for the outer hair cells (Bredberg, 1968). Because the loss of hair cells occurs in the basal coil of the tectorial membrane, hearing is impaired for high-frequency tones. The audiograms of individuals with sensory presbycusis show a sharp increase in hearing loss at high frequencies. Speech understanding is not affected, since sensitivity for the tones within the frequency range of normal speech is relatively well-preserved.

Neural Presbycusis. Atrophy of the spiral ganglia is the cause of neural presbycusis, which is called "neural" because it involves degeneration of nerve fibers. In this type of hearing loss, the sensory cells in the organ of Corti do not degenerate. One possible cause of neural presbycusis is a growth of bone tissue in the inner ear canal, so that the holes through which the fibers of the spiral ganglia exit from the tectorial membrane become blocked. As these holes close, the nerve fibers become compressed and degenerate (Krmpotic-Nemanic, 1971). The process of neural loss does not have an effect on hearing until the neurons

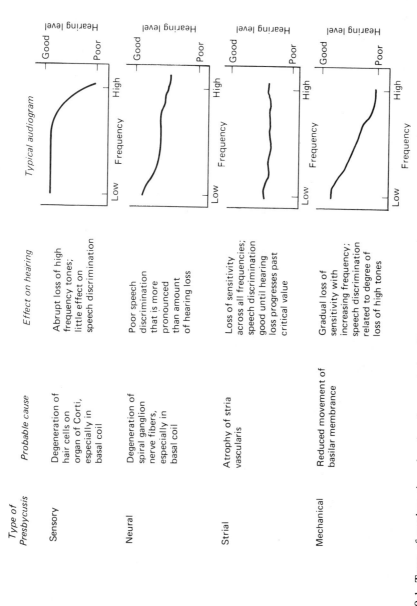

Type of Presbycusis	Probable cause	Effect on hearing	Typical audiogram
Sensory	Degeneration of hair cells on organ of Corti, especially in basal coil	Abrupt loss of high frequency tones; little effect on speech discrimination	
Neural	Degeneration of spiral ganglion nerve fibers, especially in basal coil	Poor speech discrimination that is more pronounced than amount of hearing loss	
Strial	Atrophy of stria vascularis	Loss of sensitivity across all frequencies; speech discrimination good until hearing loss progresses past critical value	
Mechanical	Reduced movement of basilar membrane	Gradual loss of sensitivity with increasing frequency; speech discrimination related to degree of loss of high tones	

FIGURE 10.1. Types of presbycusis and major characteristics. *Note.* Figure material in column 4 is based in part on "Otologic Manifestations of Aging" by R.G. Anderson and W.L. Meyerhoff, 1982, *Otolaryngologic Clinics of North America, 15,* pp. 353–370.

fall below a critical level, which usually does not happen until relatively late in adulthood. The effect on hearing is to reduce speech discrimination below the level that would be expected from the pure-tone audiogram. Presumably, this effect on speech discrimination is a consequence of its effect on the transmission of neurally coded information used in the analysis of speech patterns.

Strial Presbycusis. A slowly progressing atrophy of the stria vascularis, beginning in the 20s, results in strial presbycusis. By the 50s to 60s, the cumulative effects of this degenerative process becomes apparent. Atrophy of the strial cells changes the nature of the fluid in the scala media, which makes possible the transformation of the mechanical sound wave into a neural signal. The effect on hearing is, unlike the other forms of presbycusis, a reduction of sensitivity to all frequencies, producing a flattened audiogram with uniform hearing loss. Speech discrimination is relatively unaffected until hearing sensitivity falls below a certain threshold.

Mechanical Presbycusis. Less is known about the anatomical basis for this form of presbycusis, but it is thought that it results from some deficiency in the vibrating motion of the basilar membrane. This deficiency might be caused by loss of elasticity, atrophy of the spiral ligament, or atrophy, ruptures, and thinning of the basilar membrane. Like the audiogram of a person with sensory presbycusis, the audiogram produced in cases of mechanical presbycusis shows reduced hearing at high frequencies. However, unlike the audiogram produced by persons with sensory presbycusis, the audiogram associated with mechanical presbycusis shows a gradually descending curve moving from low to high frequencies. Speech discrimination in the case of mechanical presbycusis is relatively intact as long as there is good amplification, but is impaired when the loss of high-frequency tones progresses into the range of normal speech.

The most likely causes of the four types of presbycusis and their audiograms are summarized in Figure 10.1. As can be seen from this figure, sensitivity to tones of high frequency diminishes in most types of presbycusis. Speech discrimination is related in three out of four cases, to the degree of hearing loss.

Hearing Loss in the Aged

Audiometric curves of hearing loss by tone frequency for different age groups reveal a general pattern of greater hearing loss for high frequencies among increasingly older samples. This pattern reflects a combination of all four types of presbycusis, and the fact that it shows a downward trend at the higher tones can be attributed to the dominance of this shaped curve among the four types of presbycusis.

A curve representing average audiograms across eight cross-sectional studies for the decades of 20 through 80 is shown in Figure 10.2. The increasing degree of hearing loss at the higher frequencies for progressively older age groups beginning after the age of 50 is apparent from this figure. In addition, it can be seen that for older age groups, hearing loss progresses to the lower frequencies, so that

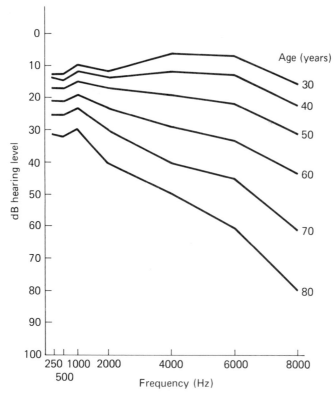

FIGURE 10.2. Hearing level frequence by age decade in adulthood. *Note.* From "The Presbycusis Component in Occupational Hearing Loss" by C.P. Lebo and R.C. Reddell, 1972, *Laryngoscope, 82*, pp. 1402–1403. Copyright 1972 by Laryngoscope Company. Adapted by permission.

hearing is impaired for a wider range of tones. Not apparent from this curve are the fairly large sex differences in hearing loss. Beginning at age 40, men have poorer hearing at frequencies of 4000 Hz and above, and by age 60, from 2000 Hz and over. It is thought that this sex difference is due to the greater exposure that men have to environmental noise in their occupations (Fisch, 1978). Since noise exposure has a detrimental impact on hearing, its cumulative effect over the working years would become apparent by middle adulthood.

In interpreting Figure 10.2, it is important to ask at what point the amount of hearing loss constitutes a deficit significant enough to impair the individual's normal everyday activities; in other words, at what point does an increase in hearing level with age become a handicap? There is a lack of agreement in the literature on aging regarding this point, with a range of from 15 dB to 40 dB regarded as a handicap. In most cases, hearing levels for high and low frequencies are regarded uniformly. Since presbycutic hearing loss differs according to frequency, the use of one hearing level irrespective of frequency as a criterion for impairment does not yield an accurate picture of the degree of handicap the individual will experience. Moreover, by placing all persons with a hearing level above a given number into one impairment category, variations are ignored in the degree of severity of handicap beyond that number. In evaluating the impact of hearing loss on a given older adult's life, then, it is important to take into account the actual amount of hearing level increase and to separate hearing levels for low and high frequencies.

The use of multiple criteria for a hearing handicap has led to a fairly wide range in published estimates of the prevalence of presbycusis in the older adult population. In one source, these estimates ranged from 12% to over 50% and led the author to conclude that approximately one-third of the over-65 population has a hearing impairment that can have "unfavourable social consequences" (Fisch, 1978, p. 283). A more conservative estimate that 13% of the 65 and older age group have hearing loss due solely to presbycusis was made by another investigator (Corso, 1977). In comparing these estimates, it is important to note that the lower number excludes other sources of hearing loss not directly related to normal aging. Secondly, the 33% estimate is based not only on actual hearing tests given to large samples of adults, but also includes figures from studies in which people were asked to estimate their own hearing impairments.

In the cases where these self-reports were used, the population estimates tend to be higher than estimates based on audiometric data. It is difficult to say which of these two sources is more valid. On the one hand, the perceived effect of hearing loss will determine much of the individual's reponse; on the other hand, audiometric data provide a firmer basis for drawing population estimates. It is probably fair to regard the differing percentages as the range of hearing impairment in the elderly due to presbycusis, with the actual figure being somewhere in between one-tenth and one-third. This range represents a large number of persons over 65 years, but is far from a majority.

Age Effects on the Higher Auditory Structures and Speech Perception

Higher Auditory Structures

The pathway carrying auditory signals to the central nervous system originates in the spiral ganglia, nerve fibers that travel from the hair cells in the cochlea to the auditory nerve through small openings in the central bony core of the cochlea out through the internal auditory canal. There are a total of 30,000 auditory nerve fibers. Since the inner hair cells feed the most information to the auditory nerve, the responses of the auditory nerve fibers closely parallel the pattern of stimulation to the inner hair cells. The auditory nerve is organized according to the point on the cochlea at which hair cells are stimulated, with fibers from the apex forming the core of the auditory nerve and fibers from the base arranged along the periphery. In addition to these afferent neural fibers, which carry information from the ears to the brain, there are also a smaller number (about 1800) of efferent fibers, which convey information from the cortex to the organ of Corti.

Auditory nerve fibers from each ear synapse at the cochlear nucleus at their own sides of the brain stem, where some preliminary auditory analyses take place. From the cochlear nucleus, most neurons travel across the trapezoidal body to the superior olivary nucleus, while others synapse on the ipsilateral side (not all neurons from the cochlear nucleus synapse at the superior olivary nucleus). Disparities between the two ears of timing and intensity of neural stimulation in the superior olivary nucleus make it possible for the individual to localize sound direction. Some neurons from the superior olivary nucleus arising ipsilaterally and contralaterally ascend through the lateral lemniscus tract to the nucleus of this tract, while other bypass this nucleus in their ascent up the system. A few fibers cross over at the commissure of Probst, providing another point of exchange between the two ears. Continuing up the lateral lemniscus tract, the next point of synapse for many neurons is in the inferior colliculus. Some neurons bypass this nucleus; others go no further than this level. Most neurons travel to the contralateral side of the midbrain here, across the inferior colliculus commisure.

Several important functions are served by the inferior colliculi, most notably, the acoustic reflex, the startle response to loud sounds, and the direction of attention to acoustic stimuli. Those fibers that pass on from the inferior colliculus synapse in the medial geniculate body of the thalamus, the final subcortical center where auditory information is sorted and recoded before passing on to the cortex. There is also an efferent pathway from the superior olivary nucleus to the cochlea. This inhibitory pathway facilitates the directing of attention onto specific features of the auditory environment.

In the primary auditory cortex, auditory stimuli become finely "tuned," since the neurons here respond to a narrower range of frequencies than do those at previous levels of the auditory system. Analysis takes place of complex tonal and sequential sound patterns, and sound direction information is interpreted. Binaural information is coded in both primary auditory areas, and there are ill-defined pathways connecting these areas to the temporal lobes on each side of the brain.

A major function of the secondary auditory cortex, the auditory association area, is the transfer of information stored in one side of the primary auditory cortex to the contralateral auditory association area. Information transferred from the right to left side of the brain underlies the conversion of tonal and sequential patterns to speech; nonverbal auditory stimuli pass from the right primary to left secondary auditory areas. From the secondary association areas, this highly refined and translated auditory information is combined with information from other senses, transferred to cortical instructions for motor responses including speech, and is processed in short- and long-term memory.

At virtually all levels of the auditory nervous system, tonotopic organization is maintained, with neurons responding to similar frequencies located adjacent to each other. Another feature of the auditory system is the variety of possibilities for bilateral stimulation, as contralateral and ipsilateral information are combined through cross-hemispheral pathways. Finally, because not all neurons synapse at all nuclei, some pathways to the cortex may have more neurons than others. As a result, information received by the ears at the same time may reach the cortex at different intervals, adding further complexity to an already elaborate aural processing system.

Speech Understanding

Hearing and comprehending oral speech depend on peripheral processes that operate in the cochlea and central processes in the brain stem, thalamus, and cortex. At the peripheral level, speech sounds are translated into neural signals on the basis of their frequencies.

The smallest unit of speech is the phoneme, which may be a single letter (such as "v") or a combination of letters that produce one sound ("th"). Each phoneme has a different frequency range. Vowel sounds generally have lower frequencies than consonants. The individual's ability to hear phonemes in normal speech can be predicted from the audiogram, which relates hearing level to sound frequency. Because of the frequency-dependent nature of the sensory processing of phonemes, the simple recognition of phonemes, words, and sentences under ideal listening conditions (no distractions or distortions of the speech signal) without attribution of meaning is considered to be based mainly on peripheral processing.

The most commonly used methods for measuring how well a person can hear speech sounds as they are combined into words are speech recognition and speech discrimination, tested under optimal listening conditions. Speech recognition is measured by presenting the listener with a list of spondee words: two-syllable words with uniform emphasis on both syllables. Examples of spondees are words such as airplane, armchair, baseball, birthday, farewell, iceberg, oatmeal, headlight. There are two measures derived from speech recognition testing. The Speech Reception Threshold (SRT) is the intensity at which the listener can identify the words 50% of the time. A somewhat more precise measure is the Performance-Intensity (P-I) function, which represents the percent of correct identifications at increasing levels of loudness. Monosyllabic words are used in speech discrimination testing. Lists of these words are called "phonetically balanced" (PB) because the phonemes in them are representative of their frequency of occurrence in the English language. The listener's speech discrimination ability is represented by a P-I function rather than a threshold.

Either of these measures of speech understanding can be made more difficult by adding interference from competing noises or competing messages, interrupting the speech signal, speeding it up, or slowing it down. When any of these changes are made, the ability to understand the spoken message becomes increasingly dependent on central processing. The auditory functions of the central nervous system enable the listener to focus attention on one set of signals and tune out others, "fill in the blanks" in interrupted or speeded speech on the basis of word meanings, and in general, take advantage of the refinement and interpretation of speech signals that occur at the level of the brain stem and above.

Separating the effects of aging on peripheral and central auditory processes is a primary focus of researchers who study speech perception as a function of age in adulthood. A reduction in hearing sensitivity for tones of higher frequency due to the peripheral changes involved in most types of presbycusis does not affect speech discrimination. This is because the majority of English phonemes have frequencies well below 2000 Hz, the first point at which age differences in sensitivity appear. However, age effects are apparent even for the speech frequencies, by ages 70 and 80, when sensitivity for these tones shows a pronounced cross-sectional decrease. Moreover, certain consonants, called sibilants, have frequencies in the upper ranges (from 3000 to over 6000 Hz). From Figure 10.2, it can be seen that there will be impaired perception of these consonants beginning in the 60s, and becoming progressively poorer through age 80. The consonants most likely to be affected are underlined in the following words: bus, zebra, shoes, azure, bench, and fudge. Since the English language is rich in sibilants, especially the s used in plurals (Singh & Singh, 1976), loss of discrimination of these tones can have a disproportionately large effect on the understanding of speech.

Research on speech understanding under ideal listening conditions by adults of different ages provides an important source of information on the effects of age on peripheral auditory processes. From the data on Speech Reception Thresholds, it appears that intelligibility of spondee words decreases cross-sectionally over the adult years (Corso, 1981; Eisdorfer & Wilkie, 1972; Goetzinger & Rousey, 1959; Jokinen, 1973). Similarly, speech discrimination of phonetically balanced lists of monosyllables diminishes systematically in adulthood, particularly after the age of 50 (Feldman & Reger, 1967; Harbert, Young, & Menduke, 1966; Jerger, 1973; Luterman, Welsh, & Melrose, 1966; Punch & McConnell, 1969). The age differences observed in both functions appear mainly to reflect the influence of aging on peripheral auditory processes, since their curves correspond to what would be predicted on the basis of auditory sensitivity scores across adulthood (Marshall, 1981; Pickett et al., 1979). This suggestion that peripheral factors determine age effects in speech understanding is reinforced by the finding that when the degree of hearing loss due to peripheral causes is controlled across age groups, older adults' speech intelligibility scores are comparable to those of younger persons (Kasden, 1970; Surr, 1977). Thus, the central auditory processes involved in interpreting speech patterns do not seem to contribute over and above the age effects on the structures within the ear to decrements in speech understanding when the signal is undistorted.

Age differences across adulthood become especially pronounced under adverse listening conditions, where central auditory processes become the predominant influences on performance. A large cross-sectional study with a longitudinal component revealed that, out of a variety of distorting conditions, sentence understanding showed the earliest (beginning in the 40s) and largest age effects when the speech signal was interrupted (Bergman, 1971; Bergman et al., 1976). Age effects also appear on tests of speech understanding when the rate of presentation is increased (Bergman et al., 1976; Calearo & Lazzaroni, 1957) or when the time frame is compressed by at least 33% so that some of the message is deleted (Calearo & Lazzaroni, 1957; Jerger, 1973; Konkle, Beasley, & Bess, 1977; Marston & Goetzinger, 1972; Sticht & Gray, 1969). Competition from background noise or competing messages is another condition that yields relatively poorer performance in older adult groups (Jerger, 1973; Jerger & Hayes, 1977; Jokinen, 1973; Orchik & Burgess, 1977; Plomp & Mimpen, 1979; Smith & Prather, 1971). Reverberation is a form of noise interference whose effect is based on the sound waves continuing to reflect off hard surfaces after the sound signal has ceased. As with other types of interference, age differences are apparent under conditions where speech is reverberated (Bergman, 1971; Bergman et al., 1976).

There is continuing debate over the relative contributions of central and peripheral factors to age differences in speech understanding under distorted conditions (cf. Marshall, 1981). Based on the small amount of anatomical evidence in addition to performance measures, it appears likely that although the cochlear nucleus appears to remain intact into old age (Konigsmark & Murphy, 1970), there is some impairment with age in higher auditory centers (Brody, 1955; Hansen & Reske-Nielsen, 1965; Henderson et al., 1980; Hinchcliffe, 1962; Kirikae, Sato, & Shitara, 1964). This conclusion is reinforced by an auditory backward masking study using letters, words, and trigrams in a dichoptic paradigm (Newman & Spitzer, 1983; see chapter 9 for a description of this paradigm). It was found that older persons required a longer interstimulus interval to reach and maintain the performance level of younger adults. The authors interpreted these findings as indicating that aging affects the processing of speech because the rate at which speech signals must be interpreted exceeds the analyzing mechanisms and channel capacity of the older adult. These findings are suggestive of central processing deficits as the source of speech perception losses associated with the normal aging process, but require replication and extension to paradigms in which central and peripheral processes are directly contrasted.

Psychological Consequences of Age Effects on Hearing

Hearing serves a number of critical adaptive functions (Ordy et al., 1981). Its adaptational value to the physical environment is to provide the individual with cues of oncoming threats that can only be heard, or that augment visual cues. These auditory cues might be the spoken words of others to watch out for danger in situations involving relatively little risk (bumping into a family member in a hallway at home) or to avoid a life-threatening situation (falling down a cliff at a riverside park). Other auditory signals of danger are fire alarms and sirens from emergency vehicles. Sounds may also be used for orienting oneself in space and for locating other people and objects. Hearing also enhances the quality of the individual's life, making it possible to enjoy music, the theater, television, and other leisure pursuits. For adaptation to the social environment, hearing plays the almost irreplaceable role of making social communication possible. Individuals are able to share information, opinions, feeling, plans, and simple observations and thoughts about their daily lives through exchanges of spoken words. The ability to hear may also influence the adult's sense of competence in being able to adapt to the physical and social environment. Not hearing a warning signal, if not dangerous, at least may be embarrassing, as can a feeling of being unable to comprehend other people's speech in face-to-face and group conversational settings.

Although there are individual variations in time of onset and degree of impairment, the average adult, between the ages of 60 and 70, begins to experience a noticeable loss of the ability to hear high-frequency sounds. This deficit will reduce the person's comprehension of words containing sibilants (of which there are many) and will be greater in listening to the voice of a woman than a man. Changes in the sensory structure of the ear with age that affect hearing sensitivity are compounded by the age effects on central auditory processes that reduce speech understanding in difficult listening situations. Most speech understanding in everyday life takes place in settings that are less than ideal (with background noise, interruptions of speech signals, variations in rate, and reverberations). Consequently, the aged individual will find it particularly difficult to communicate in a variety of social situations. Apart from these functional changes, the older adult's ears take on an altered shape and size, and may contribute to age changes elsewhere in the body to a changed self-appraisal of appearance.

It is frequently argued that loss of hearing in later life causes numerous adverse emotional reactions, such as feelings of loss of independence, social isolation, irritation, paranoia, and depression. Upon empirical investigation, though, it appears that older people with hearing loss are generally not socially maladjusted (Norris & Cunningham, 1981; Powers & Powers, 1978) nor emotionally disturbed as a result of their handicap per se. However, there may be a strain in the quality of interpersonal relationships due to the tendency of friends and relatives to attribute negative emotional consequences to hearing loss (Thomas et al., 1983) or to fail to communicate properly with the hearing-impaired older person (Eisendorfer, 1960). In addition, the older adult with hearing loss may be unable to control his or her speaking voice as a consequence of loss of auditory feedback (Fisch, 1978), so that other people find it difficult to converse with the aged individual.

Even if hearing loss does not have a direct impact on overall adjustment, the fear of making an inappropriate response to auditory stimulation may be an outcome of the older person's experience with hearing loss. When asked to report whether or not they hear a given sound in an experimental setting, aged adults are found to be more cautious about responding that they have than are younger persons (Potash & Jones, 1977; Rees & Botwinick, 1971). It may be that on prior occasions in social situations, the elderly adult with presbycutic hearing loss made errors in communicating because he or she responded to what was thought to be said rather than what was actually said. To avoid such embarrassment in the future, the individual may adopt a strategy of waiting until being absolutely sure of what was said before replying. This cautious stance may actually create other

problems in communication in that other people with whom the individual interacts must repeat themselves in order to have their message acknowledged than
would be necessary on the basis of hearing loss alone. Another outcome of the
older adult's cautious stance toward responding is that scores on tests of auditory functioning may be artificially lowered, particularly if threshold measures
are used.

It appears, then, that hearing loss does not have a direct impact on the older
person's self-concept, emotions, or even amount of social interaction, but that its
greatest effect is on feelings about communication with others: a perceived lack
of understanding and ability to communicate on the part of the non-hearing-
impaired, and avoidance of making a response that may have the unintended
effect of creating further barriers. Another outcome of hearing loss is in the
cognitive domain. There is evidence that a reduction in hearing sensitivity is
related to lower scores on tests of intellectual functioning, especially when verbal
abilities are being tested (Granick, Kleban, & Weiss, 1976; Ohta, Carlin, & Harmon, 1981; Schaie et al., 1964). It may be that hearing loss has this effect because
it closes off the older adult from cognitive stimulation created by conversation,
listening to television and radio, and general exposure to cultural influences. This
stimulation may be necessary for maintaining adequate verbal skills. Apart from
its effect on tests of cognitive functioning, this result of hearing loss may serve to
cut off the aged person even further from social stimulation, if the individual
perceives a lack of ability to contribute fully to discussions regarding substantive
issues. These feelings, in addition to the real handicap posed by hearing loss, may
serve to restrict the quality if not the quantity of stimulation the older adult
receives from the social environment.

The majority of elderly adults do not appear to have severe hearing losses. Of
those who are afflicted by this handicap, a certain proportion will be able to compensate for it, either by using hearing aids or by augmenting their auditory cues
by using their other senses, particularly vision. Knowledge of hearing loss problems and ways to overcome communication difficulties on the part of non-
hearing-impaired elderly and younger adult individuals can play a large role in
minimizing the effects of hearing loss on that percentage of the older population
with moderate to severe presbycusis.

CHAPTER 11

Somesthetic and Vestibular Systems

The somesthetic senses convey information about touch, pressure, pain, and outside temperature. This information is essential for enabling the individual to be able to judge the position and orientation of the limbs, the existence of threats to the integrity of the body's functioning from harmful agents in the environment, and also to be able to feel sensual enjoyment from physical expressions of affection with other adults and from a comfortable physical environment. The individual uses this knowledge about the state of the body to take actions to remedy an uncomfortable or dangerous situation, to continue in activities or remain in settings that are pleasurable, and to adjust the body as needed to carry out intentional movements. Unlike the other senses, the somesthetic system has multiple receptors and neural pathways, most of which convey fairly distinct features of bodily stimulation.

Adding to the information about the body's position and orientation provided by the somesthetic systems is that regarding movement and head position provided by the vestibular sense. In order to maintain an upright position when standing still and to be able to adjust the body's stance when it is moving or being moved, it is first of all necessary to be able to sense the forces of gravitational pull and acceleration applied to the body. The vestibular system contains the receptors that are sensitive to these forces as they act upon the head. These receptors provide the central nervous system with information on which it bases its instructions for the motor responses needed so the person can maintain his or her balance while moving and while standing still.

Age Effects on Somesthesis

It is of great interest to learn about the impact of the aging process on sensitivity to bodily stimulation. A common belief is that aging is associated with a greater number of "aches and pains" caused by sore joints and muscles so that the experience is more acute in the older population. On the other hand, along with decreased sensory capacities in other modalities, it might be expected that older people would have a diminished sensory experience in somesthesis, which would include a lessened awareness of touch, bodily orientation and movement, and temperature, as well as pain. Research on age differences in somesthetic sensitivity does not provide a clear-cut solution to this paradox. Instead, it appears as if there are structural and functional cross-sectional age effects but no overall deterioration across modalities in the somatic domain of perception. Moreover, despite the inherent significance of the question of how aging affects sensation of bodily stimulation, there are many thinly documented areas in the research literature on aging and others for which there is no evidence at all. Because of the relative sparseness of data, the question investigated in other sensory modalities concerning peripheral versus central age effects has just begun to be approached.

Age Effects on Peripheral Processes

Touch

In the skin are a variety of receptors that transmit neural signals when pressure from mechanical displacement is applied to them. These pressure-sensitive receptors are the structures that initiate the sensation of touch. One group of touch receptors, the Meissner's corpuscles, hair end-organs, and pacinian corpuscles, is sensitive to rapid changes in pressure, such as the pulsations of a vibrator or the movement of clothing against the skin. When a pressure is first conducted to them through the skin surface, these receptors give off a rapid burst of signals and then rapidly decrease their firing rate. High-frequency vibration is sensed by the pacinian corpuscles, while the Meissner's corpuscles respond to low-frequency vibration. The hair end-organ receptors are sensitive to light movement across hair skin surfaces, firing neural signals when the hair is moved. In skin with no hair, the Meissner's corpuscles carry out the function of responding to light movement that the hair-cell receptors perform in hairy skin. Continuous pressure is sensed by the Merkel's discs (grouped in bunches that form a dome under the epidermis), and the Ruffini end-organs. The Merkel's discs signal continuous pressure close to the surface of the skin, and the Ruffini end-organs perform this function at a deeper level within the dermis. There are regional variations across the body in the distribution of touch receptors, with the greatest density found in the lips, tongue, and fingertips.

The sensitivity of touch receptors is measured by determining their threshold, the least amount of pressure stimulation that the individual reports feeling at the site where the stimulus is applied. Another index of touch sensitivity is the two-point difference limen, the smallest distance between two sources of stimulation applied to the skin that can be detected. As the two-point difference limen increases, the degree of localization of touch sensation decreases.

Perhaps the most knowledge about age effects on somesthetic stimulation is in the area of touch. It is fairly well established, for example, that there are cross-sectional age reductions in the number of pacinian corpuscles and Meissner's corpuscles in the skin (Bolton, Winkelmann, & Dyck, 1966; Cauna, 1965). Although the pacinian corpuscles continue to regenerate throughout life, the overall number is smaller in the skin of older individuals. Moreover, the remaining touch receptors of these types are found to undergo rather striking changes in structure. In contrast, the Merkel's discs show little age-related variation in number or structure (Cauna, 1965).

Some of the function carried out by the somesthetic receptors are known to be impaired in groups of older adults. There is decreased sensitivity to touch on the skin of the hand, where sensitivity is normally high compared to the rest of the body surface (apart from the mouth). This diminished sensitivity is indicated by a higher threshold to touch and a greater two-point difference limen threshold (Axelrod & Cohen, 1961; Dyck, Schultz, & O'Brien, 1972; Gellis & Pool, 1977; Thornbury & Mistretta, 1981). Unlike the effects of age on sensitivity in smooth (nonhairy) skin, sensitivity to touch in the hair-covered parts of the body is maintained into later life. This stability corresponds to the observation that free nerve endings (Cauna, 1965) and hair end-organs (Kenshalo, 1977) do not appear to undergo age-related changes in structure. The sensitivity to vibratory stimulation, another function of the pacinian and Meissner's corpuscles, is reduced in the lower but not upper part of the body beginning at around 50 years. However, factors other than loss of receptors appear responsible for this age effect, especially decreased conduction along the peripheral neural pathways in the lower part of the body (Kenshalo, 1977).

Position

In order to monitor the location and movements of the arms and legs, it is necessary to know the degree of angular displacement between bones at the joints and also the rate at which they joints are moved by muscular contraction. Both types of information are provided by two of the pressure-sensitive touch receptors: pacinian corpuscles and Ruffini end-organs. These receptors are located deep in the tissues around joints as well as within the skin. As the degree of joint rotation changes, pressure is placed on the receptors, which then signal the onset and extent of movement. Golgi endings constitute another type of position receptor. The Golgi endings are located in the ligaments at the joint, responding to the degree of stress placed on them by the stretching of ligament that occurs when the limbs at the joint move. Each receptor covers a small angle of movement, and together all the receptors at one joint cover its entire range of movement. Information about movement and position is also processed by muscle spindle fibers, which transmit neural signals when the muscle in which they are located is stretched. Based on the multiple qualities of information it receives from the receptors near the joints regarding the body's position and movement, the central nervous system can gauge the effectiveness of the motor instructions originating from the cortex and, if necessary, initiate corrective steps so that the intended action is implemented.

The smallest degree of passive rotation of the joint that can be detected defines sensitivity of the position receptors to the arrangement of the body's limbs. The perception of effort during exercise of a limb is a measure of sensitivity indicative of the awareness of bodily movement.

The available evidence on age differences in sensations of bodily orientation and movement suggests that aging has a deleterious effect on the ability to detect the angle of rotation of some joints moved passively but not others (Laidlaw & Hamilton, 1937). There also appears to be wide individual variation in this function over adulthood (Howell, 1949). It is not known to what extent the sensitivity of muscle stretch receptors is or is not altered as a consequence of aging (Kenshalo, 1979). The perception of effort, as indexed by weight discrimination ability, does not show age differences (Landahl & Birren, 1959). If degree of perceived effort is invariant across different tasks such as this, it may be inferred that aging does not impair the perception of strain caused by muscular exertion.

Thermal Sensitivity

Detection of the outside temperature is an important feature of the autonomic mechanism that maintains homeostatic conditions in the body (see chapter 6). Another feature of thermal detection is the sensation of whether the skin (and hence the immediate environment) is warm or cold. Temperature perception serves the adaptive function of signaling the need to invoke behavioral controls to supplement the autonomic system's regulatory activities over the core body temperature.

The sensory mechanism that enables the individual to experience warmth and cold is not known. A reasonable hypothesis based on present information is that there are receptors sensitive to warmth and other receptors, which respond to cold stimuli. It seems probable that these receptors are specialized free nerve endings. It also appears that the number of cold receptors exceeds the number of warm receptors.

Within a narrow range, the individual cannot sense changes in temperature if they occur gradually. This range corresponds to temperatures slightly above and slightly below physiological zero, which is the skin temperature at which neither warmth nor cold is felt. The temperature at which physiological zero is achieved varies according to area of the body, having an average value of about 7° below the normal body core temperature.

The zone above and below physiological zero serves several important functions. Because of the relative insensitivity of thermal receptors to skin temperature in the range of physiological zero, discomfort from warmth and cold is experienced only when the temperature of the skin reflects outside conditions that can potentially threaten to lower or raise the body's core temperature below or above safe levels. The protective nature of this zone is apparent from the fact that temperature changes away from it are sensed as more intense than changes toward it. As a result, there is a greater likelihood of behavioral changes oriented toward achieving homeostasis when the possibility of risk is greater from escalating or descending temperatures relative to the risk of harm to the body's tissues.

The existence of physiological zero is due to the fact that thermal receptors adapt completely to a constant temperature maintained at this point; that is, as long as the temperature is at physiological zero, the thermal receptors do not trigger neural impulses. Complete adapatation of the receptors also takes place in a temperature range slightly above and slightly below physiological zero. Consequently, changes within the neutral range around physiological zero are not sensed if they are made so slowly that the receptors have time to adapt to one temperature increase or decrease before the next change is made. Outside the range of physiological zero, the thermal receptors show some adaptation to constant temperature but never to the same extent as when the temperature is within that zone.

The fact that physiological zero and its surrounding neutral range vary over the body probably is due partly to the differences the density of thermal receptors over the thickness of the skin. Presumably, the range in physiological zero over the body's surface reflects differences in likelihood of exposure of different body parts and in the relationship between temperature in the body's periphery and in the body's core. The forehead is most sensitive to both warming and cooling, and the abdomen the least. The forehead has, then, a higher "signal" value that the outside temperature is changing than every other part of the body. The abdomen is least likely to be exposed to the outside and its temperature is probably closer to the body's core than any other bodily region, so that it has a very low signal value, even when its temperature does change. In general, it appears that there are more areas of the body sensitive to cold than to warm, signifying perhaps greater protection against hypothermia from exposure to very cold temperatures.

There are also pain receptors that trigger neural responses at extremely high and low temperatures. The risk of tissue damage that these receptors signal should stimulate the individual to seek immediate removal of the heat or cold source from the body. A final precaution against excessive bodily damage from temperature extremes is the tendency for large body surface areas to be more sensitive to temperature changes than more circumscribed areas, leading the person to seek protection when the risk from exposure is greatest.

Absolute thresholds are defined as the temperature at which the sensations of warmth and cold are first experienced. Awareness of a change in temperature is the difference threshold. Absolute thresholds for cold and warm vary considerably according to the original or adapting temperature on the skin surface, particularly when these are very low and very high. As shown in Figure 11.1, when the person's skin is adapted to a very low temperature, a relatively large increase in temperature is needed in order to report the sensation of being "warm" and when the adapting temperature is high, a relatively large decrease is needed to make the person feel "cool." However, as can be seen in this figure, the person who is cold at the outset has a lower threshold for warmth, that is, feels warmer at a cooler temperature than a person who is warmer at the start. The person with a low skin temperature at the start also has a lower differential threshold for sensing change from cool to less cool than in the cases where skin temperature is higher at the outset. Also apparent in Figure 11.1 is the observation that at a high adapting temperature, a greater temperature decrease is needed to feel less warm than at a lower adapting temperature. Another way to look at this part of

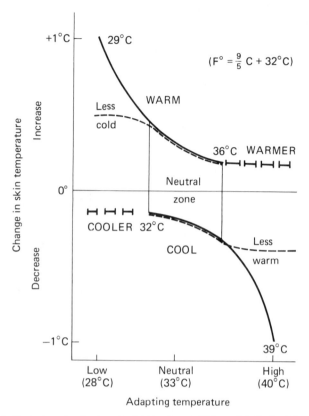

FIGURE 11.1. Variations in absolute (solid lines) and differential (dashed lines) threshold according to initial adapting temperature. The numbers on the curves represent the actual temperatures at which the thresholds were reached. The broken extensions of the solid lines represent thresholds for "cooler" and "warmer," signifying that at these temperatures, individuals were either cool or warm already because of the temperature to which they were adapted. Temperatures are in centigrade, with the conversion to Fahrenheit shown in the inset. *Note*. From "The Cutaneous Senses" by D.R. Kenshalo in L. Riggs and J.W. Kline, (Eds.), *Experimental Psychology, Vol. 1*, pp. 117–168. Copyright 1972 by Holt, Rinehart, and Winston, Inc. Reprinted by permission of CBS College Publishing.

the curve leads to the conclusion that the cold threshold at a high adapting temperature is at a much higher temperature than when the person was cool before the temperature decrease was effected. The person whose skin temperature is very warm, then, is the least sensitive to cold. Not shown in this figure is the rate of temperature change, which influences warm and cold thresholds such that the slower the temperature changes, the more the warm and cold thresholds deviate from physiological zero.

The complex relationships illustrated in Figure 11.1 demonstrate the variations in thermal sensitivity according to initial adapting temperature. These variations add to differences already discussed in thermal sensitivity according to amount and region of the body surface that is stimulated. As a consequence of these numerous interacting effects, absolute and differential thresholds of thermal sensitivity cannot be interpreted unless the exact conditions of measurement are known.

From the few studies on aging and thermal sensitivity, there would appear to be higher thresholds for warmth (Clark & Mehl, 1971) and possibly for cold in older adults, but these age effects are deemed to be of limited significance (Hensel, 1981; Kenshalo, 1979). In chapter 6, it was observed that elderly persons have diminished sensitivity to cool temperatures, and take few precautions when their core body temperature is lowered. Examining this finding in terms of Figure 11.1, it may be speculated that the elderly studied in their homes had adapted to lower temperatures necessitated by having to economize on heating bills. Small increments in temperature would therefore be perceived as warm, even though the skin temperature was actually quite low. Alternatively, it is plausible that the older persons had a lower skin temperature primarily because of defective homeostatic adjustment mechanisms, and that this intrinsic age effect makes them likely to feel warm at a lower temperature. These interpretations must remain highly speculative until thermal sensitivity data is gathered reflecting more precise specification of test conditions (initial adapting temperature, for instance).

Pain

Pain is experienced when a stimulus is applied to the skin or arises within the body that has the potential to damage bodily tissues and is felt to be unpleasant. The receptors for pain are free nerve endings located in the skin and also in the arterial walls, joint surfaces, in the skull, and in the vital organs of the body. The stimuli that can elicit pain may be mechanical, thermal, or chemical; most pain receptors are sensitive to more than one of these. Some mechanical stimuli include pressure on an artery, the stretching of a ligament, smooth muscle spasms, and distention of a hollow organ.

The threshold of pain sensitivity is the lowest intensity of stimulation at which an unpleasant feeling is reported by the observer. The highest intensity that can be withstood defines the level of pain tolerance. While there is a fairly constant pain threshold across individuals, pain tolerance shows wide variations. These variations are due to the fact that many cognitive and emotional factors influence people's tendency to report what is painful and their ability to withstand pain (Sternbach, 1978). Some individuals are more likely than others to be aware of their bodily sensations and to make fine discriminations among levels of pain stimulation. Depending on the situation, the individual may consciously or unconsciously deny or try to minimize the amount of pain being felt. Long-standing personality characteristics can also play a role. High levels of anxiety or neuroticism are linked with increased sensitivity to pain (lower tolerance) whereas social openness or gregariousness is asssociated with a greater tendency to talk about painful experiences but more stoicism as reflected in a higher pain tolerance. Finally, culturally acquired attitudes may lower or raise an individual's likelihood of reporting that a stimulus is painful.

Measures of pain sensitivity based on signal detection theory separate the contribution of these psychological factors from the purely physiological. As applied in pain research, this method involves comparing the hit rate of reporting a stimulus above zero intensity as painful to the false affirmative rate of reporting a stimulus at zero intensity as painful (Clark, 1969). The false affirmative rate serves as a control for the tendency of observers to report pain when there is no physiological basis for this response. Based on comparing the hit and false affirmative rates, separate measures can be derived for detectability (d') and for response bias (L_x), the tendency to guess.

The question concerning "aches and pains" in the aged, or more formally, whether pain thresholds and tolerance change as a function of age, has prompted a fair amount of research. However, the conclusions from this research are far from clear, with evidence existing for all possibilities of heightened, diminished, and unaltered sensitivity to pain in the later years of adulthood (Kenshalo, 1977). Part of the reason for these disparities in findings from research may be due to the effect of factors other than the intactness in the aged of pain receptors and the neural pathways that transmit pain. When signal detection theory analysis was applied to sensitivity scores obtained by putting a painfully hot stimulus on the forearm (Clark & Mehl, 1971), the older (28-67 years) women in the sample did have reduced d' scores, indicating lower pain sensitivity. However, the criteria for reporting that the stimulus was painful varied according to the participant's age and sex, and also to the intensity of the pain stimulation. Older adults, particularly men (who felt the pain as intensely as the younger people) had high L_x's, showing that they were much more likely to withhold saying they felt pain even when it must have felt quite painful. These findings are shown in Figure 11.2. Based on this type of result, it may be reasoned that differing sex compositions of samples, modalities, and levels of intensity across studies in addition to individual differences contributed to by personality, cognitive and social factors could easily account for the lack of consensus on aging effects.

Age Effects on Central Somesthetic Processes

Neural Pathways

The somatic sensations of touch, temperature, and pain are transmitted to the brain from the skin, joints, and bodily organs via two pathways, named after the portions of the spinal cord they occupy. The dorsal lemniscal pathway communicates information about what part of the body is stimulated and to what degree it is being stimulated. This information travels quickly across large myelinated axons to the ventrobasal portion of the thalamus, projecting from there to the primary somatosensory area in the cortex. There are two divisions in this pathway, one of which transmits signals about movement or changes in position (from pacinian corpuscles and Meissner's corpuscles). The other subdivision communicates steady-state, continuous information about touch and

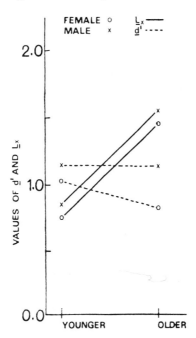

FIGURE 11.2. Values of d' and L_x based on responses to noxious stimulation. *Note*. From "Thermal Pain: A Sensory Decision Theory Analysis of the Effect of Age and Sex on d', Various Response Criteria, and 50 Percent Pain Threshold" by W.C. Clark and L. Mehl, 1971, *Journal of Abnormal Psychology*, *78*, p. 206. Copyright 1971 by the American Psychological Association. Reprinted by permission.

position (from Merkel's discs and Ruffini end-organs). These subdivisions follow parallel routes in different parts of the spinal cord.

The second major somatosensory pathway is the ventrolateral. Transmission in this system is slower than in the dorsal lemniscal, being carried over smaller nerve fibers, many of which are unmyelinated. In addition, the information transmitted in the ventrolateral pathway is cruder, since it is neither localized nor finely discriminative. This pathway passes through the reticular activating formation, and the intralaminar nuclei of the thalamus to the limbic system. The ventrolateral pathway carries the sensations of pain, thermal sensitivity, sexual feelings, tickle, and itch, and nonlocalized and poorly graded touch and pressure stimulation. Because of its connection to the autonomic and limbic systems, this pathway is thought to be responsible for linking pain and pleasurable sensations with motivations and emotions. There is some linkage between the two pathways, but for the most part, they are anatomically distinct.

The Somatosensory Cortex

The thalamic nuclei and cortical region which receive somatic information through the dorsal lemniscal pathway are organized somatotropically, that is, with different body parts represented in discretely different areas. The somatosensory cortex is also organized by modality, with separate areas representing touch and pressure, joint movement, and muscular contractions. Neurons in the primary somatosensory cortex are organized into columns, with those in the same columns having similar receptive fields and response characteristics. Because of this organization, the sources in the body of somesthetic stimulation can be identified in terms of place and type. While the sensation of pain is a function of the ventrolateral pathway, which does not have specific projections to the primary somatosensory cortex, localization of source and gradations of intensity of pain are probably dependent on the functions carried out at the cortical level. The representation of the body's surface in the primary somatosensory cortex corresponds to the distribution of receptors. The mouth and hands have the most cortical area. The trunk of the body, which has few somesthetic receptors, has proportionally the least space in this part of cortex.

In addition to its role in specifying the nature and location in the body of somesthetic stimulation, the primary somatosensory cortex appears to play a role in the functions of weight discrimination and judging the shape of objects by touch alone, called stereognosis.

The primary somatosensory cortex exerts control over the nature of incoming information through descending pathways within the dorsal lemniscal and ventrolateral systems that project downward to the thalamus, brain stem, and spinal cord. In the ventrolateral system, this descending pathway could be the basis for the influence of affective factors on the perception of noxious and sexually arousing stimuli. The descending pathway in the dorsal lemniscal system might account for the effects of knowledge from prior somatic experiences, as well as current and past sensations from other sensory systems on the way in which bodily sensations are interpreted.

There is only minimal evidence regarding the integrity of functions in the aged based on operations carried out at the central level of somesthetic processing. In one study, older men were found to be less able to recognize complex geometrical forms by touching them (Axelrod & Cohen, 1961). In another study, an increase after the age of 60 years was observed in the disruptive effects of touch on one arm to touch sensitivity on the other arm (Levin & Benton, 1973). Both of these investigations would signify some reduction in central processing, perhaps as a consequence of reduced neural numbers in the primary somatosensory area of the cortex (Henderson et al., 1980). However, other evidence from studies on somatosensory evoked potentials supports the position that there is no detrimental effect on age on central somatosensory pathways and possibly enhanced cortical functioning (Desmedt & Cheron, 1980; Kazis et al., 1983) as well as an increase in cortical cells (see chapter 8).

Psychological Consequences of Aging of the Somesthetic System

Sensations from the skin, internal organs, and joints serve critical functions in protecting the individual from harm caused by falling, cutting off of blood circulation, excessive strain on muscles, hunger, cardiovascular and respiratory stress, and excessively hot or cold environments. Touch sensitivity, in addition, serves to heighten the individual's sensual awareness as well as serving as an orienting mechanism to the immediate surroundings and as a basis for engaging in activity involving motor coordination where sensory feedback is needed. Pleasurable feelings from sexual stimulation are also communicated by the somesthetic system. All of these functions and other similar capabilities are, with varying degrees of importance, related to adaptation to the physical and social environment.

Perhaps the somesthetic function that is most critical to feelings of self-competence is the message communicated by this system about how well the body's limbs and internal organs are functioning. The somesthetic system provides essential input about the adequacy of the body's functioning. The somesthetic system provides essential input about the adequacy of the body's functioning, giving off pain signals when a joint or visceral organ is stressed. It is through the actions of this system that the aging adult becomes aware of such phenomena as muscular fatigue, joint "stiffness," cardiac stroke volume insufficiencies, ventilatory efficiency, any problems in digestion, elimination, or excretion, and loss of sexual sensitivity; in short, the age effects on all other bodily systems. If it is assumed that aging does not diminish pain sensitivity, then these signals can have a powerful impact on feeling of self-sufficiency since they can make it very evident that one's body is losing its vitality.

The very spotty evidence on aging effects on somesthetic sensitivity seems to indicate a more consistent loss of tactile functions at the peripheral level compared especially to the disparate results in pain. The area of pain sensitivity is complicated by the variation people have in their attitudes toward pain, which influence their tendency to report the presence of painful stimulation. It appears possible that older people, whether or not they experience pain, may be reluctant to report it and perhaps less likely to take therapeutic actions than they would be if they were to acknowledge pain. Apart from illustrating the effect that psychological factors play in the measurement of physiological functioning, the presence of these individual differences in willingness to admit to pain makes it hard to draw conclusions about aging.

If it is assumed that aging does not substantially reduce pain sensitivity, and in addition, if it is the case that aging does not impair the perception of effort (and possibly the higher neural functions underlying touch), then a very tentative hypothesis may be put forth regarding the effects of aging on the intensity of negative somesthetic experiences. This hypothesis is based on the assumption that the function of the perception of pain and effort is to serve as warnings to the individual of danger caused by potential destruction of bodily tissues by mechanical damage or overexertion. These warnings are protective devices that may be seen as enhancing the individual's adaptation to the environment. Touch perception may serve the same functions, to a lesser degree. Just as the autonomic system may serve as a buffer against aging of the body's vital systems, it may be hypothesized that those aspects of somesthetic function that protect the individual from danger to the body's tissue are relatively well-preserved in the elderly as an adaptive mechanism. Although awareness of reduced efficiency of the body's systems may be detrimental to the aging adult's self-esteem, this awareness is of considerable value in keeping the individual from engaging in activity that would ultimately have severely damaging effects on the body, and consequently, the individual's adaptation to the environment.

Balance

The body's orientation in space at rest and in motion is sensed by the vestibular system, which provides the central nervous system with information about the orientation of the head. From the information regarding gravitational and accelerative forces acting upon the head, knowledge of the entire body's position in space is judged.

Age Effects on the Vestibular Organs

The entire vestibular system is located in the inner ear, within the bony labyrinth in the skull that encloses the cochlea. The membrane that encloses the vestibular system and cochlea follows the the the shape of the bony labyrinth. The structures in the vestibular system are the two spherical compartments called the utricle and saccule, and the three looplike semicircular canals. On the inner walls of the utricle and saccule are the maculae, the areas containing the sensory cells. The sensory cells are hair cells, which are embedded in the maculae with the hairs projecting toward the inside of the compartment. These hair cells trigger a neural impulse when bent. Surrounding the hairs is a jellylike substance that has embedded on top of it crystals of calcium carbonate called otoconia. Movement of the head in a linear direction causes the otoconia to shift position. Because they are heavier than the tissues surrounding the hair cells, displacement of the otoconia indirectly moves the hair cells, potentially initiating a neural signal to the vestibular nerve on which they synapse. This signal, once initiated, persists until the position of the head changes again.

The receptors in the utricle and saccule provide signals that the head is being acted upon by forces that tend to pull it sideways, backward, or forward. Compensatory actions are then taken so the person maintains balance. Only linear movement stimulates the receptors in the maculae. The receptors in the semicircular canals are sensitive to movement in a curved path, as when the person moves from facing in one direction to facing in another. As the head is turned, these receptors provide the central nervous system with information about the rate and direction of rotation so that postural adjustments can be made to maintain equilibrium.

The receptors in the semicircular canals sensitive to rotational movement are hair cells, which are like those in the maculae in that they fire an impulse when bent in one direction. These hair cells are located in each of the ampullae, swellings at the ends of the three semicircular canals. The hair cells are located on the cristae, which are crest-shaped projections inside the ampullae. Covering the hair cells is a jellylike material (similar to that in the maculae) called the cupula. The semicircular canals are filled with endolymph, a thick fluid that moves at a slower rate than the structures that contain it. When the head begins to rotate, the endolymph takes several seconds to move in the same direction as the semicircular canals, causing a relative backward movement of the endolymph against the surface of the cupula. This fluid movement stimulates the hair cells to initiate their neural transmission. After a steady rate of rotation is attained (for instance, when being spun on an amusement park ride), the endolymph moves at the same rate as the semicircular canals and the hair cells decrease their firing rate. When movement ceaes and the semicircular canals have become stationary, the endolymph continues to move forward in the direction of rotation. This forward movement of the endolymph stimulates the hair cells to stop firing altogether.

The turning of the head during rotational or any other directional movement not only stimulates postural reflexes to maintain balance, but also serves to initiate nystagmus, a reflexive movement of the eyes in a direction equal and opposite to movement of the head. This reflex serves to keep the eyes fixated on the visual field that was being observed before the movement began. Without the nystagmus reflex, the person's visual images would be constantly changing each time the head moved even slightly. Stability of the visual field despite small shifts in head position allows the person to maintain constancy of visual input in such tasks as reading and engaging in conversation, where head movement occurs randomly and frequently and where continuous change of the images presented to the two eyes would be distracting.

Because the hair cells in the cristae fire signals when unusual stimulation is applied to the semicircular canals, the nytagmus reflex can be elicited to measure sensitivity of the receptors in the semicircular canals without actually rotating the person physically. A commonly used technique to stimulate the vestibular receptors is to irrigate the external ear canal with cool or warm water to produce caloric nystagmus. The temperature change alters the density of the endolymph in the nearby semicircular canals causing it to move around, which in turn stimulates the hair cells on the cristae.

The direction of either linear or rotational movement in the vestibular system is transmitted from the receptors to the central nervous system in two ways. One is by the orientations of the receptor surfaces in the two ears. The set of maculae on the two utricles and saccules and the set of cristae in the semicircular canals of both ears are each oriented on three planes such that between the two ears, a coordinate system is formed for identifying the direction of movement. The second means of transmitting information about direction through the vestibular system is based on the fact that the hair cells on both surfaces initiate an impulse only when bent in one direction. If bent in the opposite direction, the neural response is inhibited. Depending on which part of the receptor surface is stimulated, and the rate the hair cells are firing, the central nervous system can determine the orientation of the head.

Age-related processes similar to the sensory and neural forms of presbycusis appear to take place in the vestibular organs in the inner ear. There is a reduction in the number of hair cells in the maculae of both the saccule and utricle after the age of 70 and an even larger decrease in the sensory cells in the cristae of the semicircular canals (Rosenhall, 1973). This process of degeneration begins with the hairs becoming increasingly fragile and likely to form clusters. The sensory cells shrink and finally vanish altogether and are replaced by scars formed by the supporting cells (Rosenhall & Rubin, 1975). The sensory cell loss may begin as early as the 50s in the semicircular canals in some cases. Although the maculae show less severe losses of hair cells, their sensitivity is potentially reduced due to the increasing prevalences with age in defects of the layer of otoconia whose movement stimulates the hair cells. These defects occur in the saccule but not the utricle. Beginning at around 30 years, pits begin to form in the continuous covering of otoconia as the crystals of calcium carbonate break up into fragments. By the 50s and 60s, deterioration of the otoconia becomes more pronounced and may even progress to the point where the otoconia disappear entirely (Johnsson, 1971; Johnsson & Hawkins, 1972; Ross, Johnsson, Peacon, & Allard, 1976).

Neural degeneration is also evident in the vestibular nerves of older adults, a process that may begin as early as the age of 40 years (Bergström, 1973a, 1978b). As in the case of the cochlear nerve fibers, it is suggested that the loss of vestibular nerve fibers is due to the accumulation of bony material around the opening through which the nerve fibers pass (Krmpotic-Nemanic, 1969). As the holes become smaller and smaller, the vestibular nerve fibers become compressed and gradually degenerate.

Age Effects on Central Vestibular Structures

The pathway leading from the vestibular system to the central nervous system passes through the spinal cord to the brain stem and the cerebellum, where motor reflexes are initiated that maintain an upright posture. Projection of the vestibular pathway into the equilibrium center of the cortex (in the parietal lobe) appears to be responsible for the conscious sensation of balance.

Loss of neurons in the central vestibular system may contribute to the sensory and neural deterioration at the peripheral level. The brain stem nuclei involved in vestibular sensations and reflexes appear to be relatively well preserved into later life. However, the reduction of Purkinje cells in the cerebellum (Hall et al., 1975) could result in some age-related disturbances in the appraisal of vestibular information and hence the reflexive control of posture.

Age Effects on Vestibular Functioning

Dizziness and vertigo caused by vestibular malfunction (Toglia, 1975) are common disturbances experienced by older adults. Vertigo is the sensation that oneself or one's surroundings are spinning. Dizziness is the vague feeling of being unsteady, floating, and lightheaded. These feeling of disequilibrium are unpleasant to experience and they can also pose the threat of serious injury from falls caused by loss of balance. While vertigo and dizziness each may be caused by diseases of the vestibular system and pathologies elsewhere in the body, there appears to be some association of normal aging with changes in the vestibular system that may contribute to these and other alterations of vestibular functioning in later adulthood.

The question of what causes vertigo and dizziness in older adults is, as yet, only partially answered by research on the vestibular organs. The loss of otoconia in the saccules may be one source of dizziness (Corso, 1981). If the saccules are affected by the aging process more than the utricles, the ability of the vestibular system to register sensory information regarding head position on all three planes in space will be compromised. The differential reduction in sensitivity over the three spatial planes can lead to a distorted image of head position and movement and hence could be a source of the sensation of dizziness. The reduction in number of sensory cells in this structure could also account for the linear decrease in scores on tests of balance while standing still and walking forward (Fregly, Smith, & Graybiel, 1973). Since the sensory cells appear to deteriorate uniformly across the three semicircular canals, disorientation caused by differential reduction of information from the three planes in space in the maculae presumably would not occur. Instead, it would appear that the gradual deterioration of hair cells in the cristae would have the effect of lowering the sensitivity of the semicircular canals to movement of the endolymph as the head rotates. This loss of sensitivity due to sensory cell reduction would, it would seem, be augmented by decreases in the nerve fiber population.

Rather than demonstrating a linear cross-sectional decrease across the adult years, though, the sensitivity of the semicircular receptors appears to become greater in the age range of 40 to 70 years before diminishing into the 70s and beyond, as shown in Figure 11.3. There is considerable consensus that under

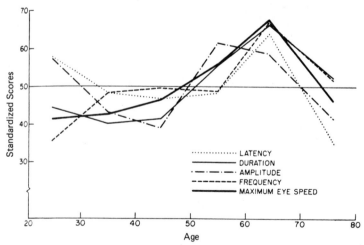

FIGURE 11.3. Measures of nystagmus response by age. Higher numbers on all measures represent better performance. The high point for latency for the 60 to 70 group represents a decreased or fast latency. The high points for the other measures represent increases. *Note.* From "Age Related Changes in Caloric Nystagmus" by A. Bruner and T. Norris, 1971, *Acta Otolaryngolica (Supplement)*, 282, p. 7. Copyright 1971, Almquist & Wiskell International. Reprinted by permission.

206 11. Somesthetic and Vestibular Systems

most conditions vestibular sensitivity, as indexed by caloric nystagmus, shows this inverted U-shaped pattern (Bruner & Norris, 1971; Karlsen, Hassanein, & Goetzinger, 1981; Mulch & Petermann, 1979). Although the heightened sensitivity before the age of 70 is difficult to explain (cf. Bruner & Norris, 1971), the decrease in nystagmus after that point may reflect the effects of cumulative loss of sensory receptors over the adult years. Poorer equilibrium and the experience of dizziness could result both from this alteration in the receptor structure and also the decrease across age groups of neuron population in the cerebellum.

It would be incorrect to convey the impression that dizziness and vertigo could be completely accounted for by these or other structural changes that are a consequence of normal aging. This is because there is considerable amount of plasticity in the vestibular system. Loss of cells in the peripheral structures of the vestibular system may not have serious effects (Kenshalo, 1979) to the extent that there is compensation by the activity of structures in other sensory systems. The positional receptors in the somesthetic system (Babin & Harker, 1982), which appear to be less vulnerable to aging effects, would be particularly important in this regard.

Psychological Consequences of Aging of the Vestibular System

Aging of the vestibular system brings with it the potential for loss of the normal mechanisms for maintaining one's posture while moving and standing still. In order to move about effectively in the physical environment, not only must the muscles and joints be able to carry out intended actions and the position receptors be supplying accurate orientation information, but the person must feel a secure sense of balance. To the extent that dizziness and vertigo reduce this security, the individual will become afraid to move around in his or her surroundings. Knowledge of the possible effects of aging on bone density can increase the older adult's fear of falling and augment real reductions in mobility caused by vestibular disturbances. Conversely, people who try to overlook transient episodes of dizziness or vertigo may be placing themselves in real danger, since their positional adjustments will be impaired if they do lose their balance.

Vestibular problems, like other motor disturbances, are also highly visible to other people. The older adult who does suffer from these sorts of dysfunctions may feel embarrassed at appearing to be disoriented in front of others, fearing that the impression is created of being inebriated or otherwise mentally confused. Apart from the real physical restrictions caused by disequilibrium, its social consequences may have additional limiting effects on the person's range of mobility.

It is true that vestibular dysfunction is neither an inevitable consequence of the aging of the vestibular apparatus nor are the changes described earlier universal in the structures of this sensory system. Compensation is possible by adapting other sensory processes to meet the individual's need for maintaining orientation during movement. The individual who reacts to any alterations in the sense of balance with neither undue alarm nor incautiousness is probably better able to take advantage of compensatory mechanisms through using kinesthetic and position cues, particularly during episodes of disequilibrium when particular attention must be paid to stance and body orientation in order to avoid falling. It is during these episodes that special precautions must be taken in order to avoid the physical and psychological harm that would otherwise occur. Repeated instances of success at overcoming vestibular disturbances may stabilize the older adult's sense of physical competence, and also lessen the physical and social risks of displaying lack of motor corrdination.

CHAPTER 12

Gustatory and Olfactory Systems

The taste receptors are the sense organs maximally sensitive to the chemical composition of foods as transmitted through direct contact. However, the eating experience represents a complex interaction of a variety of sensations in addition to taste. Food preferences, aversions, and desire reflect the joint influence of taste with the tactile, visual, and olfactory senses. In addition, eating takes place in a social milieu, and the interpersonal relations and overall "atmosphere" of this milieu can add to or detract from the enjoyment of a meal.

Effects of Age on Taste Sensitivity

Health-care practitioners are extremely concerned about the nutritional status of their elderly clients. At all ages, adequate amounts of food and water intake are essential for maintaining proper bodily stores of fluids, proteins, carbohydrates, fats, vitamins, and minerals. For the aged, though, it is especially critical that they monitor their diet, since social and economic circumstances may alter their previous eating habits established earlier in adulthood. Nutritional deficiencies may, in addition, create mental symptoms mimicking senile dementia, and so result in unnecessary suffering and deprivation for the person who could otherwise lead a normal life. If the sense of taste is impaired by the aging process, then it is even more likely that older people will lose the motivation to eat either a balanced diet, or one that is sufficient in quantity to meet their bodies' daily needs.

The Taste Bud

The sensation of taste arises from the direct contact of substances in a liquid medium with the sensory cells in the taste bud. One taste bud contains about 40 to 50 taste cells. The taste cells are the sensory cells that initiate a neural response when a substance to which they are sensitive is applied. The taste bud is spherically shaped, with its taste cells and supporting cells like the sections of an orange. The receptor surface where the neural impulse originates is on the taste hair, which extends out the pore of the taste bud onto the surface (epithelium) of the tongue. Taste cells are constantly dying, probably because of damage caused by direct exposure to harmful chemicals in the food substances with which they are in contact. Unlike other neural cells, though, taste cells do regenerate. Each taste cell lives for about 10 days, and when it dies, it is replaced by a new one grown from the epithelial cells that surround it. Another "plastic" feature of the taste receptor structure is demonstrated when a taste nerve fiber is destroyed. The taste bud will degenerate, but will re-form as new taste cells cluster around a nerve fiber growing toward the surface of the tongue.

The taste buds are found within three of the four types of papillae of the tongue. There are about 8-12 circumvallate papiillae that form a "V" toward the back of the tongue. These papillae individually look like flattened domes, and are relatively large compared to other kinds of tongue papillae. They have numerous taste buds on them and also on the walls of the troughs that surround them. In the front part of the tongue are the more rounded and small fungiform papillae. Leaf-shaped foliate papillae are situated on the sides of the tongue. All over the top surface of the tongue, giving it its rough surface, are filiform papillae. The threadlike projections of these structures do not contain taste buds. Taste buds are also located on other structures in the mouth, including the palate and tonsils.

The neural impulses leading from the taste buds travel on branches of several cranial nerves to synapse in nuclei in the brain stem, from which they pass to nuclei in the thalamus close to the projections from the facial regions of the dorsal lemniscal pathway of the somesthetic system. The pathway from the thalamus projects to the taste area of the cortex, located near the part of the primary somatic sensory area to which touch sensations from the tongue are represented.

Practical concern over how taste is affected by the aging process has stimulated a fairly large body of research on taste receptors and taste sensitivity. Most of this research is on taste detection and recognition by living subjects, but there are a few autopsy studies where the number of taste buds in different kinds of papillae were counted. From these autopsy studies, it appears as though there are few, if any, age differences in the number of taste buds. A decrease in the number of these receptors on the circumvallate papillae was noted only for persons over 75 years (Arey, Tremaine, & Monzigo, 1935). There were no age differences at all in the taste bud population in the fungiform papillae as observed in one cross-sectional study (Arvidson, 1979), and only a gradual decrease whose significance was not reported in another investigation (Moses et al., 1967). Lack of a striking age effect would be expected, considering the taste buds are continuously replenished from their surrounding epithelial cells. Unlike other neural cells, the numbers of taste buds should not be significantly depleted over the individual's lifetime (Engen, 1982).

Taste Sensitivity in Adulthood

The stimuli to which the taste cells respond fall into four categories: salty, sweet, sour, and bitter. Each taste has associated with it different molecular structures. The salty taste is caused by stimulation with a salt molecule. A variety of substances in addition to sugars account for the sweet taste, including alcohols, amino acids, and also some kinds of salts. Sourness comes from the presence of acids, and bitterness from stimulation by alkaloids or by long-chain organic molecules. Each primary taste probably does not have its own receptor. Instead, it is likely that taste receptors do vary according to which taste they are maximally stimulated by, but each taste receptor can be stimulated by each of the four primary tastes. It appears that all possible tastes can be accounted for by these four primary sensations in some combination, but the mechanism(s) by which taste mixtures are sensed is not known.

There is a tendency for different parts of the tongue to be most sensitive to one of the four primary tastes. The region of the tongue with the circumvallate papillae is most responsive to the bitter taste. The foliate papillae on the sides of the tongue seem to have maximum sensitivity to sour and salty tastes. The sweet taste seems to have the greatest stimulating effect on the front of the tongue, on which are located the fungiform papillae. It is not known to what extent this relative spatial specificity on the tongue is maintained up to higher levels in the taste pathway through the central nervous system. But it is known that different cranial nerves receive stimulation from receptors on different parts of the tongue.

Taste sensitivity is indexed in two ways. The first is by the detection threshold, which is equal to the concentration of a substance in water at which the taster can tell that the liquid being tasted is different from water. To be regarded as detected, the substance need not be identified. The concentration at which the taster can identify exactly the substance being tasted is the second taste sensitivity measure, known as the recognition threshold. These measures may be obtained either by having the taster sip solutions of various substances (sugars, salts, acids, alkaloids) or by having the examiner place drops of solutions on the person's tongue. In either case, the most reliable measures are gotten when the tasters rinse out their mouths with distilled water between each sip or drop administration.

In contrast to the relative preservation of taste bud numbers into later adulthood, there appears to be a small but progressive cross-sectional decrease with age in sensitivity to all four primary tastes as reflected by higher detection thresholds in older age groups (Balogh & Lelkes, 1961; Bourliere, Cendron, & Rapaport, 1958; Byrd & Gertman, 1959; Cooper, Bilash, & Zubek, 1959; Grzegorczyk, Jones, & Mistretta, 1979; Hermel, Schönwetter, & Samueloff, 1970; Hinchcliffe, 1958; Moore, Nielsen, & Mistretta, 1982; Murphy, 1979; Richter & Campbell, 1940; Smith & Davies, 1973). This age effect is illustrated for the salty and sweet tastes in Figures 12.1a and 12.1b, from two studies in which almost identical methods were used (including appropriate controls such as almost all nonhospitalized volunteers and having subjects rinse their mouths after each taste trial). Also evident from Figure 12.1 is the wide individual variation in taste thresholds, which is particularly pronounced in the older age range. Although the average threshold tends to be higher for persons over the age of 60, there are still a number of older individuals who have no apparent loss of sensitivity to detecting tastes.

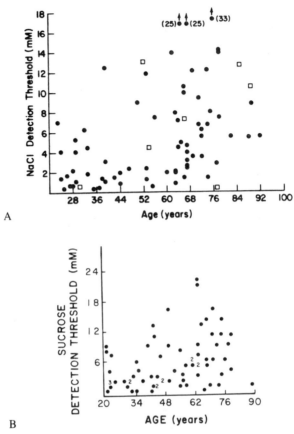

FIGURE 12.1. Comparison of age differences in thresholds for A salty and B sweet tastes. *Note.* (A) from "Age-Related Differences in Salt Taste Acuity" by P.B. Grzegoczyk, S.W. Jones, and C.M. Mistretta, 1979, *Journal of Gerontology, 34*, p. 837. Copyright 1979 by The Gerontological Society. Reprinted by permission. B from "Sucrose Taste Thresholds: Age-Related Difference" by L.M. Moore, C.R. Nielsen, and C.M. Mistretta, 1982, *Journal of Gerontology, 37*, p. 65. Copyright 1982 by The Gerontological Society. Reprinted by permission.

While anatomical evidence is not convincing regarding an overall reduction in taste buds as a function of age, then, taste threshold data show a generalized cross-sectional decline in sensitivity. What accounts for this discrepancy between the findings on taste bud numbers and taste functions? One possibility is that factors other than the integrity of the receptor structure are involved in taste detection and recognition and that these factors are impaired by aging. The most likely factors would be those relating to the processing of taste information in the

central nervous system. Although the possibility of age effects in central processing of taste information is alluded to (Hughes, 1969) in the literature, there is no
anatomical evidence bearing on this question. Another possibility is that there is
a reduction with age in the rate or efficiency with which taste buds are
regenerated that is not reflected in total counts of numbers of taste receptors.

It is also plausible that the decrease in taste sensitivity across adulthood, while
usually found to be significant in a statistical sense, is marginal in a practical
sense, so that this function is relatively well-maintained into later life. Age differences in response bias that reduce the tendency to report having tasted or
recognized a substance may create artificial age effects in threshold values. This
factor is especially relevant when it is recognized that in most everyday situations, flavors are present in amounts well above those used in threshold detection
studies (Corso, 1981). The stimulus is so much weaker than the kind of flavors
ordinarily identified outside the testing situation that the older person may be
reluctant to report its presence until it reaches a level closer to that normally
tasted. Finally, generational variations in smoking habits and use of dentures may
cause difference among age groups due to these factors rather than age (Hermel
et al., 1970; Smith & Davies, 1973).

Age Variations in Food Identification

Whatever the extent or basis for age reductions of taste sensitivity, it is likely that
there is some loss of appreciation for the flavor of food in later adulthood. As a
result, the older individual may derive less pleasure from eating, and also be less
able to identify flavors or recognize differences among foods. Moreover, food
appreciation is influenced by a variety of sensations in addition to taste which
seem to be affected by the aging process, including smell (described in the following section), vision, touch, and temperature. Color is an important feature of
food appreciation, because it influences the consumer's expectations about the
food's pleasantness before it is actually eaten, and also because color provides a
familiar cue about the food's content. Even experienced food tasters may be led
to erroneous conclusions about the flavor or freshness of food that is inappro-
priately colored or has its normal color removed. The tactile sense also provides
essential information about the texture of food, another feature that aids in its
being identified. Finally, the apparent taste of a food may vary considerably
depending on its temperature (Moskowitz, 1978).

All sensations but taste and smell were removed in a food identification task
given to adults of varying ages by pureeing the foods and serving them at a uniform temperature (Schiffman, 1977). A cross-sectional decrease was observed in
recognition of a large number of very familiar food substances, including most
kinds of fruit, vegetables, meats, and coffee. This finding might reflect the
possibility that when age effects on all the components of food appreciation and
identification are combined, the cumulative result may be to reduce even further
relative to young adults the older person's ability to discriminate among and enjoy

the variety of foods that make up a good diet. Contributing further to age effects on food appreciation could be distortions of food tastes resulting from diminished and altered salivary secretions and also dental diseases more prevalent in the elderly (Grzegorczyk et al., 1979).

Psychological Consequence of Aging of Taste Sensitivity

The effect of reduced sensitivity to flavors in food on the individual's adaptation to the environment most likely would be to create the nutritional deficits associated with a reduction in the amount and balance of solid and liquid intake. If the experience of eating were to retain its sensual pleasure, then the older adult may overcome the social and economic obstacles to maintaining an adequate diet in order to be able to enjoy the process of food consumption (see chapter 5). However, with the intrinsic motivation for eating diminished, there would be little reason to exert the necessary effort to go out of one's way to cook a variety of pleasing and also affordable meals. In addition, reductions in sensitivity to particular tastes may lead the older peron to use excessive amounts of the substances with those flavors in order to maintain previous levels of food enjoyment. Overseasoning with sugars and salts or changing to overly strong coffee, tea, or alcoholic beverages may make the adult more vulnerable to certain diseases or else exacerbate current chronic disorders.

Over and above nutritional effects, adults who normally derive a good deal of personal satisfaction out of the preparation and consumption of food and drink, a loss of discriminability would mean the loss of a major contributor to the quality of everyday life. In addition, the effect of reduced food enjoyment may spread into the adult's social spheres. There could be less incentive to entertain or be entertained at dinner parties and in restaurants, where a large proportion of socializing takes place. The older person may become a less appreciative consumer and so be less likely to be invited out for such occasions or to stimulate other friends and family to make special efforts to prepare a variety of dishes. The older person as a cook may lose some of his or her previous talent, or not develop such a proficiency if he or she begins cooking for the first time in old age (such as after the loss of a spouse who did all the cooking). The reactions of others to this lack of taste sensitivity may detract from an important source of feeling of competence for some individuals, as well as diminished social interaction.

If the process of diminution of food appreciation begins in middle adulthood, as some studies on taste imply, then adults may find themselves adapting gradually and imperceptibly over the course of several decades so that they alter their food preparation and consumption habits without sacrificing the enjoyment of food. It is probably that group of adults who most value their mealtimes who would suffer the most from the loss of food discrimination. On the other hand, it is also likely that such individuals would be most likely to try to maximize their eating pleasure by trying to enhance the flavors of the foods they consume. Flavor additives and enhancers can be used successfully in this regard, since they can have a dramatic effect on the older adult's enjoyment of food (Schiffman, Orlandi, & Erickson, 1979). Knowledge of ways to compensate for reduced discriminability without compromising one's health status in other respects by an excess of potentially harmful substances can serve to maintain an important motivator for eating a diet that is adequate in amount and variety. It may even be a boost to the older adult's feelings of competence to look for and find solutions to the problems of cooking good-tasting, nutritious, and economical foods to meet the dietary needs of oneself and one's family.

Aging and Sensitivity to Odor

Some of the enjoyment of food comes from the way that it smells when it is brought toward the mouth. The contribution that smell makes to the experience of eating is apparent by the way that one's taste seems to suffer if the nasal passages are blocked by mucus when one has a head cold. Smell has many other functions in the daily life of the individual apart from enhancing taste. Fumes from foods and liquids may serve as a warning before they are tasted that the substances contain poisonous or harmful elements. For instance, an additive to natural gas that has a distinct odor makes it possible for the homeowner to detect a leakage before it has disatrous consequences. Smoke from an accidental fire can often be smelled before it can be seen, giving the individual valuable extra time to put out the fire or to escape.

Smell also has an esthetic role in everyday life. Adults use perfume on their bodies and air fresheners in their homes to make the atmosphere around them pleasant. The large number of scents designed for men and women on the market is an indication of how important this enhancer of sexual attraction is, and also how much variation there is in odor preference. The wrong fragrance or too much of even the right one can detract noticeably from the quality of a social interaction. The opposite concern, that of eliminating and covering up socially objectionable bodily odors with soaps, deodorants, mouth rinses and mints also occupies much of the adult's grooming activities. In this case, the absence of smell, or the substitution of one smell for another is an important an objective as actually being clean. Adults who do not make efforts to smell pleasant, or at least not to smell unpleasant may find themselves left out of social and possibly business activities.

As in the case of the effects of age on taste sensitivity, persons who care for the elderly are concerned about a loss of ability to detect odors in the aged, both for reasons of safety and nutrition. An age reduction in smell sensitivity would lessen the extent to which elderly persons are able to avoid contact with dangerous substances and to escape from potentially lethal situations. Also, it seems likely that any age effects on smell might exacerbate the slight loss of taste sensitivity in old age to influence food preference (Schiffman & Covey, 1984). This would reduce below a critical level the older adult's enjoyment of food and hence motivation to eat well and regularly. These practical issues that influence older people's live have posed an impetus for research on aging and the ability to smell.

Age Effects on the Olfactory Receptors and Pathways

Olfactory Receptors

Like taste, smell is a chemical sense—it is produced by the juxtaposition of a substance onto receptors that respond differently to different molecular structures. Whereas taste results from direct contact of the substance in liquid form with taste receptors, though, the sense of smell is produced by contact of the substance with the receptor cells through the medium of air.

While it is possible to breathe only through the mouth, most breathing involves the movement of air through the nasal and oral cavities. Air can also be brought into the nose by sniffing independently of respiration, which has the effect of increasing the rate of air flow around the olfactory receptors. Air passing into the nostrils contains many different kinds of molecules. Some of these molecules possess the qualities necessary to stimulate the olfactory (smell) receptors. When they reach the receptor surface in the top part of the nostril, neural signals are transmitted through the olfactory pathway to the central nervous system. Because the receptor surface in the nostril is covered with mucus, the molecule must be soluble in water in order to provide effective stimulation to the receptors.

The olfactory receptor surface is located on the top of the nasal cavity and also along the upper portion of the septum. The conchae are bony projections within the nasal cavity that form a complex tunnel system through which air must pass before it reaches the top of the nostrils where the olfactory receptors are. Odorous vapors from foods being chewed can also reach the olfactory receptors from the pharynx, moving up behind the palate.

The receptor surface containing the sensory cells is called the olfactory epithelium. The dendrites of the olfactory cells terminate in the olfactory knobs along the mucosal surface lining the inside of the nostril. Extending into the mucus are the olfactory hairs, where molecules carried in the air probably initiate the electrochemical activities that trigger a neural impulse. The supporting cells contain granules that secrete mucus onto the surface of the epithelium. Opening onto the epithelial surface is the Bowman's gland, which also contributes mucus to the covering of the receptor surface. The basal cell in the olfactory epithelium does not participate in the olfactory sensory function. It is probable that the supply of olfactory cells is replenished by the generation of new cells from the surrounding epithelium.

If it is the case that the olfactory receptors regenerate, there should be no reduction in their numbers across the adult years. However, there are reports of decreased olfactory cell populations and abnormalities in supporting cells and Bowman's glands in progressively older age groups, beginning at as early as 30 years (Liss & Gomez, 1958; Naessen, 1971). If these results are not an artifact of sampling design or method of analysis, they can be reconciled with the existence of olfactory cell replacement if it is postulated that aging somehow impairs the growth process. Alternatively, it may be that more olfactory receptors degenerate than can be replaced due, for example, to the cumulative effects over a lifetime of damage to the olfactory epithelium. This damage may selectively destroy the neural tissue more than the supporting tissue from which the receptor cells are grown.

Olfactory Pathways and Higher Structures

Each olfactory cell sends its nonmyelinated axon to the olfactory bulb through holes in the bone of the skull over the nasal cavity. The axons of many thousands of olfactory cells synapse in the glomeruli within the olfactory bulb onto the mitral cells, whose axons form the olfactory tract that leads to the olfactory areas of the brain. There are relatively fewer mitral cells than axons entering the glomeruli, and it is probable that sensory integration of the information from the receptors takes place in the olfactory bulb. Another type of cell in the glomerulus is the tufted cell, which also sends signals along the olfactory tract.

From the olfactory bulb, the olfactory tract passes to the medial and lateral olfactory areas of the brain. Both of these areas, located near structures of the limbic system, send projections directly into the limbic system, including the hypothalamus, the brain stem, the thalamus, and an area of the prefrontal cortex believed to be the primary olfactory cortex. The subcortical interconnections are probably involved in control of feeding behavior and emotional responses to olfactory stimuli including, perhaps, connections between smell and sexual behavior. Associations of smell with sensations in other modalities such as vision and touch probably occurs as a function of the cortical projections of the olfactory tract. There are also efferent pathways extending from the cortex and brain stem through to the olfactory bulb, which seem to serve an inhibitory function to regulate the flow of olfactory stimulation.

Also within the olfactory epithelium are free nerve endings that register stimulation from irritants entering the nose. This sensory information passes along central nervous system pathways that mediate somatic sensations.

Negative effects of aging have been observed in the glomeruli and a reduction in mitral cell fibers is reported cross-sectionally as a function of age (Liss & Gomez, 1958; Smith, 1942). In the higher neural centers of the limbic system, age-related abnormalities are also reported (Tomlinson & Henderson, 1976). Even if the receptor surface remains intact by cellular regeneration, then, it is possible that the higher neural olfactory structures suffer cellular alterations that could lower the efficiency with which olfactory information is processed.

Age Effects on Olfactory Sensitivity

The way in which odors are transformed into neural signals by the receptors is not well understood. About all that can be said with any confidence is that the olfactory receptor site interacts in an unspecified way with the structure of the molecule that reaches it. The quality of the sensation, that is, which odor is perceived, may depend on the site on the olfactory epithelium that is stimulated, the type of receptor cell, or some combination of the two, so that it is the pattern of stimulation that is interpreted by the brain in differentiating among smells. Part of the difficulty in identifying the mechanism for odor identification is that there is no known scale along which to quantify or even categorize odors based on their intrinsic properties, as is done with colors according to wavelengths and sounds according to frequencies. It is believed that odors are differentiated somehow by their differing molecular structures, but so far there is no basis for making such determinations. As few as 4 and as many as 50 discrete odor qualities are defined in various classification systems. There is also a problem in relating the physical structure of odors to their perceived qualities, since people vary in their facility at describing their olfactory experiences and also in the labels for odor provided by their language.

The measurement of odor sensitivity usually involves judgments of quality rather than intensity. The mechanism underlying variations in degree of presence of odor is probably even more complicated than the mechanism underlying qualitative discrimination. Consequently, rating scales for odor intensity are even less reliable than those for odor quality.

Most cross-sectional comparisons of different age groups of adults on the ability to detect odors do yield evidence of diminished sensitivity among older adults (Anand, 1964; Doty et al., 1984; Kimbrell & Furchgott, 1963 [n-butanol only]; Murphy, 1983; Rous [cited in Corso, 1981]; Schiffman, 1979; Schiffman, Moss, & Erickson, 1976; Schiffman & Pasternak, 1979; Schiffman, Stavros, & Pasternak [cited in Schiffman, 1979]; Strauss, 1970; Venstrom & Amoore, 1968). A curve from a large cross-sectional study illustrating the age effect on smell sensitivity is shown in Figure 12.2.

There is, however, wide variation among older adults and some exceptions to the trend toward lower olfactory sensitivity among older age groups of adults

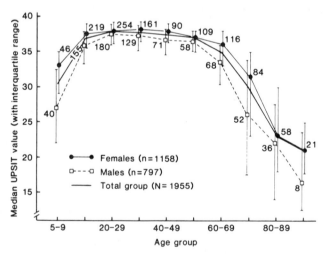

FIGURE 12.2. University of Pennsylvania Smell Identification Test (UPSIT) scores as a function of age and gender. Numbers by data points indicate sample sizes. *Note.* From "Smell Identification Ability: Changes with Age" by R.L. Doty et al., 1984, *Science, 226,* p. 1442. Copyright 1984 by AAAS. Reprinted by permission.

(Kimbrell & Furchgott, 1963 [iso-amyl acetate]; Mesolella [cited in Corso, 1981]; Rovee, Cohen, & Shlapack, 1975). Varying sample characteristics may also play a role. In Doty et al.'s (1984) data mentioned earlier, there was a confound of age and nursing-home residence. Since nursing-home residence was negatively related to smell sensitivity, the effect of this variable on cross-sectional sensitivity scores may have been to exaggerate age differences in the latter decades. Discrepancies in findings from cross-sectional age comparisons may be due to variations across studies in procedures as well as in sample composition. Some of the procedural differences might include corrections for guessing and eliminating extraneous factors, characteristics of the people being tested which lead to wide variation especially among the old who have accumulated lifetimes of olfactory experiences, and the particular odors used for testing. As is true for the other senses, aging may have differential effects on the various types of odors. However, the scaling of odors is at a very primitive level compared to quantification of stimuli in other modalities. Consequently, a way to order age effects according to the physical attributes of the odoriferous stimuli cannot be made. In addition, since the way that odors are transformed into neural signals is not known, the connection between the effect of aging on the receptors and the functional outcome on different kinds of odors cannot be made as is the case, for example, with the effects of aging on the cochlea and high-frequency tone loss in presbycusis.

The difficulties people have in labeling odors may be especially pronounced in older adults. The age difference in odor identification observed in one study comparing younger and older women (Schemper, Voss, & Cain, 1981) was reduced when the older group was provided with the correct labels for subsequent identification tasks. Older women who gave correct ("veridical") labels without help from the experimenter had identification scores that were well in the range of the scores of the younger women (see Figure 12.3). The overall difference in odor identification scores may have been due to the greater tendency of young women to apply correct labels to the odors the first time they smelled them. The age difference in odor labeling found in this study may possibly indicate an effect of aging on the processing of information at the cortical level. An effect of aging on olfactory processing in the central nervous system could, perhaps, account at least in part for age differences in olfactory identification noted in other investigations. Moreover, some of the variation among older samples in odor detection may be due to the differences in olfactory experiences that individuals accumulate over their lifetimes as a result of differences in the amount of practice they have with odorous stimuli. These olfactory experiences may influence cortical processing of olfactory information above and beyond any effects of age on the receptor or the olfactory bulb and pathways.

Psychological Consequences of Aging of the Olfactory System

Keeping in mind the differences among older individuals in their ability to detect odors, and the lack of a general aging effect for different kinds of odors, there are

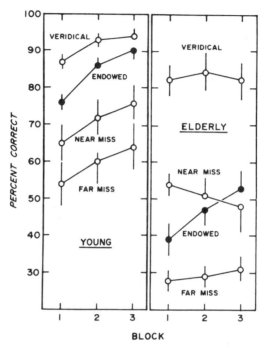

FIGURE 12.3. Odor identification scores over three blocks of trials for odors labelled by subjects (veridical = true name of substance; near miss = name reasonably close to the veridical name; far miss = name quite far from veridical label) and odor names supplied by experimenter (endowed). *Note.* From "Odor Identification in Young and Elderly Persons: Sensory and Cognitive Limitations" by T. Schemper, S. Voss, and W.S. Cain, 1981, *Journal of Gerontology, 36*, p. 449. Copyright 1981 by The Gerontological Society. Reprinted by permission.

nevertheless important consequences when an age decline does occur for a given person and for a particular odor. For instance, a larger number of people over than under 65 years of age were found to be less sensitive to the smell of natural gas (Chalke, Dewhurst, & Ward, 1958). These elderly persons would be more vulnerable to accidents caused by gas leakage and hence suffer a real danger to their lives, a dramatic consequence of diminished sensitivity to one kind of odor.

Another potential outcome of a reduced discrimination of odors could be less attentiveness to bodily odors that other people find offensive. While the individual may be perfectly well groomed in terms of cleanliness, he or she may be less likely to use the deodorizing and perfuming agents that mask these scents for hours after a bath, shower, or toothbrushing. It is conceivable that the person's quality and quantity of social interaction could suffer as a result, and he or she would believe this change was due to a personality flaw rather than a correctable sensory deficit. Similarly, the accumultion of foul odors around the home could discourage visitors until this problem was recognized by the homeowner. Of course, to the extent that older persons share an olfactory deficit, they each will be less offended by odors that others would perceive as objectionable.

Apart from these effects on the adult's social world, age effects on olfactory sensitivity would reduce the potential enjoyment of pleasant smells from natural and artificial odorants. On the other hand, the individual would be less aware of the odorous components of air pollution, and so might find living in a smog-filled urban environment more tolerable than would an individual with an acute sense of smell.

Since it seems likely that the exercising of olfactory discrimination abilities helps to preserve them (Engen, 1982), it may be argued that those smells that represent valued qualities of everyday life would be the least likely to be lost with increasing age. Whether this effect is due to maintenance of the peripheral olfactory structures or to the collection over the years of more fine-tuned cognitive labels for smells is not known. Whatever the cause, the outcome may be to ameliorate what would otherwise be a constriction of the adult's social and esthetic environment. Another compensating factor is that in the world outside the laboratory, scents that need to be recognized are typically present in levels well above the concentrations used to determine detection abilities at threshold levels. Moreover, odor identification takes place within a context that has many other sensory cues. The older person who does experience a loss of smell sensitivity may learn to use these additional cues to offset potentially dangerous or at least unpleasant effects of this age change on everyday life.

CHAPTER 13

Epilogue

This chapter is intended to serve as an integration of the numerous findings presented in preceding chapters regarding the effects of age on the human body, the theoretical meaning that these effects might have regarding the aging process, and what the psychological impact might be on the individual in whom these changes take place. In addition, some of the major gaps in the gerontological literature will be identified based on what seems to be known, what seem to be the unresolved issues, and what seem to be unasked questions regarding the physiology of human aging.

Thematic Issues Regarding Age Effects on the Human Body

Rather than summarizing the many diverse findings reviewed in previous chapters, the focus here will be upon trying to discern themes and connections among processes in specific organ systems as they adapt to the effects of time within themselves and among each other.

To begin with, the two most basic themes are loss and compensation. These themes emerged most clearly in chapter 8 regarding the controversy between the neuronal fallout and plasticity models. However, these processes of loss and compensation are consistently present in virtually every one of the organ systems and in the way many systems relate to others.

Loss and Compensation in the Aging Process

There is ample evidence that aging brings with it the eventual deterioration of basic life support functions as well as functions that contribute to the individual's quality of life. This has been the position adopted throughout the centuries regarding aging. New evidence continues to be added to support this position, evidence that becomes more sophisticated with each passing year, as more refined measurement techniques are added to the armamentarium of physiologists. What are perhaps the two most important changes with age are the loss of neurons from the central nervous system and the reduction in the heart's pumping capacity. These effects of aging seem, in turn, to underlie many other age-related decreases.

Apart from the deficits created by the loss of neurons in the central nervous system, neuronal deterioration appears to be the most plausible source of the deafferentation of muscles. Loss of muscle fibers, in turn, has been shown to cause many other changes in the body: sensory functions such as the deterioration of the iris dilator, the ciliary muscle, the muscles supporting the tympanic membrane; control mechanisms including insulin sensitivity, rates of thyroid and cortisol secretion; other critical life functions such as decreased respiratory muscle strength and lowered esophageal peristalsis; and of course, age-related losses in strength and mobility as well as contributions to loss of joint flexibility.

The second major age-related deteriorative process is the reduction of stroke volume due to diminished output by the left ventricle of the heart. This loss compounds the neural factors that cause the muscles to die, because it means that the muscles receive less aerobic support for their metabolic activity. It also has the effect of limiting the relative efficiency of the lung's ventilatory function.

In addition to these fairly massive alterations on major life processes, other dramatic losses occur in the bones, in the lung tissue, and in the kidneys which further reduce the body's ability to adapt to the demands of daily activities as well as to the stress of exertion, disease, or unusual environmental conditions. Other alterations of collagen and elastin molecules result in debilitative changes in the connective tissue throughout the body. The lowered activity of the sweat glands and the decreased adaptation that older persons are capable of making to extremely cold temperatures are two other important, and possibly life-threatening losses. The body's appearance is altered dramatically over the adult years, as is the sensitivity to sensual and sensory cues in the environment.

This portrayal of the downward spiral that constitutes what is generally recognized as the effects of the aging process does not, however, completely describe what is known about the regenerative and compensatory processes that operate in living human organisms. In chapter 2, it was suggested that aging of the joints seems to involve a "wear, tear, and repair" process based on the organic properties of living tissue. This is an important point, because it emphasizes the fact that aging occurs within an organism and not a machine. The natural tendency of an organism is to continue its life as long as possible by repairing whatever damage occurs to it. Therefore, one must look at the restorative processes within the human body in order to understand more completely how it ages.

Some of these restorative processes occur in the form of compensation. One important mechanism is the autonomic nervous system's enhanced level of activity in response to the widespread loss of tissue responsivity in the organ systems it regulates. This is the explanation that emerged from consideration of the evidence pertaining to increased levels of norepinephrine in the blood. The endocrine control system also responds by hyperactivity to overcome losses in the reproductive organs. In addition, the endocrine glands regulate their level of activity in accordance with tissue demands in the case of insulin, cortisol, thyroxine, and vasopressin. The kidney nephrons that do not suffer degenerative changes also engage in compensatory activity, thereby maintaining kidney functioning for as long as possible before the deterioration becomes too severe. Loss of vestibular sensory functioning seems to be made up, in part, by the somesthetic receptors that remain intact.

Another way that the body resists the effects of aging is through the provision of ample resources in some critical areas so that there is a reserve supply available. Usually, this surplus is accompanied by a continuous turnover of cells. Some examples in which this situation applies are in the case of the villi in the small intestine, the parenchymal cells of the liver, the photoreceptors in the retina, and various receptor cells such as the pacinian corpuscles, taste cells and, with apparently less success, the olfactory receptors.

Finally, there is plasticity in the nervous system, as described in chapter 8, and its psychological counterpart of increased knowledge gained through experience. This form of protection against the aging process is more than just reparation, it also represents growth. Plasticity signifies one means through which the individual can do more than just avoid losing function as the result of aging. In addition, plasticity in the nervous system means that the individual can also gain in ways that, if not actually life-maintaining, are nevertheless life-enhancing.

The Effects of Exercise Versus the Effects of "Bad Habits" on the Aging Process

In addition to the body's intrinsic resources for compensating for age-related losses of structure and function, there are also the steps that the aging individual can take to influence the rate at which aging affects a variety of physiological systems. It is clear that exercise activity is one readily available means that has potentially broad-ranging positive effects. Although it does not halt or reverse the aging process, exercise seems to lower the rate at which is exerts debilitative effects on critical life support functions. In particular, regular participation in aerobic exercise training has its most decisive favorable impact on cardiovascular functioning, as well as leading to enhanced respiratory, muscular, and psychomotor system activity.

In addition, there is some indication that exercise training can have beneficial effects on bone mineral content, testosterone in men, constipation, sleep patterns, growth hormone responsiveness, and insulin sensitivity. There may also be additional, unmeasured benefits of exercise training, since it might help to reverse some of the negative sequelae of cardiovascular losses and muscular degeneration described at the outset of this chapter.

The alternative picture, that losses are exaggerated by "bad habits," provides equally convincing evidence that the individual is one of the most powerful influences on his or her own rate of aging. The negative effects of these activities have been alluded to throughout the previous chapters in descriptions of, mainly cross-sectional, studies on physiological functioning. These are usually the influences that investigators control for when attempting to study the effects of aging unconfounded by extraneous factors.

One of the most common practices in which people engage that have been shown to have harmful outcomes is cigarette smoking, which seems to have widespread deleterious effects on cardiovascular functioning, the respiratory system, and taste and smell perceptions. Heavy alcohol usage is damaging to various aspects of digestive functioning, particularly the liver. Excessive exposure to the rays of the sun is damaging to the skin. Being in environments where noise levels are too high seems to accelerate the aging of the auditory system. Finally, overeating increases the risk of hypertension and other chronic cardiovascular diseases, as well as developing gastrointestinal disturbances and impaired glucose tolerance.

In addition to affecting the rate at which the individual either does or does not develop chronically impairing conditions, engaging in exercise or overindulging in harmful practices alters the description of the aging process that emerges from research. To the extent that researchers are able to determine the influence of these activities on specific physiological processes, however, important information is gained about whatever mechanisms individuals can initiate to take advantage of the compensatory processes associated with aging.

Confounding of Aging with Disease

Just as studies on human aging are inevitably biased by the habits of exercise or overindulgence engaged in by the subjects, this research is made more difficult to interpret by the inclusion of subjects in the samples who have a variety of undiagnosed disorders. In some cases, the confounding of aging with disease is more obvious, as when the subjects are selected from acute or chronic hospital wards, institutions, or out-patient clinics.

Hypertension is one of the most pervasive sources of confounds. Of course, it affects a number of cardiovascular functions, most obviously blood pressure, but also seems to have many known (and unknown) effects on other systems. Some of these effects include diminished plasma renin activity and impaired neurological functioning. Another problem presented by the case of hypertension, as well as other diseases such as osteroporosis, osteoarthritis, diabetes, and Parkinson's disease, is that it is often difficult in practice to drawn the line between normal aging and one of these pathological disorders. Similarly, diminished levels of T_3 in studies of thryoid functioning were attributed to aging, when it appears that lower levels of this hormone are associated with disease.

The result of this problem is a two-sided dilemma. One is that an older person with one of these diseases that might possibly be treatable is thought to suffer from "aging" and therefore has no alternative but to suffer the deterioration it brings. Moreover, in the case of diseases that are truly degenerative and have no cure (such as senile dementia), attribution of normal age changes to one of these diseases is unnecessarily upsetting to the afflicted individual and may bring on a chronic case of depression that aggravates the original symptoms. The other problem is that the presence of people with these disorders in empirical research distorts the findings that are arrived at, and the growth of understanding about the aging process is set back by a false lead.

In other cases, a confounding of aging in one physiological system with disease in another system or some other threat to normal functioning can exaggerate the extent of loss in what the researcher might quite confidently believe is a study of "normal" persons. Thus, in early research on testosterone levels in men, investigators claimed to have a suitable sample if the men they studied had no endocrine disorder or other related problem. However, it appears that the decline in testosterone levels identified by this type of investigation was a function of the abnormally low sexual activity levels of institutionalized men or men with chronic diseases that limited their frequency of intercourse. Another example of confounding of this nature seems to have occurred in the case of gastrointestinal functioning where the confounds are with economic and social factors as well as with other diseases that can affect the digestive organs.

An example of this type of confound can be seen in the following description of a sample taken from a frequently cited study on d-xylose absorption (Webster & Leeming, 1975, p. 109):

The 85 patients studied were selected from those admitted from the lists to a busy geriatric unit. . . . The majority had been admitted to hospital because of the results of cerebrovascular disease and a minority because of an acute exacerbation of various chronic disorders. As expected in such a group, social problems were common and had also contributed to the necessity for hospital admission. No patients with known small bowel disease or suggestive symptoms were included. . . . No acutely ill patients were studied, and urinary incontinence obviously made many patients unsuitable unless they already had an indwelling catheter, as in 16 cases.

The young sample, in comparison, included patients in an orthopedic ward admitted for treatment of sports or car accidents. Although these researchers controlled for the most obvious disease that would confound the results with age, they did not control for the other obvious differences in health, social status, and activity levels as suggested by the different life-styles of the two groups. Interestingly enough, the result of poorer d-xylose absorption in 26% of elderly is what is cited from this study (e.g., Bowman & Rosenberg, 1983) and summarized in the abstract, but not the opposite result that 16% of the elderly showed more efficient absorption than the younger adults (no significance levels were given).

Some of the confounds in studies of sensory functioning include the use of persons in the sample with lifetime exposure to noise being tested for auditory sensitivity, individuals with uncorrected visual problems in studies of visual acuity, and subjects with dental diseases having their taste sensitivity measured.

There really is no clear solution to this problem, since there invariably will be chronic diseases that neither the investigator nor the subject know about that can influence performance on any physiological measure. Instead, it is necessary for investigators to attempt to rule out systematically the effects of disease on the functions they study.

Problems With Cross-Sectional Research on Human Aging

Some of the problems regarding the confounding of aging with disease are due to the fact that, in increasingly older samples, one is more likely to find subjects who have accumulated one or more chronic diseases. On the other hand, as the sample is drawn from older and older age strata, it is increasingly likely that the group is going to be one that is composed of exceptionally healthy people who have survived their contemporaries (especially if they are taken from a community setting). Cross-sectional and longitudinal designs both will invariably suffer from this limitation. However, in the cross-sectional design, the investigator also faces the threat that what is being attributed to aging is really due to generational differences in various health practices.

For instance, while there appears to be a decrease in height across the adult years that is truly a function of changes in the bone mineral content (which causes collapse of the vertebrae), much of the difference across cohorts of adults in height seems to be due to the improved nutritional status of the young persons, who developed under conditions more favorable to optimal growth. Since some measures of lung function tend to be related to total body height, age differences in these respiratory indices will also be confounded with the same factors that influenced height differences. Other cross-sectional differences exist in body weight, and it is possible that apart from generational differences in activity level, there are other differences in nutritional habits that are confounded with age. Even so, the differences in activity level are not true age differences in physiological functions, and so represent a confound with age changes in weight. Moreover, organ weights that are directly correlated with total body weight (for instance, brain, kidney, liver) will also show cross-sectional differences for the same reasons rather than being influenced by the aging process. Finally, some age differences, such as the decrease in thyroid activity, may be confounded with age reductions in lean tissue mass that are related to generational effects on activity level.

These are all specific ways in which cross-sectional findings can present a skewed picture regarding the effects of aging on physiological functioning. In addition, there is the confound due to the fact that the same persons are not followed over time in a cross-sectional study, so what looks like age differences may be generational or other interindividual differences. The example of a lack of decrease with age in kidney functioning described in chapter 5 on the basis of the Baltimore Longitudinal Study provides an excellent illustration of how cross-sectional and follow-up studies can diverge.

While there is no real solution to the problems created by the use of cross-sectional designs, particularly in research in which autopsied cases form the sample, there needs to be continued recognition of the issues by researchers in the interpretation of their results. What would be of even greater value are investigations based upon some form of sequential design in which a longitudinal or cross-sectional study is carried out at least twice to rule out competing time of measurement or cohort effects (Baltes, 1968). In addition, research that is specifically addressed to cohort differences in, for example, activity level (e.g., Montoye, 1975; Sidney & Shephard, 1977a), nutritional habits, health practices (Belloc & Breslow, 1972) and other variables that compete with age as explanations of age group differences, has the potential to allow for the application of statistical controls if not actual experimental controls for these generational effects.

The more recent investigations of physiological functions appear to be moving in this direction with more complete sample descriptions, consideration of generational differences when they might compete with the attribution of age effects, and the inclusion of data regarding the subjects' life-styles (either retrospectively or concurrently). Moreover, long-term longitudinal data are currently becoming available from the Normative Aging Study in Boston and the Baltimore Longitudinal Study. Both of these investigations have involved collection of extensive physiological, psychological, and social data as well as sequential analyses. From such analyses it will become possible to learn more about the complex ways in which age and generational factors interact over the adult years.

The Psychological Consequences of Aging of the Body

One of the primary objectives of this book was to take a psychological approach to understanding the literature on the physiology of human aging. Accordingly, in each chapter, interpretations were offered about how the aging of the particular organ system under consideration might affect the individual in whose body these changes were taking place. In the case of some chapters, such as chapter 3 on the cardiovascular system, there was an extensive literature on which to draw, based on the research into the psychological consequences of aerobic exercise participation. The material in chapter 2, on appearance and movement, offered room for interpretation about psychological consequences because of the existence of findings based on young adults concerning body image, self-esteem, and physical attraction. In other chapters, such as chapter 5 on the digestive and excretory systems, and chapter 7 on the reproductive system, there was less empirical evidence, but a good deal of clinical work, at least in the case of digestion, from which to draw inferences. However, there were still many instances (chapter 4 on the respiratory system, chapter 6 on control mechanisms, and chapters 8 through 12 on the nervous and sensory systems) in which the psychological interpretations were much more speculative, with little if any empirical or even clinical reports on which to base statements regarding the possible effects on self-concept of changes in the body.

Since it is apparent that the approach taken to the physiological literature has been a fairly rigid empirical one here, it may appear to be incongruous to allow so much speculation to have taken place on the psychological side. However, it is unfortunately the case that psychologists have not addressed some of the critical issues regarding the effects of the aging body on the aging personality (Rossi, 1980). The model adopted in this book, based on adaptation to the environment being the source of feelings of competence and mastery and hence self-esteem, is one that has a long history in psychology (e.g., White 1963), but its specifics have not been worked out with regard to aging and its effects on adaptation at a very basic level. Another approach, advocated by Folkins and Sime (1981), would add to the notion of competence and adaptation the cognitive appraisal model of stress proposed by Lazarus (1966). This model has not been referred to earlier because it seems to have less applicability to the issues considered with regard to each specific system. However, the Lazarus model would seem to add a useful dimension to the model offered thus far.

Briefly, Lazarus proposed that the way that individuals approach new situations and whether they feel stress as a result is a function of how they appraise the situation and whether they then perceive it as threatening. The coping mechanisms that people invoke to reduce stress, suggested Lazarus, are a function of their appraisal of their resources for handling that situation and their appraisal of the situation in light of their self-evaluation of resources.

As applied to the effects of aging on the body, it may be hypothesized that part of the self-evaluation that individuals make as their bodies age includes their appraisal of themselves as capable of responding adequately to situations that tax their physical resources. Thus, the elderly woman steps out of her door on an icy winter day and appraises the situation as threatening because she knows her balance is poor and her bones have become weaker. Her real adapation to the physical environment, that is, her mobility, will be reduced to the extent that she perceives stress and stays home as a result. In addition, her negative self-appraisal of coping resources contributes to a physical self-concept of incompetence.

The stress and coping model may be a particularly useful one for predicting an individual's response to situations such as this one in later adulthood that depend on the appraisal of more than one physiological system. The concept of coping implies that the individual is attempting to draw upon as many resources as possible, and so it is a useful term to apply to the compensation model of aging in which losses in one system are overcome by drawing upon strengths in another. If the elderly woman in the example above were to cope with the situation by deciding to become involved in some kind of exercise program, to help improve her balance and flexibility, she would be illustrating this principle of compensation. Similarly, the individual who copes with hearing loss by using visual skills, such as lip reading, is compensating in this integrative fashion.

The scattering of empirical and clinical evidence cited throughout the book in specific areas of physical functioning, and the incorporation of the broad concepts of stress and coping into the competence and adaptation model presented in chapter 1 are really all in general agreement. One of the effects of the body's aging is to alter the individual's self-concept about his or her physical abilities, competence, and self-esteem, as well as adaptation to the social and physical environment. However, what is obviously needed are studies in which these questions are directly addressed.

These studies would involve asking adults (of all ages) about the physical effects of aging they have become aware of in themselves, and what the impact of these changes has been on their adaptation to their own environmental setting. The advantage of these questions would be that they would tap the individuals' phenomenological experience of aging, which is presumably as relevant to their adaptation as are the actual changes their bodies go through (Heaps, 1978). These questions could concern all bodily systems, or could be limited to one (such as changes in appearance) that the investigator is particularly interested in and could correlate with physiological measures. The second set of questions would involve asking how these changes affected the individual's feelings about his or her sense of competence and overall self-concept. These questions are probably much more difficult for people to find answers to, because feelings about the body, particularly negative ones, are not easily discussed (Rosillo & Fogel, 1971). The existing instruments measuring body image, such as the Body Cathexis scale (Secord & Jourard, 1953), do not seem to tap this quality, although a questionnaire measure of physical self-esteem might (e.g., O'Brien & Epstein, 1982).

However, what is needed is the development of interview or questionnaire measures that tap more directly the multiplicity of specific areas of physical competence, including the smooth functioning of the body's regulatory systems, and tap these in such a way that the responses can be directly correlated with physiological measurements. Ideally, a longitudinal design would be used so that the temporal sequence of changes could be more clearly understood. The knowledge that could be gained through such research would be of benefit to physiologists and psychologists. On the one hand, physiologists might learn more about the influence of personality on measures of the organ systems they are studying as well as the practical relevance of their findings. On the other hand, psychologists would learn more about the relationship of the physical self to the overall self-concept and how changes over time influence the individual's adaptation and well-being, dependent variabales that have obvious relevance to issues regarding life-span development.

Unresolved Research Issues

Throughout this book, areas have been discovered where there are heated controversies in the physiological literature, some of which have existed for many years. In addition, there are instances of wide, glaring gaps in the data base about human aging that need to be addressed. This review is intended to describe some of these controversies and gaps from which follow suggestions for future research.

Controversies in the Gerontological Literature

There are at least 35 areas that can be identified in the previous chapters where a striking difference of opinion exists between groups of investigators. Some of the more crucial ones will be briefly summarized here along with reasons for why they are important.

Reasons for Bone Loss

There are at least four competing hypotheses concerning the underlying mechanism responsible for loss of bone mineral content in adulthood: the hormonal (estrogen depletion), nutritional (deficient calcium absorption due to lowered Vitamin D), structural (continued remodeling resulting in increasing porosity and mineralization), and activity (lack of activity, which decreases the physiological stimulation for bone growth). Resolution of this issue has many implications for the preventive measures that can be taken to minimize loss of bone strength in old age.

Effects of Age on Submaximal Cardiorespiratory Measures

Due probably to lack of consistency across studies in exercise testing conditions, there are disparate findings on the effects of age on submaximal aerobic power and ventilatory rate. This issue is relevant because it has implications for the effects of age on physical work capacity.

Peripheral Versus Central Causes of Age Effects

There is a great deal of theoretical value to separating the effects of aging on the organs that control various physiological functions from the effects of aging on the effectors or peripheral afferent pathways and receptors. This issue is one that pertains to the cardiovascular system and to the visual and auditory systems. In terms of the cardiovascular system, the question is the contribution of age changes in the arteries (and if such changes would represent normal aging) versus the age-related loss in pumping capacity of the heart. In the visual and auditory systems, the question is whether age losses are greater in the central nervous system structures that process sensory information at higher, more refined levels, or whether the decrements are in the peripheral structures and pathways.

Causes of Neuron Loss and the Fallout Versus Plasticity Debate

This was reviewed in great detail in chapter 8, and is included here to reinforce its importance, given the widespread controlling influence of the nervous system on the muscles, sensory systems, autonomic control processes, and intellectual functioning. In this area of research, there are also controversies and confusion generated by the use of different indices of neuron number: packing density versus absolute counts.

Primary Aging Versus Secondary Aging

In numerous systems, there is a controversy over whether the primary loss is in the cells intrinsic to that system due to metabolic changes or other, unspecified sources of deterioration, or in the cardiovascular supply to that system. Some areas where this controversy applies are the cases of nephron loss, neuronal fallout, and muscle fiber loss. The relevance of this issue is that if the loss originates in the cardiovascular system, then prevention of age losses in the distribution of blood supply to the tissues will help offset a variety of other declines.

Effects of Aging on Digestion and Nutrition

Because of the many confounding factors that make research on the digestive system so difficult to carry out, there are a number of controversial areas, including: whether aging causes presbyesophagus and decreased gastric motility; whether or not the elderly suffer from anemia and if so, why; the effect of aging on liver function and structure; and the source of lowered vitamin D in the aged.

Issues Regarding Endocrine Functioning

Because of the historical interest in the endocrine system and the belief that the rate of aging is somehow controlled by its regulatory mechanisms, as well as the currently recognized complex interactions between tissue changes and glandular changes, there are many areas of controversy that have developed in research in this area. Some of the more heated areas of debate include whether aging affects insulin sensitivity, whether testosterone decreases or not in post climacteric men, whether any testosterone decreases are causally related to benign prostatic hypertrophy, and the effect of age on thyroid functioning.

Sensory Functions and Aging

Some isolated pockets of research controversies have formed in the literature on sensory functioning. Some of these controversies are extremely theoretical, and involve debate among cognitive psychologists as well as physiologists, and other controversies fall into the applied domain. The goal, in both situations, seems to be to learn more about the systems under study so that the results can be put to practical advantage to benefit older persons.

In the area of vision, there is disagreement among physiologists over whether there is a decrease with age in retinal cells, and if so, whether such a loss would have serious functional implications; psychologists and clinicians differ on the range of spatial frequencies in which it is claimed that contrast sensitivity is impaired in persons over 60; and gerontologists and physiological psychologists seem to be using different approaches to understanding backward masking, with the transient/sustained channel distinction receiving little attention among gerontological psychologists. In the area of hearing, clinical studies of acoustic function offer widely different estimates of the extent of hearing loss in the older population; there is also a debate about the effect of age on the acoustic reflex. In the other senses, there is a range of basic issues: whether pain sensitivity increases or decreases with age, the possible causes of dizziness, why it is that taste buds do not decrease in number with age but taste sensitivity does, and whether aging reduces olfactory sensitivity or not.

Some of these issues may be resolved by attention to confounds with age in the research designs, and some by greater communication among clinicians and researchers, as well as among researchers from different disciplines. In certain areas of physiological functioning, there are almost two separate streams of research, which neither acknowledge nor complement each other.

Areas in Need of Further Research

It is apparent that there are many unknowns in current knowledge about aging, where there are either no studies at all or there are too few to permit any even tentative conclusions. This is particularly the case when the focus is on human aging rather than comparative studies. Some of these gaps, identified in previous chapters, are summarized here.

Psychological Effects of Physiological Changes

It is apparent throughout this book as well as in the present chapter that very little attention has been paid to the whole topic of how aging of the body affects psychological functioning and vice versa.

Effects of Exercise Training on Respiratory Functioning

Given the many studies in which cardiovascular measurements are taken in systematic fashion before and after periods of aerobic exercise training, it is surprising that there has been so little data collected on the respiratory system. Some of the reasons for this lack of attention to respiratory functioning were identified in chapter 4. Unlike some of the other areas needing research, however, this area does not lack previous investigations. Instead, those that exist have included a disparate array of variables. One of the more intriguing questions not well addressed in this literature is why exercise training does not improve ventilatory efficiency, and whether the limitation is respiratory or cardiovascular.

Effects of Exercise on Kidney Functions

Although there is reason to suspect that exercise might affect the urine concentrating mechanism in the aged, this issue has not been addressed either in terms of discovering the effects of aging on the kidney's response to exercise, or the effects of exercise training on the adequacy of this process.

Effects of Aging on Digestive Functions

Apart from the areas of controversy discussed earlier, there is a lack of research on the effects of aging on segmentation in the small intestine, and little if any information on protein absorption in old age.

Unresolved Areas of Research in the Nervous and Sensory Systems

There is very little information on most of the higher pathways and structures of the sensory systems regarding age effects. In addition, the limbic system (apart from the hippocampus) has received almost no attention in humans, yet because of the complex relationships between cognition and emotion, particularly in later life, this would seem to be an important area of study.

Concluding Remarks

This review of issues and areas of needed research is not an exhaustive one, but is an attempt to highlight some of the critical issues that face the physiological study of aging. Research in virtually all areas of physiological functioning is growing at a very rapid rate and increasing design and instrumentation sophistication are adding to the quality of the available information. It is hoped that by identifying some of the critical issues and gaps, as well as highlighting the relationship between physiological and psychological processes in aging, both disciplines will add productively to knowledge of how the body changes and how it adapts to these changes.

References

Abend, S. A. (1974). Problems of identity: Theoretical applications. *Psychoanalytic Quarterly, 43*, 606-637.

Adams, G. M., deVries, H. A., Girandola, R. M., & Birren, J. E. (1977). The effect of exercise training on systolic time intervals in elderly men. *Medicine and Science in Sports and Exercise, 9*, 68.

Adams, P., Davies, G. T., & Sweetname, P. (1970). Osteoporosis and the effect of aging on bone mass in elderly men and women. *Quarterly Journal of Medicine, 39*, 601-615.

Adrian, M. J. (1981). Flexibility in the aging adult. In E. L. Smith & R. C. Serfass (Eds.), *Exercise and aging: The scientific basis* (pp. 45-58). Hillside, NJ: Enslow.

Aloia, J. F., Cohn, S. H., Ostuni, J. A., Cane, R., & Ellis, K. (1978). Prevention of involutional bone loss by exercise. *Annals of Internal Medicine, 89*, 356-358.

Anand, M. P. (1964). Accidents in the home. In W. F. Anderson & B. Isaacs (Eds.), *Current achievements in geriatrics* (pp. 239-245). London: Cassell.

Anderson, B., & Palmore, E. (1974) Longitudinal evaluation of ocular function. In E. Palmore (Ed.), *Normal aging II* (pp. 24-32). Durham, NC: Duke University Press.

Anderson, K. L., & Hermansen, L. (1965). Aerobic work capacity in middle-aged Norwegian men. *Journal of Applied Physiology, 20*, 432-436.

Anderson, R. G., & Meyerhoff, W. L. (1932). Otologic manifestations of aging. *Otolaryngologic Clinics of North America, 15*, 353-370.

Anderson, T. W., & Shephard, R. J. (1969). Normal values for single breath diffusing capacity: The influences of age, body size, and smoking habits. *Respiration, 26*, 1-7.

Anderson, W. F., & Cowan, N. R. (1966). Handgrip pressure in older people. *British Journal of Preventive Social Medicine, 20*, 141-147.

Andres, R. (1971). Aging and diabetes. *Medical Clinics of North America, 55*, 835-845.

Andres, R., & Tobin, J. D. (1977). Endocrine systems. In C. E. Finch & L. Hayflick (Eds.), *Handbook of the biology of aging*. New York: Van Nostrand Reinhold.

Aniansson, A., Grimby, G., Nygaard, E., & Saltin, B. (1980). Muscle fiber composition and fibre area in various age groups. *Muscle and Nerve, 2*, 271-272.

Aniansson, A., & Gustafsson, E. (1981). Physical training in elderly men with special reference to quadriceps muscle strength and morphology. *Clinical Physiology, 1*, 87-98.

Anthonisen, N. R., Danson, J., Robertson, P. C., & Ross, W. R. D. (1969-70). Airway closure as a function of age. *Respiratory Physiology, 8*, 58-65.

Arden, G. B. (1978). The importance of measuring contrast sensitivity in cases of visual disturbance. *British Journal of Ophthalmology, 62*, 198-209.

Arden, G. B., & Jacobson, J. J. (1978). A simple grating test for contrast sensitivity: Preliminary results indicate value for screening in glaucoma. *Investigative Ophthalmology and Visual Science*, *17*, 23-32.

Arey, L. B., Tremaine, M. J., & Monzingo, F. L. (1935). The numerical and topographical relations of taste buds to human circumvallate papillae throughout the life span. *Anatomical Record*, *64*, 9-25.

Arundale, K. (1978). An investigation into the variation of human contrast sensitivity with age and ocular pathology. *British Journal of Ophthalmology*, *62*, 213-215.

Arvidson, K. (1979). Location and variation in number of taste buds in human fungiform papillae. *Scandinavian Journal of Dental Research*, *87*, 435-442.

Asmussen, E. (1981). Aging and exercise. In S. M. Horvath & M. K. Yousef (Eds.), *Environmental physiology: Aging, heat and altitude* (pp. 419-428). New York: Elsevier.

Asmussen, E., Fruensgaard, K., & Norgaard, S. (1975). A follow-up longitudinal study of selected physiologic functions in former physical education students—after 40 years. *Journal of the American Geriatrics Society*, *23*, 442-450.

Asmussen, E., & Mathiasen, P. (1962). Some physiological functions in physical education students reinvestigated after twenty-five years. *Journal of the American Geriatrics Society*, *10*, 379-387.

Åstrand, I., Åstrand, P. O., Hallback, I., & Kilbom, Å. (1973). Reduction in maximal oxygen uptake with age. *Journal of Applied Physiology*, *35*, 649-654.

Avioli, L. V. (1982). Aging, bone, and osteoporosis. In S. G. Korenman (Ed.), *Endocrine aspects of aging*. New York: Elsevier Biomedical.

Axelrod, S., & Cohen, L. D. (1961). Senescence and embedded-figure performance in vision and touch. *Perceptual and Motor Skills*, *12*, 283-288.

Azizi, F., Vagenakis, A. G., Portnay, G. I., Rapoport, B., Ingbar, S. H., & Braverman, L. E. (1979). Pituitary-thyroid responsiveness to intramuscular thyrotropin-releasing hormone based on analyses of serum thyrozine, tri-iodothyronine, and thyrotropin concentrations. *New England Journal of Medicine*, *292*, 273-277.

Babin, R. W., & Harker, L. A. (1982). The vestibular system in the elderly. *Otolaryngologic Clinics of North America*, *15*, 387-393.

Back, K. W., & Gergen, K. J. (1968). The self through the latter span of life. In C. Gordon & K. J. Gergen (Eds.), *The self in social interaction* (pp. 241-250). New York: Wiley.

Bahrke, M. S. (1981). Alterations in anxiety following exercise and rest. In F. J. Nagle & H. J. Montoye (Eds.) *Exercise in health and disease* (pp. 291-298). Springfield, IL: C C Thomas.

Baker, H., Jaslow, S. P., & Frank, O. (1978). Severe impairments of dietary folate utilization in the elderly. *Journal of the American Geriatrics Society*, *26*, 218-221.

Bakke, J. L., Lawrence, N., Knudtson, K. P., Roy, S., & Needman, G. H. (1964). A correlative study of the content of thyroid-stimulating hormone (TSH) and cell morphology of the human adenophypophysis. *American Journal of Clinical Pathology*, *41*, 576-588.

Balazs, E. A. (1977). Intercellular matrix of connective tissue. In C. E. Finch & L. E. Hayflick (Eds.) *Handbook of the biology of aging*. New York: Van Nostrand Reinhold.

Balazs, E. A., & Denlinger, J. L. (1982). Aging changes in the vitreus. In R. Sekuler, L. P. Hutman, & C. Owsley (Eds.) *Aging and human visual function*. New York: Alan R. Liss.

Baldwin, E. de F., Cournand, A., & Richards, D. W., Jr. (1948). Pulmonary insufficiency: Physiological classification, clinical methods of analysis, standard values in normal subjects. *Medicine*, *27*, 243-278.

Ball, M. J. (1977). Neuronal loss, neurofibrillary tangles and granuovacuolar degeneration in the hippocampus with ageing and dementia. *Acta Neuropathologica, 37,* 111-118.

Balogh, K., & Lelkes, K. (1961). The tongue in old age. *Gerontologia Clinica, 3,* 38-54.

Baltes, P. B. (1968). Longitudinal and cross-sectional sequences in the study of age and generation effects. *Human Development, 11,* 145-171.

Baltes, P. B., & Schaie, K. W. (1976). On the plasticity of intelligence in adulthood and old age. *American Psychologist, 31,* 720-725.

Barnes, R. F., Raskind, M., Gumbrecht, G., & Halter, J. B. (1982). The effects of age on the plasma catecholamine response to mental stress in man. *Journal of Clinical Endocrinology and Metabolism, 54,* 64-69.

Baron, J. H. (1963). Studies of basal and peak acid output with an augmented histamine test. *Gut, 4,* 136-144.

Barry, A. J., Daly, J. W., Pruett, E. D. R., Steinmetz, J. R., Page, H. F., Birkhead, N. C., & Rodahl, K. (1966). The effects of physical conditioning on older individuals. *Journal of Gerontology, 21,* 182-191.

Barzel, U. S. (1983). Common metabolic disorders of the skeleton in aging. In W. Reichel (Ed.), *Clinical aspects of aging* (3rd ed., pp. 360-370). Baltimore: Williams & Wilkins.

Bassey, E. J. (1978). Age, inactivity, and some physiological responses to exercise. *Gerontology, 24,* 66-77.

Beausoleil, N. I., Sparrow, D., Rowe, J. W., & Silbert, J. E. (1980). Longitudinal analysis of the influence of age on bone loss in men. *Gerontologist, 20,* (No. 5, Part 2), 63 (abstract).

Begin, R., Renzetti, A. D., Bigler, A., & Watanabe, S. (1975). Flow and age dependence of airway closure and dynamic compliance. *Journal of Applied Physiology, 38,* 199-207.

Bell, B., Wolf, E., & Bernholtz, C. D. (1972). Depth perception as a function of age. *Aging and Human Development, 3,* 77-81.

Bell, G. H., Dunbar, O., & Beck, J. S. (1967). Variations in strength of vertebrae with age and their relation to osteoporosis. *Calcified Tissue Research, 1,* 75-86.

Belloc, N. B., & Breslow, L. (1972). Relationship of physical health status and health practices. *Preventive Medicine, 1,* 409-421.

Benestad, A. M. (1965). Trainability of old men. *Acta Medica Scandinavica, 178,* 321-327.

Bengsston, C., Vedin, J. A., Grimby, G., & Tibblin, G. (1978). Maximal work performance test in middle-aged women: Results from a population study. *Scandinavian Journal of Clinical and Laboratory Investigations, 38,* 181-188.

Benjamin, B. J. (1981). Frequency variability in the aged voice. *Journal of Gerontology, 36,* 722-726.

Benjamin, B. J. (1982). Phonological performance in gerontological speech. *Journal of Psycholinguistic Research, 11,* 159-167.

Berger, D., Crowther, R. C., Floyd, J. C. Jr., Pek, S., & Fajans, S. S. (1978). Effects of age on fasting plasma levels of pancreatic hormones in man. *Journal of Clinical Endocrinology and Metabolism, 47,* 1183-1189.

Berglund, E., Birath, G., Bjure, J., Grimby, G., Kjellmer, I., Sandqvist, L., & Söderholm, B. (1963). Spirometric studies in normal subjects. I. Forced expirograms in subjects between 7 and 70 years of age. *Acta Medica Scandinavica, 173,* 185-192.

Bergman, M. (1971). Hearing and aging. *Audiology, 10,* 164-171.

Bergman, M., Blumenfeld, V. G., Cascardo, D., Dash, B., Levitt, H., & Margulies, M. K. (1976). Age-related decrement in hearing for speech: Sampling and longitudinal studies. *Journal of Gerontology, 31*, 533-538.

Bergström, B. (1973a). Morphology of the vestibular nerve: II. The number of myelinated vestibular nerve fibers in man at various ages. *Acta Otolaryngolica, 76*, 173-179.

Bergström, B. (1973b). Morphology of the vestibular nerve: III. Analysis of the myelinated vestibular nerve fibers in man at various ages. *Acta Otolaryngolica, 76*, 331-338.

Bernier, J. J., Vidon, N., & Mignon, M. (1973). The value of a cooperative multicenter study for establishing a table of normal values for gastric acid secretion and as a function of sex, age, and weight. *Biologie et Gastro-Enterologie (Paris), 6*, 287-296.

Berscheid, E., Walster, E., & Bohrnstedt, G. (1973). Body image. The happy American body: A survey report. *Psychology Today, 7*(6), 119-131.

Bhanthumnavin, K., & Schuster, M. M. (1977). Aging and gastrointestinal function. In C. E. Finch & L. Hayflick (Eds.), *Handbook of the biology of aging*. New York: Van Nostrand Reinhold.

Birren, J. E. (1959). Principles of research on aging. In J. E. Birren (Ed.), *Handbook of the psychology of aging*. Chicago: University of Chicago Press.

Björntorp, P., Berchtold, P., & Tibblin, G. (1971). Insulin secretion in relation to adipose tissue in men. *Diabetes, 20*, 65-70.

Björntorp, P., Fahlén, M., Grimby, G., Gustafson, A., Holm, J., Renström, P., & Scherstén, T. (1972). Carbohydrate and lipid metabolism in middle-aged, physically well-trained men. *Metabolism, 21*, 1037-1044.

Blichert-Toft, M. (1975). Secretion of corticotrophin and somatotrophin by the senescent adenohypophysis in man. *Acta Endodrinologica (Supp.), 195*, 15-154.

Blichert-Toft, M., Hummer, L., & Dige-Petersen, H. (1975). Human serum thyrotrophin level and response to thyrotrophin-releasing hormone in the aged. *Gerontologica Clinica, 17*, 191-203.

Blinkov, S. M., & Glezer, I. I. (1968). *The human brain in figures and tables*. New York: Plenum and Basic Books.

Block, E. (1952). Quantitative morphological investigations of the follicular system in women. Variations at different ages. *Acta Anatomy, 14*, 108-123.

Blumenthal, J. A., Schocken, D. D., Needels, T. L., & Hindle, P. (1982). Psychological and physiological effects of physical conditioning on the elderly. *Journal of Psychosomatic Research, 26*, 505-510.

Bode, R. J., Dosman, J., Martin, R. R., Ghezzo, H., & Macklem, P. T. (1976). Age and sex difference in lung elasticity and in closing capacity in nonsmokers. *Journal of Applied Physiology, 41*, 129-135.

Bolton, C. F., Winkelmann, R. K., & Dyck, P. J. (1966). A quantitative study of Meissner's corpuscles in man. *Neurology, 16*, 1-9.

Bondareff, W. (1981). The neurobiological basis of age-related changes. In J. L. McGaugh & S. B. Kiesler (Eds.), *Aging: Biology and behavior*. New York: Academic Press.

Boone, D. C., & Azen, S. P. (1979). Normal range of motion of joints in male subjects. *Journal of Bone and Joint Surgery, 61A*, 756-759.

Borkan, G. A., & Norris, A. H. (1980). Assessment of biological age using a profile of physical parameters. *Journal of Gerontology, 25*, 177-184.

Bortz, W. M. II. (1982). Disuse and aging. *Journal of the American Medical Association, 248*, 1203-1208.

Bottcher, J. (1975). Morphology of the basal ganglia in Parkinson's disease. *Acta Neurologica Scandinavia*, *52*, Supplement 62.

Botwinick, J. (1977). Intelligence and aging. In J. E. Birren & K. W. Schaie (Eds.), *Handbook of the psychology of aging*. New York: Van Nostrand Reinhold.

Botwinick, J. (1978). *Aging and behavior* (2nd ed.). New York: Springer.

Bourlière, F., Cendron, H., & Rapaport, A. (1958). Modification avec l' age des seuils gustatifs de perception et de reconnaissance aux saveurs salée et sucrée, chez l'homme. [Change with age in smell perception and recognition of salty and sweet taste in man.] *Gerontologia*, *2*, 104-112.

Bowman, B. B., & Rosenberg, I. H. (1983). Digestive function and aging. *Human Nutrition: Clinical Nutrition*, *37C*, 75-89.

Boyarsky, R. (1976). Sexuality. In F. U. Steinberg (Ed.), *Cowdry's the care of the geriatric patient*. St. Louis: C. V. Mosby.

Brandes, D., & Garcia-Bunuel, R. (1978). Aging of the male sex accessory organs. In E. L. Schneider (Ed.), *Aging: Vol. 4. The aging reproductive system*. New York: Raven Press.

Brandfonbrener, M., Landowne, M., & Shock, N. W. (1955). Changes in cardiac output with age. *Circulation*, *12*, 557-566.

Brandstetter, R. D., & Kazemi, H. (1983). Aging and the respiratory system. *Medical Clinics of North America*, *67*, 419-431.

Brauer, P. M., Slavin, J. L., & Marlett, J. A. (1981). Apparent digestibility of neutral detergent fiber in elderly and young adults. *American Journal of Clinical Nutrition*, *34*, 1061-1070.

Bredburg, C. (1968). Cellular pattern and nerve supply for the human organ of corti. *Acta Otolaryngolica Supplement*, *236*, 1-135.

Breitmeyer, B. G. (1980). Unmasking visual masking: A look at the "why" behind the veil of the "how." *Psychological Review*, *87*, 52-69.

Breitmeyer, B. G., & Ganz, L. (1976). Implications of sustained and transient channels for theories of visual pattern masking, saccadic suppression, and information processing. *Psychological Review*, *83*, 1-36.

Brewer, B. J. (1979). Aging of the rotator cuff. *American Journal of Sports Medicine*, *7*, 102-110.

Brizzee, K. R. (1975). Gross morphometric analyses and quantitative histology of the aging brain. In J. M. Ordy & K. R. Brizzee (Eds.), *Neurobiology of aging*. New York: Plenum.

Brizzee, K. R., Klara, P. & Johnson, J. E. (1975). Changes in microanatomy, neurocytology and fine structure with aging. In J. M. Ordy & K. R. Brizzee (Eds.) *Neurology of aging*. New York: Plenum.

Brocklehurst, J. C. (1978). The large bowel. In J. C. Brocklehurst (Ed.), *Textbook of geriatric medicine and gerontology* (2nd ed.). New York: Churchill Livingstone.

Brody, A. W., Johnson, J. R., Townley, R. G., Herrera, H. R., Snider, D., & Campbell, J. C. (1974). The residual volume. Predicted values as a function of age. *American Review of Respiratory Disease*, *109*, 98-105.

Brody, H. (1955). Organization of cerebral cortex: III. A study of aging in the human cerebral cortex. *Journal of Comparative Neurology*, *102*, 511-556.

Brody, H. (1970). Structural changes in the aging nervous system. In H. T. Blumental (Ed.), *Interdisciplinary topics in gerontology* (Vol. 7). New York: Karger.

Brody, H. (1982). Age changes in the nervous system. In F. I. Caird (Ed.), *Neurological disorders in the elderly*. Bristol: John Wright & Sons.

Brody, H., & Vijayashankar, N. (1977). Anatomical changes in the nervous system. In C. E. Finch & L. Hayflick (Eds.), *Handbook of the biology of aging*. New York: Van Nostrand Reinhold.

Brooks, G. A., & Fahey, T. D. (1984). *Exercise physiology: Human bioenergetics and its applications*. New York: Wiley.

Bruner, A., & Norris, T. (1971). Age related changes in caloric nystagmus. *Acta Otolaryngologica (Supplement), 282*.

Brunner, D., & Meshulam, N. (1970). Physical fitness of trained elderly people. In D. Brunner & E. Jokl (Eds.), *Medicine and Sport: Vol. 4. Physical activity and aging* (pp. 80-88). Baltimore: University Park Press.

Buccola, V., & Stone, W. J. (1975). Effects of jogging and cycling program on physiological and personality variables in aged men. *Research Quarterly, 46*, 134-139.

Buell, S. J. (1982). Golgi-Cox and rapid Golgi methods as applied to autopsied human brain tissue: Widely disparate results. *Journal of Neuropathology and Experimental Neurology, 41*, 500-507.

Buell, S. J., & Coleman, P. D. (1979). Dendritic growth in the aged human brain and failure of growth in senile dementia. *Science, 206*, 854-856.

Buell, S. J., & Coleman, P. D. (1981). Individual differences in dendritic growth in human aging and senile dementia. In D. Stein (Ed.), *The psychobiology of aging: Problems and perspectives*. Amsterdam: Elsevier North Holland.

Bugiani, C., Salvariani, S., Perdelli, F., Mancardi, G. L., & Leonardi, A. (1978). Nerve cell loss with aging in the putamen. *European Neurology, 17*, 286-291.

Buist, A. A., & Ross, B. B. (1973). Predicted values for closing volumes using a modified single breath nitrogen test. *American Review of Respiratory Diseases, 107*, 744-752.

Burgess, R., & Huston, T. (Eds.), *Social exchange in developing relationships*. New York: Academic Press.

Burstein, A. H., Reilly, D. T., & Martens, M. (1976). Aging of bone tissue: Mechanical properties. *Journal of Bone and Joint Surgery (A), 58*, 82-86.

Butler, R. N. (1978). Psychosocial aspects of reproductive agir E. L. Schneider (Ed.), *Aging, Vol. 4, The aging reproductive system*. New York: Raven Press.

Byrd, E., & Gertman, S. (1959). Taste sensitivity in aging persons. *Geriatrics, 14*, 381-384.

Calearo, C., & Lazzaroni, A. (1957). Speech intelligibility in relation to the speed of the message. *Laryngoscope, 67*, 410-419.

Calloway, N. O., & Merrill, R. S. (1965). The aging adult liver. I. Bromsulphalein and bilirubin clearances. *Journal of the American Geriatrics Society, 13*, 594-598.

Campbell, E. J., & Lefrak, S. S. (1978). How aging affects the structure and function of the respiratory system. *Geriatrics, 33* (6), 68-78.

Campbell, M. J., McComas, A. J., & Petito, F. (1973). Physiological changes in aging muscles. *Journal of Neurology, Neurosurgery and Psychiatry, 36*, 174-182.

Carlson, H. E., Gillin, J. C., Gorden, P., & Snyder, F. (1972). Absence of sleep-related growth hormone peaks in aged normal subjects and in acromegaly. *Journal of Clinical Endocrinology and Metabolism, 34*, 1102-1105.

Carlsson, A., Adolfsson, R., Aquilonius, S.-M., Gottfries, C.-G., Oreland, L., Svennerholm, L., & Winblad, B. (1980). Biogenic amines in human brain in normal aging, senile dementia, and chronic alcoholism. In M. Goldstein, D. B. Calne, A. Lieberman, & M. O. Thorner (Eds.), *Advances in biochemical psychopharmacology: Vol. 23. Ergot compounds and brain function: Neuroendocrine and neuropsychiatric aspects*. New York: Raven Press.

Carr, R. D., Smith, M. J., & Keil, P. G. (1960). The liver in the aging process. *Archives of Pathology, 70,* 15-18.

Carter, J. H. (1982a). The effects of aging on selected visual functions: Color vision, glare sensitivity, field of vision, and accommodation. In R. Sekuler, L. P. Hutman, & C. Owsley (Eds.), *Aging and human visual function.* New York: Alan R. Liss.

Carter, J. H. (1982b). Predicting visual responses to increasing age. *Journal of the American Optometric Association, 53,* 31-36.

Cartlidge, N. E. F., Black, M. M., Hall, M. R. P., & Hall, R. (1970). Pituitary function in the elderly. *Gerontologia Clinica, 12,* 65-70.

Cauna, N. (1965). The effects of aging on the receptor organs of the human dermis. In W. Montagna (Ed.), *Advances in biology of skin: Volume 6. Aging* (pp. 63-96). New York: Pergamon Press.

Chalke, H. D., Dewhurst, J. R., & Ward, C. W. (1958). Loss of sense of smell in old people. *Public Health, 72*(6), 223-230.

Chapman, E. A., deVries, H. A., & Swezey, R. (1972). Joint stiffness: Effects of exercise on old and young men. *Journal of Gerontology, 27,* 218-221.

Chien, S., Peng, M. T., Chen, K. P., Huang, T. F., Chang, C., & Fang, H. S. (1975). Longitudinal studies on adipose tissues and its distribution in human subjects. *Journal of Applied Physiology, 39,* 825-830.

Christensen, N. J. (1973). Plasma noradrenaline and adrenaline in patients with thyrotoxicosis and myxoedema. *Clinical Science and Molecular Medicine, 45,* 163-171.

Chung, E. B. (1966a). Aging in human joints: I. Articular cartilage. *Journal of the National Medical Association, 58,* 87-95.

Chung, E. B. (1966b). Aging in human joints: II. Joint capsule. *Journal of the National Medical Association, 58,* 87-95.

Clark, W. C. (1969). Sensory-decision theory analysis of the placebo effect on the criterion for pain & thermal sensitivity. *Journal of Abnormal Psychology, 74,* 363-371.

Clark, W. C., & Mehl, L. (1971). Thermal pain: A sensory decision theory analysis of the effect of age and sex on d^1, various response criteria, and 50 percent pain threshold. *Journal of Abnormal Psychology, 78,* 202-212.

Clarke, H. H. (1977). Exercise and aging. *Physical Fitness Research Digest,* Series 7, No. 2, 1-27.

Clarkson, P. M. (1978). The effect of age and activity level on simple and choice fractionated response time. *European Journal of Applied Physiology, 40,* 17-25.

Cohen, T., Gitman, L., & Lipshutz, E. (1960). Liver function studies in the aged. *Geriatrics, 15,* 824-836.

Cohn, J. E., Carroll, D. G., Armstrong, B. W., Shepard, R. H., & Riley, R. L. (1954). Maximal diffusing capacity of the lungs. *Journal of Applied Physiology, 6,* 588-597.

Cohn, S. H., Vaswani, A., Zanzi, I., & Ellis, K. J. (1976). Effect of aging on bone mass in adult women. *American Journal of Physiology, 230,* 143-148.

Cole, G. M., Segall, P. E., & Timiras, P. S. (1982). Hormones during aging. In A. Vernandakis (Ed.), *Hormones in development and aging* (pp. 477-550). New York: Spectrum.

Cole, T. V. (1974). The influence of height on the decline in ventilatory function. *International Journal of Epidemiology, 3,* 145-152.

Colebatch, H. J. H., Greaves, I. A., & Ng, C. K. Y. (1979). Exponential analysis of elastic recoil and aging in healthy males and females. *Journal of Applied Physiology, 47,* 683-691.

Coleman, R. M., Miles, L. E., Guilleminault, C. C., Zarcone, V. P., van den Hoed, J., & Dement, W. C. (1981). Sleep-wake disorders in the elderly: A polysomnographic analysis. *Journal of the American Geriatrics Society, 29,* 289-296.

Collingwood, T. R. (1972). The effect of physical training upon behavior and self-attitude. *Journal of Clinical Psychology, 28,* 583-585.

Collingwood, T. R., & Willett, L. (1971). The effects of physical training upon self-concept and body attitude. *Journal of Clinical Psychology, 27,* 411-412.

Collins, K. J., Dore, C., Exton-Smith, A. N., Fox, R. H., MacDonald, I. C., & Woodward, P. M. (1977). Accidental hypothermia and impaired temperature homeostasis in the elderly. *British Medical Journal, 1,* 353-356.

Collman, R. D., & Stoller, A. (1962). A survey of mongoloid births in Victoria, Australia, 1942-1957. *American Journal of Public Health, 52,* 813-829.

Connor, C. L., Walsh, R. P., Litzelman, D. K., & Alvarez, M. G. (1978). Evaluation of job applicants. The effects of age vs. success. *Journal of Gerontology, 33,* 246-252.

Conway, J., Wheeler, R., & Sannerstedt, R. (1971). Sympathetic nervous activity during exercise in relation to age. *Cardiovascular Research, 5,* 577-581.

Cooley, C. H. (1902). *Human nature and the social order.* New York: Scribner's.

Cooper, R. M., Bilash, I., & Zubek, J. P. (1959). The effect of age on taste sensitivity. *Journal of Gerontology, 14,* 56-58.

Cornes, J. S. (1965). Number, size, and distribution of Peyer's patches in the human small intestine. II. The effect of age on Peyer's patches. *Gut, 6,* 230-233.

Corso, J. F. (1977). Presbycusis, hearing aids, and aging. *Audiology, 16,* 146-163.

Corso, J. F. (1981). *Aging sensory systems and perception.* New York: Praeger.

Costa, P. T., Jr., & McCrae, R. R. (1980). Somatic complaints in males as a function of age and neuroticism: A longitudinal analysis. *Journal of Behavioral Medicine, 3,* 245-258.

Cotes, J. E., Hall, A. M., Johnson, G. R., Jones, P. R. M., & Knibbs, A. V. (1973). Decline with age of cardiac frequency during submaximal exercise on healthy women. *Journal of Physiology, 238,* 24-45P.

Cotlier, E. (1981). The lens. In R. A. Moses (Ed.), *Adler's physiology of the eye.* St. Louis: C. V. Mosby.

Coulombe, P., Dussault, J. A., & Walker, P. (1977). Catecholamine metabolism in thyroid tissue. II. Norepinephrine secretion rate in hyperthryoidism and hypothyroidism. *Journal of Clinical Endocrinology and Metabolism, 44,* 1185-1189.

Coyne, A. C. (1981). Age differences and practice in forward visual masking. *Journal of Gerontology, 36,* 730-732.

Cragg, B. G. (1975). The density of synapses and neurons in normal, mentally defective, and aging human brains. *Brain, 98,* 81-90.

Cramer, S., Kietzman, M. L., & Laer, J. Van. (1982). Dichoptic backward masking of letters, words, and trigrams in old and young subjects. *Experimental Aging Research, 8,* 103-108.

Croft, L. H. (1982). *Sexuality in later life: A counseling guide for physicians.* Boston, MA: John Wright.

Cumming, G. R. (1967). Current levels of fitness. *Canadian Medical Journal, 96,* 868-877.

Cunningham, D. A., & Hill, J. S. (1975). Effect of training on cardiovascular response to exercise in women. *Journal of Applied Physiology, 39,* 891-895.

Cunningham, D. A., Rechnitzer, P. A., Pearce, M. E., & Donner, A. P. (1982). Determinants of self-selected walking pace across ages 19 to 66. *Journal of Gerontology, 37,* 560-564.

Curcio, C. A., Buell, S. J., & Coleman, P. D. (1982). Morphology of the aging central nervous system: Not all downhill. In J. A. Mortimer, F. J. Pirozzola, & G. I. Maletta

(Eds.), *Advances in neurogerontology: The aging motor system*. New York: Praeger.

Cureton, T. K. (1969). *The physiological effects of exercise programs on adult men*. Springfield, IL: C C Thomas.

Cureton, T. K. (1963). Improvement of psychological states by means of exercise fitness programs. *Journal of the Association for Physical and Mental Rehabilitation, 17*, 14-18; 25.

Currey, J. D. (1979). Changes in the impact energy absorption of bone with age. *Journal of Biomechanics, 12*, 459-469.

Currey, J. D. (Feb. 13, 1984). Effects of differences in mineralization on the mechanical properties of bone. *Philosophical Transactions of the Royal Society of London (Biology), 304* (1121), 509-518.

Daly, J. W., Barry, A. J., & Birkhead, N. C. (1968). The physical working capacity of older individuals. *Journal of Gerontology, 23*, 134-139.

Damon, A. (1972). Predicting age from body measurements and observations. *Aging and Human Development, 3*, 169-174.

Damon, A., Seltzer, C. C., Stoudt, H. W., & Bell, B. (1972). Age and physique in healthy white veterans at Boston. *Journal of Gerontology, 27*, 202-208.

Darmady, E. M., Offer, J., & Woodhouse, M. A. (1973). The parameters of the aging kidney. *Journal of Pathology, 109*, 195-207.

Davidson, A. J., Talner, L. B., & Downs, W. M. (1969). A study of the angiographic appearance of the kidney in an aging normotensive population. *Radiology, 92*, 975-983.

Davidson, J. M., Chen, J. J., Crapo, L., Gray, G. D., Greenleaf, W. J., & Catania, J. A. (1983). Hormonal changes and sexual function in aging men. *Journal of Clinical Endocrinology and Metabolism, 57*, 71-77.

Davidson, M. B. (1979). The effect of aging on carbohydrate metabolism: A review of the English literature and a practical approach to the diagnosis of diabetes mellitus in the elderly. *Metabolism, 28*, 688-705.

Davidson, M. B. (1982). The effect of aging on carbohydrate metabolism: A comprehensive review and a practical approach to the clinical problem. In S. G. Korenman (Ed.), *Endocrine aspects of aging*. New York: Elsevier Biomedical.

Davies, D. F., & Shock, N. W. (1950). Age changes in glomerular filtration rate, effective renal plasma flow, and tubular excretory capacity in adult males. *Journal of Clinical Investigation, 29*, 496-507.

Davies, I., & Fotheringham, A. P. (1981). Lipofuscin—does it affect cellular performance? *Experimental Gerontology, 16*, 119-125.

DeFronzo, R. A. (1979). Glucose intolerance and aging: Evidence for tissue sensitivity to insulin. *Diabetes, 28*, 1095-1101.

DeFronzo, R. A. (1982). Glucose intolerance and aging. In M. E. Reff and E. L. Schneider (Eds.), *Biological markers of aging* (pp. 98-119). (NIH Publication No. 82-2221)

DeFronzo, R. A., Tobin, J. D., & Andres, R. A. (1979). Glucose clamp technique: A method for quantifying insulin secretion and resistance. *American Journal of Physiology, 237*, E214-E223.

Dekaban, A. S., & Sadowsky, D. (1978). Changes in brain weights during the span of human life: Relation of brain weights to body heights and body weights. *Annals of Neurology, 4*, 345-356.

Dekoninck, W. J., Jacquy, J., Jocquet, P., & Noel, G. (1976). Cerebral blood flow and metabolism in senile dementia. In J. S. Meyer, H. Lechner, & M. Reivich (Eds.), *Cerebral vascular disease*. Stuttgart: Thieme.

Dement, W. C., Miles, L. E., & Bliwise, D. L. (1982). Physiological markers of aging: Human sleep pattern changes. In M. E. Reff & E. L. Schneider (Eds.), *Biological markers of aging* (pp. 177-187). (NIH Publication No. 82-2221)

Denolin, H., Messin, R., Degre, S., Vandermoten, P., & de Coster, A. (1970). Influence of age on the behaviour of normal subjects during exercise. In D. Brunner & E. Jokl (Eds.), *Medicine and sport: Vol. 4, Physical activity and aging* (pp. 309-315). Baltimore: University Park Press.

Derefeldt, G., Lennerstrand, G., & Lindh, B. (1979). Age variations in normal human contrast sensitivity. *Acta Ophthalmologica, 57,* 679-690.

Desmedt, J. E., & Cheron, G. (1980). Aging of the somatosensory system in man. In D. G. Stein (Ed.), *The psychobiology of aging: Problems and perspectives* (pp. 387-394). New York: Elsevier.

Devaney, K. O., & Johnson, H. A. (1980). Neuron loss in the aging visual cortex of man. *Journal of Gerontology, 35,* 836-841.

deVries, H. A. (1970). Physiological effects of an exercise training regimen upon men aged 52-88. *Journal of Gerontology, 25,* 325-336.

deVries, H. A. (1980). *Physiology of exercise for physical education and athletics* (3rd ed.). Dubuque, IA: Wm. C. Brown.

deVries, H. A., & Adams, G. M. (1972a). Comparison of exercise responses in old and young men. *Journal of Gerontology, 27,* 344-348.

deVries, H. A., & Adams, G. M. (1972b). Comparison of exercise responses in old and young men: II. Ventilatory mechanics. *Journal of Gerontology, 27,* 349-352.

deVries, H. A., & Adams, G. M. (1977). Effect of the type of exercise upon the work of the heart in older men. *Journal of Sports Medicine and Physical Fitness, 17,* 41-48.

Dill, D. B., Yousef, M. K., Vitez, T. S., Goldman, A., & Patzer, R. (1982). Metabolic observations on caucasian men and women aged 17 to 88 years. *Journal of Gerontology, 37,* 565-571.

Donevan, R. E., Palmer, W. H., Varvis, C. J., & Bates, D. V. (1955). Influence of age on pulmonary diffusing capacity. *Journal of Applied Physiology, 14,* 483-492.

Donohue, S. (1977). The correlation between physical fitness, absenteeism, and work performance. *Canadian Journal of Public Health, 68,* 201-203.

Doty, R. L., Shaman, P., Applebaum, S. L., Giberson, R., Sikosorski, L., & Rosenberg, L. (1984). Smell identification ability: Changes with age. *Science, 226,* 1441-1443.

Dowell, L. J., Badgett, J. L. Jr., & Landiss, C. W. (1968). A study of the relationship between selected physical attributes and self-concept. In G. S. Kenyon (Ed.), *Contemporary psychology of sport* (pp. 657-672). Chicago: Athletic Institute.

Drachman, D. A. (1977). Memory and cognitive function in man: Does the cholinergic system have a specific role? *Neurology, 27,* 783-790.

Drachman, D. A., & Leavitt, J. (1974). Human memory and the cholinergic system: A relationship to aging? *Archives of Neurology, 30,* 113-121.

Drachman, D. A., Noffsinger, D., Sahakian, B. J., Kurdziel, S., & Fleming, P. (1980). Aging, memory and the cholinergic system: A study of dichotic listening. *Neurobiology of Aging, 1,* 39-43.

Drachman, D. A., & Sahakian, B. M. (1980). Memory, aging and pharmacosystems. In D. G. Stein (Ed.), *The psychobiology of aging: Problems and perspectives* (pp. 347-368). New York: Elsevier.

Dressler, M., & Rassow, B. (1981). Neural contrast sensitivity measurement with a laser interference system for clinical screening application. *Investigative Ophthalmology and Visual Science, 21,* 737-744.

Drinkwater, B. L., Bedi, J. F., Loucko, A. B., Roche, S., Horvath, S. M. (1982). Sweating sensitivity in capacity of women in relation to age. *Journal of Applied Physiology, 53*, 671-676.

Drinkwater, B. L., Horvath, S. M., & Wells, C. L. (1975). Aerobic power of females ages 10 to 68. *Journal of Gerontology, 30*, 385-394.

Dudl, R. J., & Ensinck, J. W. (1972). The role of insulin, glucagon, and growth hormone in carbohydrate homeostasis during aging. *Diabetes, 21* (Suppl. 1), 357.

Dudl, R. J., & Ensinck, J. W. (1977). Insulin and glucagon relationships during aging in man. *Metabolism, 26*, 33-41.

Dudl, R. J., Ensinck, J. W., Palmer, H. E., & Williams, R. H. (1973). Effect of age on growth hormone secretion in man. *Journal of Clinical Endocrinology and Metabolism, 37*, 11-16.

Dunnill, M. S., & Halley, W. (1973). Some observations on the quantitative anatomy of the kidney. *Journal of Pathology, 110*, 113-121.

Durin, J. V. G. A., & Womersley, J. (1974). Body fat assessed from total body density and its estimation from skinfold thickness: Measurements on 481 men and women aged from 16 to 72 years. *British Journal of Nutrition, 32*, 77-97.

Dyck, P. J., Schultz, P. W., & O'Brien, P. C. (1972). Quantitation of touch-pressure sensation. *Archives of Neurology, 26*, 465-473.

Eastwood, H. D. H. (1972). Bowel transit studies in the elderly. Radio opaque markers in the investigation of constipation. *Gerontologica Clinica, 14*, 154-159.

Edelhauser, H. E., VanHorn, D. L., & Records, R. E. (1979). Cornea and sclera. In R. E. Records. *Physiology of the human eye and visual system*. New York, NY: Harper & Row.

Edelman, N. H., Mittman, C., Norris, A. H., & Shock, N. W. (1968). Effects of respiratory pattern on age differences in ventilation uniformity. *Journal of Applied Physiology, 24*, 49-53.

Edwards, A. E., & Husted, J. R. (1976). Penile sensitivity, age, and sexual behavior. *Journal of Clinical Psychology, 32*, 697-700.

Eisdorfer, C., Nowlin, J., & Wilkie, F. (1970). Improvement of learning in the aged by modification of autonomic nervous system activity. *Science, 170*, 1327-1329.

Eisdorfer, C., & Wilkie, F. (1972). Auditory changes in the aged: A follow-up study. *Journal of the American Geriatrics Society, 20*, 377-382.

Eisendorfer, C. (1960). Rorschach rigidity and sensory decrement in a senescent population. *Journal of Gerontology, 15*, 188-190.

Elahi, E., Muller, D. C., Tzankoff, S. P., Andres, R., & Tobin, J. D. (1982). Effect of age and obesity on fasting levels of glucose, insulin, glucagon, and growth hormone in man. *Journal of Gerontology, 37*, 385-391.

Ellis, F. P., Exton-Smith, A. N., Foster, K. G., & Weiner, J. S. (1976). Eccrine sweating and mortality during heat waves in very young and very old persons. *Israel Journal of Medical Science, 12*, 815-817.

Ellis, K. J., Shukla, K. K., Cohn, S. H., & Pierson, R. N. Jr. (1974). A predictor for total-body potassium based on height, weight, sex, and age: Application in medical disorders. *Journal of Laboratory and Clinical Medicine, 83*, 716-727.

Engen, T. (1982). *The perception of odors*. New York: Academic Press.

Epstein, M. (1979). Effects of aging on the kidney. *Federation Proceedings, 38*, 168-172.

Epstein, M., & Hollenberg, N. K. (1976). Age as a determinant of renal sodium conservation in normal man. *Journal of Laboratory and Clinical Medicine, 87*, 411-417.

Epstein, S. (1973). The self-concept revisited: Or a theory of a theory. *American Psychologist, 28*, 404-416.

Erdelyi, M. H. (1974). A new look at the New Look: Perceptual defense and vigilance. *Psychological Review*, *81*, 1-25.

Ericcson, P., & Irnell, L. (1969). Effect of 5 years aging on ventilatory capacity and physical work capacity in elderly people. *Acta Medica Scandinavica*, *185*, 193-199.

Eriksen, C. W., Hamlin, R. M., & Breitmeyer, B. G. (1970). Temporal factors in visual perception as related to aging. *Perception and Psychophysics*, *7*, 354-356.

Esler, M., Skews, H., Leonard, P., Jackman, G., Bobik, A., & Korner, P. (1981). Age-dependence of noradrenaline kinetics in normal subjects. *Clinical Science*, *60*, 217-219.

Etholm, B., & Belal, A., Jr. (1974). Senile changes in the middle ear joints. *Annals of Otology, Rhinology, and Laryngology*, *83*, 49-54.

Evans, F. G. (1973). *Mechanical properties of bone*. Springfield, IL: C C Thomas.

Evans, F. G. (1976). Mechanical properties and histology of cortical bone from younger and older men. *Anatomical Record*, *185*, 1-12.

Evans, M. A., Triggs, E. J., Cheung, M., Broe, G. A., & Creasey, H. (1981). Gastric emptying rate in the elderly: Implications for drug therapy. *Journal of the American Geriatrics Society*, *29*, 201-205.

Evered, D. C., Tunbridge, W. M. G., Hall, R., Appleton, D., Bruvis, M., Clark, F., Manuel, P., & Young, E. (1972). Thyroid hormone concentrations in a large scale community survey. Effect of age, sex, and medication. *Clinica Chimica Acta*, *83*, 223-229.

Everitt, A. V. (1976a). The female climacteric. In A. V. Everitt & J. A. Burgess (Eds.), *Hypothalamus, pituitary, and aging*. Springfield, IL: C C Thomas.

Everitt, A. V. (1976b). The thyroid gland, metabolic rate and aging. In A. V. Everitt & J. A. Burgess (Eds.), *Hypothalamus, pituitary and aging*. Springfield, IL: C C Thomas.

Everitt, A. V., & Burgess, J. A. (1976). Growth hormone and aging. In A. V. Everitt & A. Burgess (Eds.), *Hypothalamus, pituitary and aging*. Springfield, IL: C C Thomas.

Fardy, P. S. (1971). Left ventricle time component changes in middle-aged men following a twelve week physical training intervention program. *Journal of Sports Medicine and Physical Fitness*, *13*, 212-225.

Feifel, H. (1957). Judgment of time in younger and older men. *Journal of Gerontology*, *12*, 71-74.

Feinberg, I. (1974). Changes in sleep cycle patterns with age. *Journal of Psychiatric Research*, *10*, 283-306.

Feinberg, I., Koresko, R. L., & Heller, N. (1967). EEG sleep patterns as a function of normal and pathological aging in man. *Journal of Psychiatric Research*, *5*, 107-144.

Feldman, J. M., & Plonk, J. W. (1976). Effect of age on intravenous glucose tolerance and insulin secretion. *Journal of the American Geriatrics Society*, *24*, 1-3.

Feldman, R. M., & Reger, S. N. (1967). Relations among hearing, reaction time, and age. *Journal of Speech and Hearing Research*, *10*, 479-495.

Fikry, M. E. (1965). Gastric secretory functions in the aged. *Gerontologica Clinca*, *7*, 216-226.

Finch, C. E. (1977). Neuroendocrine and autonomic aspects of aging. In C. E. Finch & L. Hayflick (Eds.), *Handbook of the biology of aging*. New York: Van Nostrand Reinhold.

Finch, C. E. (1982). Rodent models for aging processes in the human brain. In S. Corkin, K. L. Davis, J. H. Growdon, E. Usdin, & R. J. Wurtman (Eds.), *Aging: Vol. 19, Alzheimer's disease: A report of progress*. New York: Raven Press.

Fink, R. I., Kolterman, O. G., Griffin, J., & Olefsky, J. M. (1983). Mechanisms of insulin resistance in aging. *Journal of Clinical Investigation, 71*, 1523-1525.

Finkelstein, J. W., Roffwarg, H. P., Boyar, R. M., Kream, J., & Hellman, L. (1972). Age-related changes in 24-hour spontaneous secretion of growth hormone. *Journal of Clinical Endocrinology and Metabolism, 35*, 665-670.

Fisch, L. (1978). Special senses: The aging auditory system. In J. C. Brocklehurst (Ed.), *Textbook of geriatric medicine and gerontology* (p. 283). New York: Churchill Livingstone.

Fisher, R. F. (1969). Elastic constants of human lens capsule. *Journal of Physiology, 201*, 1-19.

Fisher, S., & Cleveland, S. E. (1958). *Body image and personality* (2nd ed.). Princeton, NJ: D. Van Nostrand.

Folkins, C. H., Lynch, S., & Gardner, M. M. (1973). Psychological fitness as a function of physical fitness. *Archives of Physical Medicine and Rehabilitation, 53*, 503-508.

Folkins, C. H., & Sime, W. E. (1981). Physical fitness training and mental health. *American Psychologist, 36*, 373-389.

Foote, N. N., & Cottrell, L. S. (1955). *Identity and interpersonal competence.* Chicago: University of Chicago Press.

Forbes, G. B., & Reina, J. C. (1970). Adult lean body mass declines with age: Some longitudinal observations. *Metabolism, 19*, 653-663.

Foster, K. G., Ellis, F. P., Dore, C., Exton-Smith, A. N., & Weiner, J. S. (1976). Sweat responses in the aged. *Age and Ageing, 5*, 91-101.

Fox, R. H., MacGibbon, R., Davies, L., & Woodward, P. M. (1973). Problem of the old and the cold. *British Medical Journal, 1*, 21-24.

Fozard, J. L., Wolf, E., Bell, B., McFarland, R. A., & Podolsky, S. (1977). Visual perception and communication. In J. E. Birren & K. W. Schaie (Eds.), *Handbook of the psychology of aging.* New York: Van Nostrand Reinhold.

Frackowiak, R. S. J., Jones, T., Lenzi, G. L., & Heather, J. D. (1980). Regional cerebral oxygen utilization and blood flow in normal man using oxygen-I5 and positron emission tomography. *Acta Neurologica Scandinavica, 62*, 336-344.

Frackowiak, R. S. J., Lenzi, G. L., Jones, T., & Heather, J. D. (1980). The quantitatiave measurement of regional cerebral blood flow and oxygen metabolism in man using oxygen-I5 and positron emission tomography: Theory, procedure, and normal values. *Journal of Computer Assisted Tomography, 4*, 727-736.

Fraisse, P. (1963). *The psychology of time.* New York: Harper & Row.

Franco-Morselli, R., Elghozi, J. L., Joly, E., DiGiulio, S., & Meyer, P. (1977). Increased plasma adrenaline concentrations in benign essential hypertension. *British Medical Journal, 2*, 1251-1254.

Franks, D. D., & Marolla, J. (1976). Efficacious action and social approval as interacting dimensions of self-esteem: A tentative formulation through construct validation. *Sociometry, 39*, 324-340.

Fregly, A., Smith, M., & Graybiel, A. (1973). Revised normative standards of performance of men on quantitative ataxia test battery. *Acta Otolarygngolica, 75*, 10-16.

Frekany, G. A, & Leslie, D. K. (1975). Effects of an exercise program on selected flexibility measurements of senior citizens. *Gerontology, 15*, 182-183.

Frolkis, V. V. (1982). *Aging and life-prolonging processes.* Vienna: Springer-Verlag.

Frolkis, V. V., & Bezrukov, V. V. (1979). *Aging of the central nervous system: Vol. 11. Interdisciplinary topics in human aging.* New York: Karger.

Gacek, R. R. (1975). Degenerative hearing loss in aging. In W. S. Fields (Ed.), *Neurological and sensory disorders in the elderly*. New York: Stratton.

Galbo, H., Richter, E. A., Hilstead, J., Holst, J.-J., Christensen, N. J., & Henriksson, J. (1977). Hormonal regulation during prolonged exercise. *Annals of the New York Academy of Science, 301*, 72-80.

Gallagher, J. C., Riggs, B. L., Eisman, J., Hamstra, A., Arnaud, S. B., & DeLuca, H. F. (1979). Intestinal calcium absorption and serum vitamin D metabolites in normal subjects and osteoparotic patients. *Journal of Clinical Investigation, 64*, 729-736.

Garn, S. M. (1975). Bone loss and aging. In R. Goldman & M. Rockstein (Eds.), *The physiology and pathology of aging* (pp. 39-57). New York: Academic Press.

Garn, S. M., Rohmann, C. G., & Wagner, B. (1967). Bone loss as a general phenomenon in man. *Federation Proceedings, 26*, 1729-1736.

Gelfand, G. A., & Piper, N. (1981). Acoustic reflex thresholds in young and elderly subjects with normal hearing. *Journal of the Acoustical Society of America, 69*, 295-297.

Gellis, M., & Pool, R. (1977). Two-point discrimination distances in the normal hand and forearm. *Plastic and Rconstructive Surgery, 59*, 57-63.

Gergen, K. J. (1977). The social construction of self-knowledge. In T. Mischel (Ed.), *The self: Psychological and philosophical issues* (pp. 140-169). Oxford: Blackwell.

Gershberg, H. (1957). Growth hormone content and metabolic actions of human pituitary glands. *Endocrinology, 61*, 160-165.

Gerstenblith, G. (1980). Noninvasive assessment of cardiovascular function in the elderly. In M. L. Weisfeldt (Ed.), *Aging: Vol. 12. The aging heart: Its function and response to stress*. New York: Raven Press.

Getchell, L. H., & Moore, J. C. (1975). Physical training: Comparative responses of middle-aged adults. *Archives of Physical Medicine and Rehabilitation, 56*, 250-254.

Gibson, G. J., Pride, N. B., O'Cain, C., & Quagliato, R. (1976). Sex and age differences in pulmonary mechanics in normal nonsmoking subjects. *Journal of Applied Physiology, 41*, 20-25.

Goetzinger, C., & Rousey, C. (1959). Hearing problems in later life. *Medical Times, 87*, 771-780.

Goldberg, B., & Folkins, C. (1974). Relationship of body image to negative emotional attitudes. *Perceptual and Motor Skills, 39*, 1053-1054.

Goldman, R. (1977). Aging of the excretory system: Kidney and bladder. In C. E. Finch & L. Hayflick (Eds.), *Handbook of the biology of aging*. New York: Van Nostrand Reinhold.

Gollnick, P. D., Armstrong, R. B., Saubert, C. V. IV, Piehl, K., & Saltin, B. (1972). Enzyme activity and fiber composition in skeletal muscle of untrained and trained men. *Journal of Applied Physiology, 33*, 312-319.

Goyal, V. K. (1982). Changes with age in the human kidney. *Experimental Gerontology, 17*, 321-331.

Granath, A., Jonsson, B., & Strandell, T. (1970). Circulation in healthy old men studied by right-heart catheterization at rest and during exercise in supine and sitting position. In D. Brunner & E. Jokl (Eds.), *Medicine and Sport: Vol. 4. Physical activity and aging* (pp. 48-79). Baltimore: International Universities Press.

Granick, S., Kleban, M. H., & Weiss, A. D. (1976). Relationships between hearing loss and cognition in normally hearing aged persons. *Journal of Gerontology, 31*, 434-440.

Green, E. J., Greenough, W. T., & Schlumpf, B. E. (1983). Effects of complex or isolated environments on cortical dendrites of middle-aged rats. *Brain Research, 264*, 233-240.

Greenough, W. T., & Green, E. J. (1981). Experience and the changing brain. In J. L. McGaugh & S. B. Kiesler (Eds.), *Aging: Biology and behavior*. New York: Academic.

Gregerman, R. I., & Bierman, E. L. (1974). Aging and hormones. In R. H. Williams (Ed.), *Textbook of endocrinology* (5th ed.). Philadelphia: W. B. Saunders.

Gregerman, R. I., Gaffney, G. W., Shock, N. W., & Crowder, S. E. (1962). Thyroxine turnover in euthyroid man with special reference to changes with age. *Journal of Clinical Investigations, 41*, 2065-2074.

Greifenstein, F. E., King, R. M., Latch, S. S., & Comroe, J. H. Jr. (1952). Pulmonary function studies in healthy men and women 50 years and older. *Journal of Applied Physiology, 4*, 641-648.

Gribbin, B., Pickering, T. G., Sleight, P., & Peto, R. (1971). Effect of age and high blood pressure on baroreflex sensitivity in man. *Circulation Research, 29*, 424-431.

Griest, J. H., Klein, M. H., Eischens, R. R., Faris, J., Gurman, A. S., & Morgan, W. P. (1979). Running as a treatment for depression. *Comprehensive Psychiatry, 20*, 41-54.

Grimby, G., Danneskiold-Samsoe, B., Hvid, K., & Saltin, B. (1982). Morphology and enzymatic capacity in arm and leg muscles in 78-82 year old men and women. *Acta Physiologica Scandinavica, 115*, 124-134.

Grimby, G., Nilsson, N. J., & Saltin, B. (1966). Cardiac output during submaximal and maximal exercise in active middle-aged athletes. *Journal of Applied Physiology, 21*, 1150-1156.

Grimby, G., & Saltin, B. (1966). Physiological analysis of physically well-trained middle-aged and old athletes. *Acta Medica Scandinavica, 179*, 513-526.

Grimby, G., & Saltin, B. (1983). The aging muscle. *Clinical Physiology, 3*, 209-218.

Grzegorczyk, P. B., Jones, S. W., & Mistretta, C. M. (1979). Age-related differences in salt taste acuity. *Journal of Gerontology, 34*, 834-840.

Guardo, C. J. (1968). Self revisited: The sense of self identity. *Journal of Humanistic Psychology, 8*, 137-142.

Guth, P. H. (1968). Physiologic alterations in small bowel function with age: The absorption of d-xylose. *American Journal of Digestive Diseases, 13*, 565-571.

Gutmann, E. (1977). Muscle. In C. E. Finch & L. Hayflick (Eds.), *Handbook of the biology of aging*. New York: Van Nostrand Reinhold.

Gutmann, E., & Hanzlikova, V. (1976). Fast and slow motor units in aging. *Gerontology, 22*, 280-300.

Haber, R. N., & Standing, L. G. (1969). Direct measures of short-term visual storage. *Quarterly Journal of Experimental Psychology, 21*, 43-54.

Hall, T. C., Miller, A. K. H., & Corsellis, J. A. N. (1975). Variations in the human Perkinje cell poulation according to age and sex. *Neuropathology and Applied Neurobiology, 1*, 267-292.

Hammer, W. M., & Wilmore, J. H. (1973). An exploratory investigation in personality measures and physiological alterations during a 10-week jogging program. *Journal of Sports Medicine and Physical Fitness, 13*, 238-247.

Hammett, V. B. P. (1967). Psychological changes with physical fitness training. *Canadian Medical Asociation Journal, 96*, 764-769.

Hanley, T. (1974). "Neuronal fall-out" in aging brain: a critical review of the quantitative data. *Age and Aging, 3*, 133-151.

Hansen, C. C., & Reske-Nielson, E. (1965). Pathological studies in presbycusis. *Archives of Otolaryngology, 82*, 115-132.

Hanson, J. S., & Nedde, W. H. (1974). Long-term physical training effect in sedentary females. *Journal of Applied Physiology, 37*, 112-116.

Hanson, J. S., Tabakin, B. S., Levy, A. M., & Nedde, W. (1968). Long-term physical training and cardiovascular dynamics in middle-aged men. *Circulation, 38,* 783-799.

Harbert, F., Young, I. M., & Menduke, H. (1966). Audiological findings in presbycusis. *Journal of Auditory Research, 6,* 297-312.

Harbitz, T. B. (1973). Testis weight and the histology of the prostate in elderly men. Analysis in an autopsy series. *Acta Pathologica Microbiologica Scandinavica, 81A,* 148-158.

Harman, S. M. (1978). Clinical aspects of the male reproductive system. In E. L. Schneider (Ed.), *Aging: Vol. 4. The aging reproductive system.* New York: Raven Press.

Harman, S. M., & Tsitouras, P. D. (1980). Reproductive hormones in aging men. I. Measurement of sex steroids, basal luteinizing hormone and Leydig cell response to human chorionic gonadotropin. *Journal of Clinical Endocrinology and Metabolism, 51,* 35-40.

Hartley, G. H., & Farge, E. J. (1977). Personality and physiological traits in middle-aged runners and joggers. *Journal of Gerontology, 32,* 541-548.

Hartley, L. H., Grimby, G., Kilbom, Å., Nilsson, N. J., Åstrand, I., Bjure, J., Ekblom, B., & Saltin, B. (1969). Physical training in sedentary middle-aged and older men. III. Cardiac output and gas exchange at submaximal and maximal exercise. *Scandinavian Journal of Clinical Laboratory Investigations, 24,* 335-344.

Hartman, D. E., & Danhauer, J. L. (1976). Perceptual features of speech for males in four perceived age decades. *Journal of the Acoustical Society of America, 59,* 713-715.

Haskell, W. L. (1984). The influence of exercise on the concentrations of triglyceride and cholestrol in human plasma. *Exercise and Sports Sciences Reviews, 12,* 205-244.

Hass, G. M. (1943). Studies of cartilage (iv). A morphological and clinical analysis of aging human costal cartilage. *Archives of Pathology, 35,* 275-284.

Hass, G. M. (1956). Pathological calcification. In G. H. Bourne (Ed.), *The biochemistry and physiology of bone* (chap. 24). New York: Academic Press.

Hayashi, Y., & Endo, S. (1982). All-night sleep polygraphic recordings of healthy aged persons: REM and slow-wave sleep. *Sleep, 5,* 277-283.

Heaney, R. P. (1982). Age-related bone loss. In M. E. Reff & E. L. Schneider (Eds.), *Biological markers of aging* (pp. 161-167). (NIH Publication No. 82-2221).

Heaney, R. P., Gallagher, J. C., Johnston, C. C., Neer, R., Parfitt, A. M., & Whedon, G. D. (1982). Calcium nutrition and bone health in the elderly. *American Journal of Clinical Nutrition, 36,* 986-1013.

Heaps, R. A. (1978). Relating physical and psychological fitness: A psychological point of view. *Journal of Sports Medicine and Physical Fitness, 18,* 399-408.

Heath, G. W., Hagberg, J. M., Ehseni, A. A., & Holloszy, J. O. (1981). A physiological comparison of young and old endurance athletes. *Journal of Applied Physiology, 51,* 634-640.

Hegedus, L., Perrild, H., Poulson, L. R., Anderson, J. R., Holm, B., Schnohr, P., Jensen, G., & Hansen, J. M. (1983). The determination of thyroid volume by ultrasound and its relationship to body weight, age, and sex in normal subjects. *Journal of Clinical Endocrinology and Metabolism, 56,* 260-263.

Hegstad, R., Brown, R. D., Jiang, N. S., Kao, P., Weinshilboum, R. M., Strong, C., & Wisgerhof, M. (1983). Aging and aldosterone. *The American Journal of Medicine, 74,* 442-448.

Heikkenen, E. (1978). Studies on aging, physical fitness and health. In F. Landry & W. A. R. Orban (Eds.), *Sports medicine* (Vol. 5, pp. 331-337). Miami: Symposia Specialists.

Heinzelman, F. (1973). Social and psychological factors that influence the effectiveness of exercise programs. In J. Naughton & H. Hellerstein (Eds.), *Exercise testing and exercise training in coronary heart disease*. New York: Academic Press.

Heinzelman, F., & Bagley, R. W. (1970). Response to physical activity programs and their effects on health behavior. *Public Health Reports, 85*, 905-911.

Heiss, G., Tamir, I., Davis, C. E., Tyroler, H. A., Rifkind, B. M., Schonfeld, G., Jacobs, D., & Frantz, I. D., Jr. (1980). Lipoprotein-cholesterol distributions in selected North American populations: The Lipid Research Clinics Program Prevalence Study. *Circulation, 61*, 302-315.

Helderman, J. H., Vestal, R. E., Rowe, J. W., Tobin, J. D., Andres, R., & Robertson, G. L. (1978). The response of arginine vasopressin to intravenous ethanol and hypertonic saline in man: The impact of aging. *Journal of Gerontology, 33*, 39-47.

Hellerstein, H. K. (1973). Exercise therapy in coronary disease. Rehabilitation and secondary prevention. In J. H. de Haas, H. C. Hemker, & H. A. Snellen (Eds.), *Ischaemic heart disease* (pp. 406-429). Baltimore: Williams & Wilkins.

Hellon, R. F., & Lind, A. R. (1956). Activity of the sweat glands with reference to the influence of aging. *Journal of Physiology, 133*, 132-144.

Hellon, R. F., Lind, A. R., & Weiner, J. S. (1956). The physiological reactions of men of two age groups to a hot environment. *Journal of Physiology, 133*, 118-131.

Henderson, G., Tomlinson, B., & Gibson, P. H. (1980). Cell counts in human cerebral cortex in normal adults throughout life using an image analysing computer. *Journal of the Neurological Sciences, 46*, 113-136.

Hendricks, C. D., & Hendricks, J. (1976). Concepts of time and temporal construction among the aged, with implications for research. In J. F. Gubrium (Ed.), *Time, roles, and self in old age*. New York: Human Sciences Press.

Henschel, A., Cole, M. B., & Lyczkowskyj, O. (1968). Heat tolerance of elderly persons living in a subtropical climate. *Journal of Gerontology, 23*, 17-22.

Hensel, H. (1981). *Thermoreception and temperature regulation*. New York: Academic Press.

Hermel, J., Schönwetter, S., & Samueloff, S. (1970). Taste sensation and age in man. *Journal of Oral Medicine, 25*, 39-42.

Herrmann, J., Rusche, H., Kroll, H. J., Hilger, P., & Kruskemper, G. (1974). Free triiodothyronine (T_3) and thyroxine (T_4) serum levels in old age. *Hormone and Metabolic Research, 6*, 239-240.

Hesch, R.-D., Gatz, J., Juppner, H., & Stubbe, P. (1977). TBG-dependency of age related variations of thyroxine and triiodothyronine. *Hormone and Metabolic Research, 9*, 141-146.

Higatsberger, M. R., Budka, H., & Bernheimer, H. (1982). Neurochemical investigations of aged human brain cortex. In S. Hoyer (Ed.), *The aging brain: Physiological and pathophysiological aspects*. New York: Springer-Verlag.

Hilyer, J. C., & Mitchell, W. (1979). Effect of physical fitness training combined with counseling on the self-concept of college students. *Journal of Counseling Psychology, 26*, 427-436.

Hinchcliffe, R. (1958). Clinical quantitative gustometry. *Acta Otolaryngolica, 49*, 453-466.

Hinchcliffe, R. (1962). The anatomical locus of presbycusis. *Journal of Speech and Hearing Disorders, 27*, 301-310.

Hodgson, J. L., & Buskirk, E. R. (1977). Physical fitness and age, with emphasis on cardiovascular function in the elderly. *Journal of the American Geriatrics Society, 25*, 385-392.

Hodkinson, H. M., Round, P., Stanton, B. R., & Morgan, C. (1973). Sunlight, vitamin D, and osteomalacia in the elderly. *Lancet, 1*, 910-912.

Hofstetter, H. W., & Bertsch, J. D. (1976). Does stereopsis change with age? *American Journal of Optometry and Physiological Optics, 53*, 644-667.

Holland, J., Milic-Emili, J., Macklem, R. T., & Bates, D. V. (1968). Regional distribution of pulmonary ventilation and perfusion in elderly subjects. *Journal of Clinical Investigations, 47*, 81-92.

Hollander, N., & Hollander, V. P. (1958). The microdetermination of testosterone in human spermatic vein blood. *Journal of Clinical Endocrinology and Metabolism, 18*, 966-970.

Hollenberg, N. K., Adams, D. F., Solomon, H. S., Rashid, A., Abrams, H. L., & Merrill, J. P. (1974). Senescence and the renal vasculature in normal man. *Circulation Research, 34*, 309-316.

Hollis, J. B., & Castell, D. O. (1974). Esophageal function in old men: A new look at "presbyesophagus." *Annals of Internal Medicine, 80*, 371-374.

Horn, J. L., & Cattell, R. B. (1982). Whimsy and misunderstandings of G_f-G_c theory: A comment on Guilford. *Psychological Bulletin, 91*, 623-633.

Horn, J. L., & Donaldson, G. (1980). Cognitive development in adulthood. In J. Kagan & O. G. Brim, Jr. (Eds.), *Constancy and change in development*. Cambridge, MA: Harvard University Press.

Horsman, A., & Currey, J. D. (1983). Estimation of mechanical properties of the distal radius from bone mineral content and cortical width. *Clinical Orthopedics and Related Research, 176*, 298-304.

Hossack, K. F., & Bruce, R. A. (1982). Maximal cardiac function in sedentary normal men and women: Comparison of age-related changes. *Journal of Applied Physiology, 53*, 799-804.

Howell, T. H. (1949). Senile deterioration of the central nervous system: Clinical study. *British Medical Journal, 1*, 56-58.

Hoyer, W. J., & Plude, D. J. (1980). Attentional and perceptual processes in the study of cognitive aging. In L. W. Poon (Ed.), *Aging in the 1980s: Psychological issues*. Washington, DC: American Psychological Association.

Hoyer, W. J., & Plude, D. J. (1982). Aging and the allocation of attentional resources in visual information processing. In R. Sekuler, L. P. Hutman, & C. Owsley (Eds.), *Aging and human visual function*. New York: Alan R. Liss.

Hsu, J. M., & Smith, J. C. Jr. (1984). B-Vitamins and ascorbic acid in the aging process. In J. M. Ordy, D. Harman, & R. B. Alfin-Slater (Eds.), *Aging: Vol. 26. Nutrition in gerontology* (pp. 87-118). New York: Raven Press.

Hughes, G. (1969). Changes in taste sensitivity with advancing age. *Gerontologia Clinica, 11*, 224-230.

Hunziker, O., Abdel'Al, S., & Schulz, U. (1979). The aging human cerebral cortex: A stereological characterization of changes in the capillary net. *Journal of Gerontology, 34*, 345-350.

Huston, T. L., & Levinger, G. (1978). Interpersonal attraction and relationships. *Annual Review of Psychology, 29*, 115-156.

Huttenlocher, P. R. (1979). Synaptic density in human frontal cortex—developmental changes and effects of aging. *Brain Research, 163*, 195-205.

Hyams, D. E. (1978). The liver and biliary system. In J. Brocklehurst (Ed.), *Textbook of geriatric medicine and gerontology* (2nd ed.). New York: Churchill-Livingstone.

Inokuchi, S., Ishikawa, H., Iwamato, S., & Kimura, T. (1975). Age related changes in the histological composition of the rectus abdominis muscle of the adult human. *Human Biology, 47*, 231-249.

Ismail, A. H., & Young, R. J. (1977). Effects of chronic exercise on the personality of adults. *Annals of the New York Academy of Sciences, 301*, 958-969.

Jackson, R. A., Blix, P. M., Matthews, J. A., Hamling, J. B., Din, B. M., Brown, D. C., Belin, J., Rubenstein, A. H., & Nabarro, J. D. N. (1982). Influence of aging on glucose homeostasis. *Journal of Clinical Endocrinology and Metabolism, 55*, 840-848.

Jensen, H. K., & Blichert-Toft, M. (1971). Serum corticotrophin, plasma cortisol and urinary excretion of 17-ketogenic steroids in the elderly (age group: 66-94 years). *Acta Endocrinologica, 66*, 25-34.

Jerger, J. (1973). Audiological findings in aging. *Advances in Oto-Rhino-Laryngology, 20*, 115-124.

Jerger, J., & Hayes, D. (1977). Diagnostic speech audiometry. *Archives of Otolaryngology, 103*, 216-222.

John, E. R., & Schwartz, E. L. (1978). The neurophysiology of information processing and cognition. *Annual Review of Psychology, 29*, 1-29.

Johnsson, L. G. (1971). Degenerative changes and anomalies of the vestibular system in man. *Laryngoscope, 81*, 1682-1694.

Johnsson, L. G., & Hawkins, J. E., Jr. (1972). Sensory and neural degeneration with aging, as seen in micro-dissections of the human inner ear. *Annals of Otolaryngology, Rhinology, and Laryngology, 81*, 179-193.

Jokinen, K. (1973). Presbycusis VI. Masking of speech. *Acta Oto-Laryngolica, 76*, 426-430.

Judd, H. L., Judd, G. E., Lucas, W. E., & Yen, S. S. C. (1974). Endocrine function of the postmenopausal ovary: Concentrations of androgens and estrogens in ovarian and peripheral vein blood. *Journal of Clinical Endocrinology and Metabolism, 39*, 1020-1024.

Judd, H. L., & Korenman, S. G. (1982). Effects of aging on reproductive function in women. In S. G. Korenman (Ed.), *Endocrine aspects of aging*. New York: Elsevier Biomedical.

Julius, S., Amery, A., Whitlock, L. S., & Conway, J. (1967). Influence of age on the hemodynamic response to exercise. *Circulation, 36*, 222-230.

Kalant, N., Leiborici, D., Leibovici, T., & Fukushima, N. (1980). Effect of age on glucose utilization and responsiveness to insulin in forearm muscle. *Journal of the American Geriatrics Society, 28*, 304-307.

Kalk, W. J., Vinik, A. I., Pimstone, B. L., & Jackson, W. P. U. (1973). Growth hormone response to insulin hypoglycemia in the elderly. *Journal of Gerontology, 28*, 431-434.

Kamocka, D. (1970). Cytological studies of parotid glands secretion in people over 60 years of age. *Excerpta Medica*, Section 20, *13*, 412.

Kampmann, J. P., Sinding, J., & Møller-Jørgensen, I. (1975). Effect of age on liver function. *Geriatrics, 30*(8), 91-95.

Kannel, W. B., & Hubert, H. (1982). Vital capacity as a biomarker of aging. In M. E. Reff & E. L. Schneider (Eds.), *Biological markers of aging* (pp. 145-160). (NIH Publication No. 62-2221)

Kanstrup, I. L., & Ekblom, B. (1978). Influence of age and physical activity on central hemodynamics and lung function in active adults. *Journal of Applied Physiology, 45*, 709-717.

Karacan, I., Williams, R. L., Thornby, J. I., & Salis, P. J. (1975). Sleep-related penile tumescence as a function of age. *American Journal of Psychiatry, 132*, 932-937.

Karlsen, E., Hassanein, R., & Goetzinger, C. (1981). The effects of age, sex, hearing loss and water temperature on caloric nystagmus. *Laryngoscope, 91*, 620-627.

Kasch, F. W. (1976). The effects of exercise on the aging process. *The Physician and Sportsmedicine, 4*, 64-68.

Kasch, F. W., & Kulberg, J. (1981). Physiological variables during 15 years of endurance exercise. *Scandinavian Journal of Sports Sciences, 3*, 59-62.

Kasch, F. W., & Wallace, J. P. (1976). Physiological variables during 10 years of endurance exercise. *Medicine and Science in Sports and Exercise, 8*, 5-8.

Kasden, S. (1970). Speech discrimination in two age groups matched for hearing loss. *Journal of Auditory Research, 10*, 210-212.

Kastenbaum, R., Derbin, V., Sabatini, P., & Artt, S. (1972). "The ages of me": Toward personal and interpersonal definitions of functional aging. *Aging and Human Development, 3*, 197-211.

Katsuki, S., & Masuda, M. (1969). Physical exercise for persons of middle and elder age in relation to their physical activity. *Journal of Sports Medicine and Physical Fitness, 9*, 193-199.

Kausler, D. H. (1982). *Experimental psychology and human aging.* New York: Wiley.

Kavanagh, T., & Shephard, R. J. (1978). The effects of continued training on the aging process. *Annals of the New York Academy of Science, 301*, 656-670.

Kazis, A., Vlaikidis, N., Pappa, P., Papanastasiou, J., Vlaheveis, G., & Routsonis, K. (1983). Somatosensory and visual evoked potentials in human aging. *Electromyography and Clinical Neurophysiology, 23*, 49-59.

Kendall, M. J. (1970). The influence of age on the xylose absorption test. *Gut, 11*, 498-501.

Kendall, M. J., Woods, K. L., Wilkins, M. R., & Worthington, D. J. (1982). Responsiveness to B-adrenergic receptor stimulation: The effects of age are cardioselective. *British Journal of Clinical Pharmacology, 14*, 821-826.

Kenney, R. A. (1982). *Physiology of aging: A synopsis.* Chicago: Yearbook Medical.

Kenshalo, D. R. (1972). Cutaneous senses. In L. Riggs & J. W. Kling (Eds.), *Experimental psychology* (Vol. 1, pp. 117-168). New York: Holt, Rinehart, & Winston.

Kenshalo, D. R. (1977). Age changes in touch, vibration, temperature, kinesthesis, and pain sensitivity. In J. E. Birren & K. W. Schaie (Eds.), *Handbook of the psychology of aging.* New York: Van Nostrand Reinhold.

Kenshalo, D. R., Sr. (1979). Changes in the vestibular and somesthetic systems as a function of age. In J. M. Ordy & K. Brizzee (Eds.), *Aging: Vol. 10. Sensory systems and communication in the elderly* (pp. 269-282). New York: Raven Press.

Kent, J. Z., & Acone, A. B. (1966). Plasma androgens and aging. In A. Vermeulen & D. Exley (Eds.), *Androgens in normal and pathological conditions.* Excerpta Medica International Congress Series, *101*, Amsterdam.

Keys, A., Taylor, H. L., & Grande, F. (1973). Basal metabolism and age of adult man. *Metabolism, 22*, 579-587.

Khan, T. A., Shragge, B. W., Crispin, J. S., & Lind, J. F. (1977). Esophageal motility in the elderly. *American Journal of Digestive Diseases, 22*, 1049-1054.

Kiessling, K.-H., Pilström, L., Bylund, A.-Ch., Saltin, B., & Piehl, K. (1974). Enzyme activities and morphometry in skeletal muscle of middle-aged men after training. *Scandinavian Journal of Clinical and Laboratory Investigation, 33*, 63-69.

Kilbom, Å. (1971a). Physical training with submaximal intensities in women. III. Effect on adaptation to professional work. *Scandinavian Journal of Clinical and Laboratory Investigations, 28*, 331-343.

Kilbom, Å. (1971b). Physical training and submaximal intensities in women. I. Reaction to exercise and orthotasis. *Scandinavian Journal of Clinical and Laboratory Investigations, 28*, 141-161.

Kimbrell, G. McA., & Furchgott, E. (1963). The effect of aging on olfactory threshold. *Journal of Gerontology, 18*, 364-365.

Kimmerling, G., Javorski, W. C., & Reaven, G. M. (1977). Aging and insulin resistance in a group of nonobese male volunteers. *Journal of the American Geriatrics Society, 25*, 349-353.

Kino, M., Lance, V. Q., Shamatpour, A., & Spodick, D. (1975). Effects of aging on response to isometric exercise. *American Heart Journal, 90*, 575-581.

Kirikae, I., Sato, R., & Shitara, T. (1964). A study of hearing in advanced age. *Laryngoscope, 74*, 205-220.

Kline, D. W., & Birren, J. E. (1975). Age differences in backward dichoptic masking. *Experimental Aging Research, 1*, 17-25.

Kline, D. W., Ikeda, D. M., & Schieber, F. J. (1982). Age and temporal resolution in color vision: When do red and green make yellow. *Journal of Gerontology, 37*, 705-709.

Kline, D. W., & Schieber, F. J. (1981). Visual aging: A transient/sustained shift? *Perception and Psychophysics, 29*, 181-182.

Kline, D. W., & Schieber, F. J. (1982). Visual persistence and temporal resolution. In R. Sekuler, D. Kline, & K. Dismukes (Eds.), *Aging and human visual function*. New York: Alan R. Liss.

Kline, D. W., Schieber, F., Abusamra, L. C., & Coyne, A. C. (1983). Age, the eye, and the visual channels: Contrast sensitivity and response speed. *Journal of Gerontology, 38*, 211-216.

Kline, D. W., & Szafran, J. (1975). Age differences in backward monoptic visual noise masking. *Journal of Gerontology, 30*, 307-311.

Knudson, R. J., & Kaltenborn, W. T. (1981). Evaluation of lung elastic recoil by exponential curve analysis. *Respiration Physiology, 46*, 29-42.

Koff, R. S., Garvey, A. J., Burney, S. W., & Bell, B. (1973). Absence of an age effect on sulfobromopthalein retention in healthy men. *Gastroenterology, 65*, 300-302.

Konigsmark, B. W., & Murphy, E. A. (1970). Neuronal populations in the human brain. *Nature, 228*, 1335-1336.

Konkle, D. F., Beasley, D. S., & Bess, F. M. (1977). Intelligibility of time-altered speech in relation to chronological aging. *Journal of Speech and Hearing Research, 20*, 108-115.

Kovacs, K., Ryan, N., Horvath, E., Penz, G., & Ezrin, C. (1977). Prolactin cells of the human pituitary gland in old age. *Journal of Gerontology, 32*, 534-540.

Kram, D., & Schneider, E. L. (1978). An effect of reproductive aging: Increased risk of genetically abnormal offspring. In E. L. Schneider (Ed.), *Aging: Vol. 4. The aging reproductive system*. New York: Raven Press.

Krmpotic-Nemanic, J. (1969). Presbycusis and retrocochlear structures. *International Audiology, 8*, 210-220.

Krmpotic-Nemanic, J. (1971). A new concept of the pathogenesis of presbycusis. *Archives of Otolaryngology, 93*, 161-166.

Kronenberg, R. S., & Drage, C. W. (1973). Attenuation of the ventilatory and heart rate responses to hypoxia and hypercapnia with aging in normal men. *Journal of Clinical Investigation, 52*, 1812-1819.

Kuta, I., Pařizková, J., & Dýcka, J. (1970). Muscle strength and lean body mass in old men of different physical activity. *Journal of Applied Physiology*, *29*, 168-171.

Kuwabara, T. (1977). Age-related changes of the eye. In S. S. Han & D. H. Coons (Eds.), *Special senses in aging*. Ann Arbor, MI: Institute of Gerontology, University of Michigan.

Laidlaw, R. W., & Hamilton, M. A. (1937). A study of thresholds in apperception of passive movement among normal control subjects. *Bulletin of the Neurological Institute*, *6*, 268-273.

Lakatta, E. G. (1979). Alterations in the cardiovascular system that occur in advanced age. *Federation Proceedings*, *38*, 163-167.

Lakatta, E. G. (1980). Age-related alterations in the cardiovascular response to adrenergic mediated stress. *Federation Proceedings*, *39*, 3173-3177.

Lake, C. R., Ziegler, M. G., Coleman, M. D., & Kopin, I. J. (1977). Age-adjusted plasma norepinephrine levels are similar in normotensive and hypertensive subjects. *New England Journal of Medicine*, *296*, 208-209.

Lake, C. R., Ziegler, M. G., & Kopin, I. J. (1976). Use of plasma norepinephrine for evaluation of sympathetic neuronal function in man. *Life Sciences*, *18*, 1315-1326.

Landahl, H. D., & Birren, J. E. (1959). Effects of age on the discrimination of lifted weights. *Journal of Gerontology*, *14*, 48-55.

Landowne, M., Brandfonbrener, M., & Shock, N. W. (1955). The relation of age to certain measures of performance of the heart and circulation. *Circulation*, *12*, 567-576.

Larsson, L. (1982). Physical training effects on muscle morphology in sedentary men at different ages. *Medicine and Science in Sports and Exercise*, *14*, 203-206.

Larsson, L., Grimby, G., & Karlsson, J. (1979). Muscle strength and speed of movement in relation to age and muscle morphology. *Journal of Applied Physiology*, *46*, 451-456.

Larsson, L., & Karlsson, J. (1978). Isometric and dynamic endurance as a function of age and skeletal muscle characteristics. *Acta Physiologica Scandinavica*, *104*, 129-136.

Larsson, L., Sjödin, B., & Karlsson, J. (1978). Histochemical and biochemical changes in human skeletal muscle with age in sedentary males. *Acta Physiologica Scandinavica*, *103*, 31-39.

Lasada, K., & Roberts, P. (1974). Variation in the morphometry of the normal human thyroid in growth and aging. *Journal of Pathology*, *112*, 161-168.

Lavker, R. M., Kwong, F., & Kligman, A. M. (1980). Changes in skin surface patterns with age. *Journal of Gerontology*, *35*, 348-354.

Lavy, S., Melamed, E., Bentin, S., Cooper, G., & Rinot, Y. (1978). Bihemispheric decreases of regional blood flow in dementia: Correlation with age-matched normal controls. *Annals of Neurology*, *4*, 445-450.

Lawton, M. P. (1977). The impact of the environment on aging and behavior. In J. E. Birren & K. W. Schaie (Eds.), *Handbook of the psychology of aging*. New York: Van Nostrand Reinhold.

Lawton, M. P., & Nahemow, L. (1973). Ecology and the aging process. In C. Eisdorfer & M. P. Lawton (Eds.), *The psychology of adult development and aging*. Washington, DC: The American Psychological Association.

Lazarus, L., & Eastman, C. J. (1976). Assessment of hypothalamic pituitary function in old age. In A. V. Everitt & J. A. Burgess (Eds.), *Hypothalamus, pituitary and aging*. Springfield, IL: C C Thomas.

Lazarus, R. S. (1966). *Psychological stress and the coping process*. New York: McGraw-Hill.

Learoyd, B. M., & Taylor, M. G. (1966). Alterations with age in the viscoelastic properties of human arterial walls. *Circulation Research, 18*, 278-292.

LeBlanc, P., Ruff, F., & Milic-Emili, J. (1970). Effects of age and body position on 'airway
closure' in man. *Journal of Applied Physiology, 28*, 448-451.

Lebo, C. P., & Reddell, R. C. (1972). The presbycusis component in occupational hearing
loss. *Laryngoscope, 82*, 1399-1409.

L'Ecuyer, R. (1981). The development of the self-concept through the life span. In M. D.
Lynch, A. A. Norem-Hebeisen, & K. J. Gergen (Eds.), *Self-concept: Advances in
theory and research* (pp. 203-218). Cambridge, MA: Ballinger.

Leonardson, G. R. (1977). Relationship between self-concept and perceived physical
fitness. *Perceptual and Motor Skills, 44*, 62.

Lesser, M. (1978). The effects of rhythmic exercise on the range of motion in older adults.
American Corrective Therapy Journal, 32, 118-122.

Levin, H. S., & Benton, A. L. (1973). Age and susceptibility to tactile masking effects.
Gerontologia Clinica, 15, 1-9.

Lexell, J., Henriksson-Larsson, K., Winblad, B., & Sjöstrom, M. (1983). Distribution of
different fibre types in human skeletal muscle. 3. Effects of aging on m. vastus lateralis
studied in whole muscle cross-sections. *Muscle and Nerve, 6*, 588-595.

Lieberman, M., & Coplan, A. S. (1970). Distance from death as a variable in the study
of aging. *Developmental Psychology, 2*, 71-84.

Lind, A. R., Humphreys, P. W., Collins, K. J., Foster, K., & Sweetland, K. F. (1970).
Influence of age and daily duration of exposure on responses of men to work in the
heat. *Journal of Applied Physiology, 28*, 50-56.

Lindblad, L. E. (1977). Influence of age on sensitivity and effector mechanisms of the
carotid baroreflex. *Acta Physiologica Scandinavica, 101*, 43-49.

Lindemann, R. D. (1975). Age changes in renal function. In R. Goldman & M. Rockstein
(Eds.), *The physiology and pathology of human aging*. New York: Academic Press.

Lindeman, R. D., van Buren, H. C., & Raisz, L. G. (1960). Osmolar renal concentrating
ability in healthy young men and hospitalized patients without renal disease. *New
England Journal of Medicine, 262*, 1306-1309.

Lippa, A. S., Pelham, R. W., Beer, B., Critchett, D. J., Dean, R. L., & Bartus, R. T.
(1980). Brain cholinergic dysfunction and memory in aged rats. *Neurobiology of
Aging, 1*, 13-19.

Liss, L., & Gomez, F. (1958). The nature of senile changes of the human olfactory bulb
and tract. *Archives of Otolaryngology, 67*, 167-171.

Lockett, M. F. (1976). Aging of the adenophypophysis in relation to renal aging. In A. V.
Everitt & J. A. Burgess (Eds.), *Hypothalamus, pituitary and aging*. Springfield, IL: C
C Thomas.

Loewenfeld, I. E. (1979). Pupillary changes related to age. In H. S. Thompson (Ed.),
Topics in neuro-optholmology. Baltimore: Williams & Wilkins.

Ludwig, F., & Smoke, M. E. (1980). The measurement of biological age. *Experimental
Aging Research, 6*, 497-522.

Luterman, D. M., Welsh, O. L., & Melrose, J. (1966). Responses of aged males to time-
altered speech stimuli. *Journal of Speech and Hearing Research, 9*, 226-230.

Lynch, G., & Gerling, S. (1981). Aging and brain plasticity. In J. L. McGaugh & S. B.
Kiesler (Eds.), *Aging: Biology and behavior*. New York: Academic Press.

Lynch, S. R., Finch, C. A., Monsen, E. R., & Cook, J. D. (1982). Iron status of elderly
Americans. *American Journal of Clinical Nutrition, 36*, 1032-1045.

Lynne-Davies, P. (1977). Influence of age on the respiratory system. *Geriatrics, 32,* 57-60.

Malina, R. M. (1969). Quantification of fat, muscle and bone in man. *Clinical Orthopaedics and Related Research, 65,* 9-38.

Mankovsky, N. B., Mints, A. Y., & Lisenyuk, V. P. (1982). Age peculiarities of human motor control in aging. *Gerontology, 28,* 314-322.

Mann, D. N. A., & Yates, P. O. (1974). Lipoprotein pigments—their relationship to aging in the human nervous system. I. The lipofuscin content of nerve cells. *Brain, 97,* 481-488.

Marmor, M. F. (1977). The eye and vision in the elderly. *Geriatrics, 32,* 63-67.

Marmor, M. F. (1980). Clinical physiology of the retina. In G. A. Reyman, D. R. Sanders, & M. F. Goldberg (Eds.), *Principles and practice of opthalmology* (Vol. 2, pp. 823-856). Philadelphia: Saunders.

Marmor, M. F. (1982). Aging of the retina. In R. Sekuler, D. Kline, & K. Dismukes (Eds.), *Aging and human visual function.* New York: Alan R. Liss.

Marshall, B. E., & Wyche, M. Q. (1972). Hypoxemia during and after anesthesia. *Anesthesiology, 37,* 178-209.

Marshall, L. (1981). Auditory processing in aging listeners. *Journal of Speech and Hearing Disorders, 46,* 226-240.

Marston, L. E., & Goetzinger, C. P. (1972). A comparison of sensitized words and sentences for distinguishing non-peripheral auditory changes as a function of aging. *Cortex, 8,* 213-223.

Martin, C. E. (1977). Sexual activity in the aging male. In J. Money & H. Mousafh (Eds.), *Handbook of sexology.* Amsterdam: ASP Biological and Medical Press.

Marx, J. J. M. (1979). Normal iron absorption and decreased red cell iron uptake in the aged. *Blood, 53,* 204-211.

Massie, J. F., & Shephard, R. J. (1971). Physiological and psychological effects of training. *Medicine and Science in Sports, 3,* 110-117.

Masters, W. H., & Johnson, V. E. (1966). Human sexual response. Boston: Little, Brown.

Masters, W. H., & Johnson, V. E. (1970). *Human sexual inadequacy.* Boston, Little, Brown.

Mauderly, J. L. (1978). Effect of age on pulmonary structure and function of immature and adult animals and man. *Federation Proceedings, 39,* 173-177.

Mazess, R. B. (1982). On aging bone loss. *Clinical Orthopaedics and Related Research, 165,* 239-252.

McArdle, W. D., Katch, F. I., & Katch, V. L. (1981). *Exercise physiology: Energy, nutrition, and human performance.* Philadelphia: Lea & Febiger.

McConnell, J. G., Buchanan, K. D., Ardill, J., & Stout, R. W. (1982). Glucose tolerance in the elderly: The role of insulin and its receptor. *European Journal of Clinical Investigation, 12,* 55-61.

McDonald, R. K., Solomon, D. H., & Shock, N. W. (1951). Aging as a factor in the renal hemodynamic changes induced by a standardized pyrogen. *Journal of Clinical Investigations, 30,* 457-462.

McDonough, J. R., Kusumi, F., & Bruce, R. A. (1970). Variations in maximal oxygen intake with physical activity in middle-aged men. *Circulation, 41,* 743-751.

McFarland, R. A., Domey, R. G., Warren, A. B., & Ward, D. C. (1960). Dark adapation as a function of age: I. A statistical analysis. *Journal of Gerontology, 15,* 149-154.

McGandy, R. B., Barrows, C. H., Jr., Spanias, A., Meredith, A., Stone, J. L., & Norris, A. H. (1966). Nutrient intake and energy expenditure in men of different ages. *Journal of Gerontology, 21,* 581-587.

McGeer, E. G., & McGeer, P. L. (1975). Age changes in the human for some enzymes associated with metabolism of catecholamines, GABA, and acetylcholine. In J. M. Ordy & K. R. Brizzee (Eds.), *Neurobiology of aging*. New York: Plenum Press.

McGeer, E. G., & McGeer, P. L. (1976). Neurotransmitter metabolism in aging brain. In R. D. Terry & S. Gershon (Eds.), *Aging: Vol. 3, Neurobiology of aging*. New York: Raven Press.

McGeer, E. G., & McGeer, P. L. (1980). Aging and neurotransmitter systems. In M. Goldstein, D. B. Calne, A. Lieberman, & M. O. Thorner (Eds.), *Advances in biochemical psychopharmacology: Vol. 23. Ergot compounds and brain function: Neuroendocrine and neuropsychiatric aspects*. New York: Raven Press.

McGrath, C., & Morrison, J. D. (1980). Age-related changes in spatial frequency perception. *Journal of Physiology, 310*, 52 P.

McGrath, J., & O'Hanlon, J. F. (1968). Relationships among chronological age, intelligence, and rate of subjective time. *Perceptual and Motor Skills, 26*, 1083-1088.

McGrath, M. W., & Thomson, M. L. (1959). The effect of age, body size, and lung volume change on alveolar-capillary permeability and diffusing capacity in man. *Journal of Physiology, 146*, 572-582.

McGuire, E. A., Tobin, J. D., Berman, M., & Andres, R. (1979). Kinetics of native insulin in diabetic, obese, and aged men. *Diabetes, 28*, 110-120.

McLachlan, M. S. F. (1978). The aging kidney. *Lancet, 2*, 143-146.

McLachlan, M. S. F., Guthrie, J. C., Anderson, C. K., & Fulker, M. J. (1977). Vascular and glomerular changes in the aging kidney. *Journal of Pathology, 121*, 65-78.

McPherson, B. D. (1980). Social factors to consider in fitness programming and motivation: Different strokes for different groups. In R. R. Danielson & K. F. Danielson (Eds.), *Fitness motivation* (pp. 8-17). Toronto: Orcol Publications.

McPherson, B. D., Paivio, A., Yuhasz, M. S., Rechnitzer, P. A., Pickard, H. A., & Lefcoe, N. M. (1967). Psychological effects of an exercise program for post-infarct and normal adult men. *Journal of Sports Medicine and Physical Fitness, 7*, 95-102.

Meier-Ruge, W., Hunziker, O., Iwangoff, P., Reichlmeier, K., & Schulz, U. (1980). Effect of age on morphological and biochemical parameters of the human brain. In D. G. Stein (Ed.), *The psychobiology of aging: Problems and perspectives* (pp. 197-318). New York: Elsevier.

Meindok, H., & Dvorsky, R. (1970). Serum folate and vitamin-B_{12} levels in the elderly. *Journal of the American Geriatrics Society, 18*, 317-326.

Melmed, S., & Hershman, J. M. (1982). The thyroid and aging. In S. G. Korenman (Ed.), *Endocrine aspects of aging*. New York: Elsevier Biomedical.

Melmgaard, K. (1966). The alveolar-arterial oxygen difference: Its size and components in normal man. *Acta Physiologica Scandinavica, 67*, 10-20.

Meyer, J. S., Deshmukh, V. D., Mathew, N. T., Naritomi, H., Ishihara, N., Sakai, F., Hsu, M.-Ch., Pollack, P., Bishop, L., Perez, F. I., & Gedye, J. L. (1976). Non-invasive 133-Xe inhalation measurements of regional cerebral blood flow in dementia. In J. S. Meyer, H. Lechner, & M. Reivich (Eds.), *Cerebral vascular disease*. Stuttgart: Thieme.

Miller, J. H., McDonald, R. K., & Shock, N. W. (1951). The renal extraction of p-aminohippurate in the age individual. *Journal of Gerontology, 6*, 213-216.

Miller, J. H., McDonald, R. K., & Shock, N. W. (1952). Age changes in the maximal rate of renal tubular resorption of glucose. *Journal of Gerontology, 7*, 196-200.

Miller, J. H., & Shock, N. W. (1953). Age differences in the renal tubular response to antidiuretic hormone. *Journal of Gerontology, 8*, 446-450.

Minaker, K. L., & Rowe, J. L. (1982). Gastrointestinal system. In J. W. Rowe & R. W. Besdine (Eds.), *Health and disease in old age*. Boston: Little, Brown.

Mittman, C., Edelman, N. H., Norris, A. H., & Shock, N. W. (1965). Relationship between chest wall and pulmonary compliance and age. *Journal of Applied Physiology*, *20*, 1211-1216.

Moatamed, F. (1966). Cell frequencies in the human inferior olivary complex. *Journal of Comparative Neurology*, *128*, 109-116.

Möller, W.-D., & Wolschendorf, K. (1978). The dependence of cerebral blood flow and age. *European Neurology*, *17*, 276-279.

Monagle, R. D., & Brody, H. (1974). The effects of age upon the main nucleus of the inferior olive in the human. *Journal of Comparative Neurology*, *155*, 61-66.

Montagna, W. (1965). Morphology of the aging skin: The cutaneous appendages. In W. Montagna (Ed.), *Aging: Vol. 6. Advances in biology of skin*. New York: Pergamon Press.

Montemayor, R. (1978). Men and their bodies: The relationships between body type and behavior. *Journal of Social Issues*, *34*, 48-64.

Montoye, H. J. (1975). *Physical activity and health: An epidemiologic study of an entire community*. Englewood Cliffs, NJ: Prentice-Hall.

Montoye, H. J. (1982). Age and oxygen utilization during submaximal treadmill exercise in males. *Journal of Gerontology*, *37*, 396-402.

Montoye, H. J., & Lamphiear, D. E. (1977). Grip and arm strength in males and females, age 10 to 69. *Research Quarterly*, *48*, 109-120.

Moore, J. G., Tweedy, C., Christian, P. E., & Datz, F. L. (1983). Effect of age on gastric emptying of liquid-solid meals in man. *Digestive Diseases and Sciences*, *28*, 340-344.

Moore, L. M., Nielsen, C. R., & Mistretta, C. M. (1982). Sucrose taste thresholds: Age-related differences. *Journal of Gerontology*, *37*, 64-69.

Moore, R. A. (1952). Male secondary sexual organs. In A. I. Lansing (Ed.), *Cowdry's problems of aging*. Baltimore, MD: Williams & Wilkins.

Morgan, D. B., Spiers, F. W., Pulvertaft, C. N., & Fourman, P. (1967). The amount of bone in the metacarpal and the phalanx according to age and sex. *Clinical Radiology*, *18*, 101-108.

Morgan, W. P. (1981). Psychological benefits of physical activity. In F. J. Nagle & H. J. Montoye (Eds.), *Exercise in health and disease* (pp. 299-314). Springfield, IL: C C Thomas.

Morgan, W. P., & Pollock, M. L. (1978). Physical activity and cardiovascular health: Psychological aspects. In F. Landry & W. Orban (Eds.), *Physical activity and human well being* (pp. 163-181). Miami: Symposium Specialists.

Morgan, W. P., Roberts, J. A., & Feinerman, A. D. (1971). Psychologic effect of acute physical activity. *Archives of Physical Medicine and Rehabilitation*, *52*, 420-425.

Morgan, Z. R., & Feldman, M. (1957). The liver, biliary tract and pancreas in the aged: An anatomic and laboratory evaluation. *Journal of the American Geriatrics Society*, *5*, 59-65.

Moritani, T., & deVries, H. A. (1980). Potential for gross muscle hypertrophy in older man. *Journal of Gerontology*, *35*, 672-682.

Morris, A. F., & Husman, B. F. (1978). Life quality changes following an endurance conditioning program. *American Corrective Therapy Journal*, *32*, 3-6.

Morris, J. F., Koski, A., & Johnson, L. C. (1971). Spirometric standards for healthy non-smoking adults. *American Review of Respiratory Disease*, *103*, 57-67.

Moses, R. A. (1981). Accommodation. In R. A. Moses (Ed.), *Adler's physiology of the eye*. St. Louis: C. V. Mosby.

Moses, S. W., Rotem, Y., Jagoda, N., Talmon, N., Eichorn, F., & Levin, S. (1967). A clinical, genetic and biochemical study of familial dysautonomia in Israel. *Israel Journal of Medical Sciences, 3,* 358-371.

Moskowitz, H. R. (1978). Taste and food technology: Acceptability, aesthetics and preferences. In C. Carterette & M. P. Friedman (Eds.), *Handbook of perception: Vol. VIA. Tasting and smelling*. New York: Academic Press.

Mouritzen Dam, A. (1979). The density of neurons in the human hippocampus. *Neuropathology and Applied Neurobiology, 5,* 249-264.

Muiesan, G., Sorbini, C. A., & Grassi, V. (1971). Respiratory function in the aged. *Bulletin de Physio-Pathologie Respiratoire, 7,* 973-1009.

Mulch, G., & Petermann, W. (1979). Influence of age on results of vestibular function tests. *Annals of Otology, Rhinology, and Laryngology, (Supplement 56), 88,* 1-17.

Munn, K. (1981). Effects of exercise on the range of motion in elderly subjects. In E. Smith & R. Serfass (Eds.), *Exercise and aging: The scientific bases* (pp. 167-186). Hillside, NJ: Enslow.

Murono, E. P., Nankin, H. R., Lin, T., & Osterman, J. (1982). The aging Leydig cell. V. Diurnal rhythms in aged men. *Acta Endocrinologica, 99,* 619-623.

Murphy, C. (1979). The effect of age on taste sensitivity. In S. S. Hay & D. H. Coons (Eds.), *Special senses in aging: A current biological assessment* (pp. 21-33). Ann Arbor: University of Michigan Press.

Murphy, C. (1983). Age-related effects on the threshold, psychophysical function, and pleasantness of menthol. *Journal of Gerontology, 38,* 217-222.

Murphy, G. (1947). *Personality: A biosocial approach*. New York: Harper & Brothers.

Naessen, R. (1971). An inquiry on the morphological characteristics of possible changes with age in the olfactory region of man. *Acta Otolaryngolica, 71,* 49-62.

Nakamaru, M., Ogihara, T., Hata, T., Maruyama, A., Mikami, H., Naka, T., Iwanaga, K., & Kumahara, Y. (1981). The effect of age on active and cryoactivatable inactive plasma renin in normal subjects and patients with essential hypertension. *Japanese Circulation Journal, 45,* 1231-1235.

Newman, C. W., & Spitzer, J. B. (1983). Prolonged auditory processing time in the elderly: Evidence from a backward recognition-masking paradigm. *Audiology, 22,* 241-252.

Niinimaa, V., & Shephard, R. J. (1978a). Training and oxygen conductance in the elderly. II. The cardiovascular system. *Journal of Gerontology, 33,* 362-367.

Niinimaa, V., & Shephard, R. J. (1978b). Training and oxygen conductance in the elderly. I. The respiratory system. *Journal of Gerontology, 33,* 354-361.

Nordberg, A., Adolfsson, R., Marcusson, J., & Winblad, B. (1982). Cholinergic receptors in the hippocampus in normal aging and dementia of the Alzheimer type. In E.Giacobini, G. Filogano, & A. Vernadakis (Eds.), *Aging: Vol. 20. The aging brain: Cellular and molecular mechanisms of aging in the nervous system*. New York: Raven Press.

Norris, A. H., Shock, N. W., Landowne, M., & Falzone, J. A., Jr. (1956). Pulmonary function studies: Age differences in lung volumes and bellows function. *Journal of Gerontology, 11,* 379-387.

Norris, A. H., Shock, N. W., & Yiengst, M. J. (1953). Age changes in heart rate and blood pressure responses to tilting and standardized exercise. *Circulation, 8,* 521-526.

Norris, A. H., Shock, N. W., & Yiengst, M. J. (1955). Age differences in ventilatory and gas exchange responses to graded exercise in males. *Journal of Gerontology, 10*, 145-155.

Norris, M. L., & Cunningham, D. R. (1981). Social impact of hearing loss in the aged. *Journal of Gerontology, 36*, 727-729.

Noth, R. H., Lassman, M. N., Tan, S. Y., Fernandez-Cruz, A., Jr., & Mulrow, P. J. (1977). Age and the renin-aldosterone system. *Archives of Internal Medicine, 137*, 1414-1417.

Novak, L. P. (1972). Aging, total body potassium, fat-free mass, and cell mass in males and females between 18 and 85 years. *Journal of Gerontology, 27*, 438-443.

Nunn, J. F. (1977). *Applied respiratory physiology* (2nd ed.). London: Butterworths.

O'Brien, E. J., & Epstein, S. (1982). *Sources of self-esteem inventory*. Unpublished manuscript, University of Massachusetts at Amherst.

Ohta, R. J., Carlin, M. F., & Harmon, B. M. (1981). Auditory acuity and performance on the Mental Status Questionnaire in the elderly. *Journal of the American Geriatrics Society, 29*, 476-478.

Olsen, T., Laurberg, P., & Weeke, J. (1978). Low serum triiodothyronine and high serum reverse triiodothyronine in old age: An effect of disease not age. *Journal of Clinical Endocrinology and Metabolism, 47*, 1111-1115.

Orchik, D. J., & Burgess, J. (1977). Synthetic sentences identification as a function of age of the listener. *Journal of the American Audiology Society, 3*, 42-46.

Ordway, G. A., & Wekstein, D. R. (1979). Effect of age on cardiovascular responses to static (isometric) exercise. *Proceedings of the Society for Experimental Biology and Medicine, 161*, 189-192.

Ordy, J. M. (1975). The nervous system, behavior, and aging: An interdisciplinary approach. In J. M. Ordy & K. R. Brizzee (Eds.), *Neurobiology of aging*. New York: Plenum Press.

Ordy, J. M. (1981). Neurochemical aspects of aging in humans. In H. M. van Praag, M. H. Lader, O. J. Rafaelson, & E. J. Sachar (Eds.), *Handbook of biological psychiatry*. New York: Dekker.

Ordy, J. M., Brizzee, K. R., & Johnson, H. A. (1982). Cellular alterations in visual pathways and the limbic system: Implications for vision and short-term memory. In R. Sekuler, D. Kline, & K. Dismukes (Eds.), *Aging and human visual function*. New York: Alan R. Liss.

Örlander, J., Kiessling, K.-H., Larsson, L., Karlsson, J., & Aniansson, A. (1978). Skeletel muscle metabolism and ultrastructure in relation to age in sedentary men. *Acta Physiologica Scandinavica, 104*, 249-261.

Örlander, J., & Aniansson, A. (1980). Effects of physical training on skeletal muscle metabolism and ultrastructure in 70-75 year old men. *Acta Physiologica Scandinavica, 109*, 149-154.

O'Sullivan, J. B., Mahan, C. M., Freedlender, A. E., & Williams, R. F. (1971). Effect of age on carbohydrate metabolism. *Journal of Clinical Endocrinology and Metabolism, 33*, 619-623.

Owsley, C., Sekuler, R., & Boldt, C. (1981). Aging and low-contrast vision: Face perception. *Investigative Ophthalmology and Visual Science, 21*, 362-365.

Owsley, C., Sekuler, R., & Siemsen, D. (1983). Contrast sensitivity throughout the lifespan. *Vision Research, 23*, 689-699.

Palmer, G. J., Ziegler, M. G., & Lake, C. R. (1978). Response of norepinephrine and blood pressure to stress increases with age. *Journal of Gerontology, 33*, 482-487.

Panek, P. E., Barrett, G. V., Sterns, H. L., & Alexander, R. A. (1977). A review of age changes in perceptual information ability with regard to driving. *Experimental Aging Research*, *3*, 387-449.

Pařizková, J. (1974). Body composition and exercise during growth and development. In G. L. Rarick (Ed.), *Physical activity: Human growth and devlopment* (pp. 98-104). New York: Academic Press.

Paterson, C. A. (1979). Crystalline lens. In R. E. Records (Ed.), *Physiology of the human eye and visual system*. New York: Harper & Row.

Pedersen, E. B., & Christensen, N. J. (1975). Catecholamines in plasma and urine in patients with essential hypertension determined by double-isotope derivative techniques. *Acta Medica Scandinavica*, *198*, 373-377.

Perry, E. K., Perry, R. H., Gibson, P. H., Blessed, G., & Tomlinson, B. E. (1977). A cholinergic connection between normal aging and senile dementia in the human hippocampus. *Neuroscience Letters*, *6*, 85-89.

Persky, H., Smith, K. D., & Basu, G. K. (1971). Relation of psychologic measures of aggression and hostility to testosterone production in man. *Psychosomatic Medicine*, *33*, 265-277.

Petersen, D. D., Pack, A. I., Silage, D. A., & Fishman, A. P. (1981). Effects of aging on ventilatory and occlusion pressure responses to hypoxia and hypercapnia. *American Review of Respiratory Diseases*, *124*, 387-391.

Petrofsky, J. S., Burse, K. L., & Lind, A. R. (1975). Comparison of physiological responses of women and men to isometric exercise. *Journal of Physiology*, *38*, 863-868.

Petrofsky, J. S., & Lind, A. R. (1975). Aging, isometric strength and endurance, and cardiovascular responses to static effort. *Journal of Applied Physiology*, *38*, 91-95.

Petrushevskii, I. I. (1966). Effects of fitness training on general psychological functioning, job performance and WAIS scores on telegraph operators. *Psychological Abstracts*, *40*, 767 (abstract).

Pfeifer, M. A., Halter, J. B., Wilkie, F., Cook, D. L., Brodsky, J., & Porte, D., Jr. (1980). Autonomic nervous system function and age-related increases blood pressure and heart rate in man. *Clinical Research*, *28*, 335 (abstract).

Pfeiffer, E., & Davis, G. C. (1974). Determinants of sexual behavior in middle and old age. In E. Palmore (Ed.), *Normal aging II* (pp. 251-262). Durham, NC: Duke University Press.

Pfeiffer, E., Verwoerdt, A., & Davis, G. C. (1974). Sexual behavior in middle life. In E. Palmore (Ed.), *Normal aging II* (pp. 243-251). Durham, NC: Duke University Press.

Pfeiffer, E., Verwoerdt, A., & Wang, H. S. (1969). Sexual behavior in aged men and women. In E. Palmore (Ed.), *Normal aging* (pp. 299-303). Durham, NC: Duke University Press.

Pickett, J. M., Bergman, M., & Levitt, H. (1979). Aging and speech understanding. In J. M. Ordy & K. Brizzee (Eds.), *Aging: Vol. 10. Sensory systems and communication in the elderly* (pp. 167-186). New York: Raven Press.

Pierce, J. A., & Ebert, R. V. (1958). The elastic properties of the lungs in the aged. *Journal of Laboratory and Clinical Medicine*, *51*, 63-71.

Pirke, K. M., & Doerr, P. (1975). Age-related changes in free plasma testosterone, dihydrotestosterone, and oestradiol. *Acta Endocrinologica*, *80*, 171-178.

Pitkanën, P., Westermack, P., Cornwell, C. G., III, & Murdoch, W. (1983). Amyloid of the seminal vesicles: A distinctive and common localized form of senile amyloidosis. *American Journal of Pathology*, *110*, 64-69.

266 References

Pitts, D. G. (1982a). The effects of aging on selected visual functions: Dark adaptation, visual acuity stereopsis, and brightness contrast. In R. Sekuler, D. Kline, & K. Dismukes (Eds.), *Aging and human visual function*. New York: Alan R. Liss.

Pitts, D. G. (1982b). Visual acuity as a function of age. *Journal of the American Optometric Association, 53*, 117-124.

Plato, C. C., & Norris, A. H. (1979a). Osteoarthritis of the hand: Age specific joint-digit and prevalence rates. *American Journal of Epidemiology, 109*, 169-180.

Plato, C. C., & Norris, A. H. (1979b). Osteoarthritis of the hand: Longitudinal studies. *American Journal of Epidemiology, 110*, 740-746.

Plemons, J. K., Willis, S. L., & Baltes, P. B. (1978). Modifiability of fluid intelligence in aging: A short-term longitudinal training approach. *Journal of Gerontology, 33*, 224-231.

Pliska, A., & Gilchrest, B. A. (1983). Growth factor responsiveness of cultured human fibroblasts declines with age. *Journal of Gerontology, 38*, 513-518.

Plomp, R., & Mimpen, A. M. (1979). Speech reception threshold for sentences as a function of age and noise level. *Journal of the Acoustical Society of America, 66*, 1333-1342.

Plowman, S. A., Drinkwater, B. L., & Horvath, S. (1979). Age and aerobic power in women: A longitudinal study. *Journal of Gerontology, 34*, 512-520.

Pollock, M. L. (1974). Physiological charcteristics of older champion athletes. *Research Quarterly, 45*, 363-373.

Pollock, M. L., Dawson, G. A., Miller, H. S., Ward, A., Cooper, D., Headley, W., Linnerud, A. C., & Nomeir, M.-M. (1976). Physiologic responses of men 49 to 65 years of age to endurance training. *Journal of the American Geriatrics Society, 24*, 97-104.

Pollock, M. L., Miller, H. S. Jr., Janeway, R., Linnerud, A. C., Robertson, B., & Valentino, R. (1971). Effects of walking on body composition and cardiovascular function of middle aged men. *Journal of Applied Physiology, 30*, 126-130.

Port, S., Cobb, F. R., Coleman, R. E., & Jones, R. H. (1980). Effect of age on the response of the left ventricular ejection fraction to exercise. *New England Journal of Medicine, 303*, 1133-1137.

Potash, M., & Jones, B. (1977). Aging and decision criteria for the detection of tones in noise. *Journal of Gerontology, 32*, 436-440.

Powell, A. H., Jr., Eisdorfer, C., & Bogdanoff, M. D. (1964). Physiologic response patterns observed in a learning task. *Archives of General Psychiatry, 10*, 192-195.

Powers, J. K., & Powers, E. A. (1978). Hearing problems of elderly persons: Social consequences and prevalence. *ASHA, 20*, 79-83.

Prinz, P. N. (1977). Sleep patterns in the healthy aged: Relationship with intellectual function. *Journal of Gerontology, 32*, 179-186.

Prinz, P. N., & Halter, J. (1983). Sleep disturbances in the elderly: Neurohormonal correlates. In E. D. Weitzman & M. Chase (Eds.), *Advances in sleep research* (Vol. 8, pp. 463-501). New York: Spectrum.

Prinz, P. N., Halter, J., Benedetti, C., & Raskind, M. (1979). Circadian variation of plasma catecholamines in young and old men: Relation to rapid eye movement and slow wave sleep. *Journal of Clinical Endocrinology and Metabolism, 49*, 300-304.

Prinz, P. N., Vitiello, M. V., Smallwood, R. G., Schoene, R. B., & Halter, J. B. (1984). Plasma norepinephrine in normal young and aged men: Relationship with sleep. *Journal of Gerontology, 39*, 561-567.

Prinz, P. N., Weitzman, E. D., Cunningham, G. R., & Karacan, I. (1983). Plasma growth hormone during sleep in young and aged men. *Journal of Gerontology, 38*, 519-524.

Profant, G. R., Early, R. G., Nilson, K. L., Kusumi, F., Hofer, V., & Bruce, R. A. (1972). Responses to maximal exercise in healthy middle-aged women. *Journal of Applied Physiology*, *33*, 595-599.

Proper, R., & Wall, F. (1972). Left ventricular stroke volume measurements not affected by chronologic aging. *American Heart Journal*, *83*, 843-845.

Ptacek, P. H., & Sander, E. K. (1966). Age recognition from voice. *Journal of Speech and Hearing Research*, *9*, 273-277.

Ptacek, P. H., Sander, E. K., Maloney, W. H., & Jackson, C. C. R. (1966). Phonatory and related changes with advancing age. *Journal of Speech and Hearing Research*, *9*, 353-360.

Punch, J., & McConnell, F. (1969). The speech discrimination function of elderly adults. *Journal of Auditory Research*, *9*, 159-166.

Rabbitt, P. M. A. (1980). A fresh look at changes in reaction times in old age. In D. G. Stein (Ed.), *The psychobiology of aging: Problems and perspectives* (pp. 425-442). New York: Elsevier.

Ramig, L. A., & Ringel, R. L. (1983). Effects of physiological aging on selected acoustic characteristics of voice. *Journal of Speech and Hearing Research*, *26*, 22-30.

Ratzmann, K. P., Witt, S., Heinke, P., & Shulz, B. (1982). The effect of ageing on insulin sensitivity and insulin secretion in non-obese healthy subjects. *Acta Endocrinologica*, *100*, 543-549.

Raven, P. B., & Mitchell, J. (1980). The effect of aging on the cardiovascular response to dynamic and static exercise. In M. L. Weisfeldt (Ed.), *Aging: Vol. 12. The aging heart: Its function and response of stress*. New York: Raven.

Ravens, J. R., & Calvo, W. (1966). Neurological changes in the senile brain. In F. Luthy & A. Bischoff (Eds.), *Proceedings of the Fifth International Congress of Neuropathology*. New York: Excerpta Medica Foundation, International Congress Series No. 100.

Reading, V. M. (1968). Disability glare and age. *Vision Research*, *8*, 207-214.

Rees, J. N., & Botwinick, J. (1971). Detection and decision factors in auditory behavior of the elderly. *Journal of Gerontology*, *26*, 133-136.

Reid, L. (1967). The aged lung. In *The Pathology of Emphysema*. London: Lloyd-Luke.

Richards, O. W. (1977). Effects of luminance and contrast on visual acuity ages 16 to 90 years. *American Journal of Optometry and Physiological Optics*, *54*, 178-184.

Richter, C. P., & Campbell, K. H. (1940). Sucrose taste thresholds of rats and humans. *American Journal of Physiology*, *128*, 291-297.

Riggs, B. L., Wahner, W. H., Dunn, W. L., Mazess, R. B., Offord, K. P., & Milton, L. J. (1981). Differential changes in bone mineral density of the appendicular and axial skeleton with aging. *Journal of Clinical Investigation*, *67*, 328-335.

Robinson, J. K. (1983). Skin problems of aging. *Geriatrics*, *38*, 57-65.

Robinson, P. K. (1983). The sociological perspective. In R. B. Weg (Ed.), *Sexuality in the later years* (pp. 81-103). New York: Academic Press.

Robinson, S., Belding, H. S., Consolazio, S., Horvath, S. M., & Turrell, E. S. (1965). Acclimatization of older men to work in the heat. *Journal of Applied Physiology*, *20*, 583-586.

Robinson, S., Dill, D. B., Ross, J. C., Robinson, R. D., Wagner, J. A., & Tzankoff, S. P. (1973). Training and physiological aging in man. *Federation Proceedings*, *32*, 1628-1634.

Robinson, S., Dill, D. B., Tzankoff, S. P., Wagner, J. A., & Robinson, R. D. (1975). Longitudinal studies of aging in 37 men. *Journal of Applied Physiology*, *38*, 263-267.

Rogers, J. Silver, M. A., Shoemaker, W. J., & Bloom, F. E. (1980). Senescent changes in a neurobiological model system: Cerebellar Purkinje cell electrophysiology and correlative anatomy. *Neurobiology of Aging*, *1*, 3-11.

Root, A. W., & Oski, F. A. (1969). Effects of human growth hormone in elderly males. *Journal of Gerontology*, *24*, 97-104.

Rose, C. L., & Cohen, M. L. (1977). Relative importance of physical activity for longevity. *Annals of the New York Academy of Sciences*, *301*, 671-697.

Rosenberg, I. H., Bowman, B. B., Cooper, B. A., Halsted, C. H., & Lindenbaum, J. (1982). Folate nutrition in the elderly. *American Journal of Clinical Nutrition*, *36*, 1060-1066.

Rosenberg, I. R., Friedland, N., Janowitz, H. D., & Dreiling, D. A. (1966). The effect of age and sex upon human pancreatic secretion of fluid and bicarbonate. *Gastroenterology*, *50*, 191-194.

Rosenhall, U. (1973). Degenerative patterns in the aging human vestibular neuroepithelia. *Acta Otolaryngolica*, *76*, 208-220.

Rosenhall, U., & Rubin, W. (1975). Degenerative changes in the human vestibular sensory epithelia. *Acta Otolaryngolica*, *79*, 67-80.

Rosillo, R. H., & Fogel, M. L. (1971). Correlation of psychologic variables and progress in physical rehabilitation: IV. The relation of body image to success in physical rehabilitation. *Archives of Physical Medicine and Rehabilitation*, *52*, 182-186.

Ross, M. D., Johnsson, L. G., Peacor, D., & Allard, L. T. (1976). Observations on normal and degenerating human otoconia. *Annals of Otology, Rhinology, and Laryngology*, *85*, 310-326.

Rossi, A. S. (1980). Aging and parenthood in the middle years. In P. B. Baltes (Ed.), *Lifespan development and behavior* (Vol. 3, pp. 137-205). New York: Academic.

Rossi, A. S., & Rossi, P. E. (1980). Body time and social time. Mood patterns by menstrual cycle phase and day of week. In J. E. Parsons (Ed.), *The psychobiology of sex differences in sex roles* (pp. 269-304). Washington, DC: Hemisphere.

Rossman, I. (1977). Anatomic and body composition changes with age. In C. E. Finch & L. Hayflick (Eds.), *Handbook of the biology of aging*. New York: Van Nostrand Reinhold.

Rossman, I. (1979). *Clinical geriatrics*. Philadelphia: Lippincott.

Rost, R., Dreisback, W., & Hollmann, W. (1978). Hemodynamic changes in 50–70-year-old men due to endurance training. In F. Landry & W. A. R. Orban (Eds.), *Sports medicine* (Vol. 5, pp. 121-124). Miami: Symposia Specialists.

Rovee, C. K., Cohen, R. Y., & Shlapack, W. (1975). Lifespan stability in olfactory sensitivity. *Developmental Psychology*, *11*, 311-318.

Rowe, J. W. (1982). Renal function and aging. In M. E. Reff & E. L. Schneider (Eds.), *Biological markers of aging* (pp. 228-236). (NIH Publication No. 82-2221)

Rowe, J. W., Andres, R. A., Tobin, J. D., Norris, A. H., & Shock, N. W. (1976). The effect of age on creatinine clearance in man: A cross-sectional and longitudinal study. *Journal of Gerontology*, *31*, 155-163.

Rowe, J. W., Minaker, K. L., Pallotta, J. A., & Flier, J. S. (1983). Characterization of the insulin resistance of aging. *Journal of Clinical Investigation*, *71*, 1581-1587.

Rowe, J. W., Shock, N. W., & DeFronzo, R. A. (1976). The influence of age on the renal response to water deprivation in man. *Nephron*, *17*, 270-278.

Rowe, T. W., & Troen, B. R. (1980). Sympathetic nervous system and aging in man. *Endocrinological Review*, *1*, 167-179.

Rubenstein, H. A., Butler, V. P., & Werner, S. C. (1973). Progressive decrease in serum triiodothyronine concentrations with human aging: Radioimmuno-assay following extraction of serum. *Journal of Clinical Endocrinology and Metabolism, 37*, 247-253.

Rubin, S., Tack, M., & Cherniak, N. S. (1982). Effect of aging on respiratory responses to CO_2 and inspiratory resistive loads. *Journal of Gerontology, 37*, 306-312.

Saar, N., & Gordon, R. D. (1979). Variability of plasma catecholamine levels: Age, duration of posture and time of day. *British Journal of Clinical Pharmacology, 8*, 353-358.

Salthouse, T. (1982). *Adult cognition.* New York: Springer-Verlag.

Salthouse, T. A., Wright, R., & Ellis, C. L. (1979). Adult age and the rate of an internal clock. *Journal of Gerontology, 34*, 53-57.

Saltin, B., & Grimby, G. (1968). Physiological analysis of middle-aged and old former athletes. *Circulation, 38*, 1104-1115.

Saltin, B., Hartley, L. H., Kilbom, Å, Åstrand, I. (1969). Physical training in sedentary middle-aged and older men. II. Oxygen uptake, heart rate, and blood lactate concentration at submaximal and maximal exercise. *Scandinavian Journal of Clinical Laboratory Investigations, 24*, 323-334.

Sandoz, P., & Meier-Ruge, W. (1977). Age-related loss of nerve cells from the human inferior olive, and unchanged volume of its gray matter. *IRCS Medical Science, 5*, 376.

Sandqvist, L., & Kjellmer, I. (1960). Normal values for the single breath nitrogen elimination test in different age groups. *Scandinavian Journal of Clinical Laboratory Investigations, 12*, 131-135.

Sato, I., Hasegawa, Y., Takahashi, N., Hirata, Y., Shimomura, K., & Hotta, K. (1981). Age-related changes of cardiac control function in man. With special reference to heart rate control at rest and during exercise. *Journal of Gerontology, 36*, 564-572.

Sawin, C. T., Chopra, D., Azizi, F., Mannix, J. E., & Bacharach, P. (1979). The aging thyroid: Increased prevalence of elevated serum thyrotropin levels in the elderly. *Journal of the American Medical Association, 242*, 247-250.

Schaffner, F., & Popper, H. (1959). Non-specific reactive hepatitis in aged and infirm people. *American Journal of Digestive Diseases, 4*, 389-399.

Schaie, K. W., & Baltes, P. B. (1975). On sequential strategies in developmental research. *Human Development, 18*, 384-390.

Schaie, K. W., Baltes, P., & Strother, C. R. (1964). A study of auditory sensitivity in advanced age. *Journal of Gerontology, 19*, 453-457.

Schalekamp, M. A. D. H., Krauss, X. H., Schalekamp-Kuyken, M. P. A., Kolsters, G., & Birkenhager, W. H. (1971). Studies on the mechanism of hypernatriuresis in essential hypertension in relation to measurements of plasma renin concentration, body fluid compartments, and renal function. *Clinical Science, 41*, 219-231.

Scheibel, A. B. (1979). Aging in human motor control systems. In J. M. Ordy & K. Brizzee (Eds.), *Aging: Vol. 10. Sensory systems and communication in the elderly.* New York: Raven Press.

Scheibel, A. B. (1982). Age-related changes in the human forebrain. *Neurosciences Research Progress Bulletin, 20*, 577-583.

Scheibel, M. E., Lindsay, R. D., Tomiyasu, U., & Scheibel, A. B. (1975). Progressive dendritic changes in aging human cortex. *Experimental Neurology, 47*, 392-403.

Scheibel, M. E., Lindsay, R. D., Tomiyasu, U., & Scheibel, A. B. (1976). Progressive dendritic changes in the aging human limbic systems. *Experimental Neurology, 53*, 420-430.

Scheibel, M. E., & Scheibel, A. B. (1975). Structural changes in the aging brain. In H. Brody, D. Harman, & J. M. Ordy (Eds.), *Aging: Vol. 1. Clinical, morphological, and neurochemical aspects of aging in the central nervous system*. New York: Raven Press.

Scheibel, M. E., Tomiyasu, U., & Scheibel, A. B. (1977). The aging human Betz cell. *Experimental Neurology, 56*, 598-609.

Scheie, H. G., & Albert, D. M. (1977). *Textbook of ophthalmology* (9th ed.). Philadelphia, PA: Saunders.

Schemper, T., Voss, S., & Cain, W. S. (1981). Odor identification in young and elderly persons: Sensory and cognitive limitations. *Journal of Gerontology, 36*, 446-452.

Schiff, I., & Wilson, E. (1978). Clinical aspects of aging of the female reproductive system. In E. L. Schneider (Ed.), *Aging: Vol. 4. The aging reproductive system*. New York: Raven Press.

Schiffman, S. (1977). Food recognition by the elderly. *Journal of Gerontology, 32*, 586-592.

Schiffman, S. (1979). Changes in taste and smell with age: Psychophysical aspects. In J. M. Ordy & K. Brizzee (Eds.), *Aging: Vol. 10. Sensory systems and communication in the elderly* (pp. 227-246). New York: Raven Press.

Schiffman, S., Orlandi, M., & Erickson, R. P. (1979). Changes in taste and smell with age: Biological aspects. In J. M. Ordy & K. Brizzee (Eds.), *Aging: Vol. 10. Sensory systems and communication in the elderly* (pp. 247-268). New York: Raven Press.

Schiffman, S. S., & Covey, E. (1984). Changes in taste and smell with age: Nutritional aspects. In J. M. Ordy, D. Harman, & R. B. Alfin-Slater (Eds.), *Aging: Vol. 26. Nutrition in Gerontology* (pp. 43-64). New York: Raven Press.

Schiffman, S. S., Moss, J., & Erickson, R. P. (1976). Thresholds of food odors in the elderly. *Experimental Aging Research, 2*, 389-398.

Schiffman, S. S., & Pasternak, M. (1979). Decreased discrimination of food odors in the elderly. *Journal of Gerontology, 34*, 73-79.

Schilder, P. (1935/1950). *The image and appearance of the human body*. New York: International Universities Press.

Schmidt, R. F. (1978). *Fundamentals of neurophysiology (2nd ed.)*. New York: Springer-Verlag.

Schneider, D. J., Hastorf, A. H., & Ellsworth, P. C. (1979). *Person perception* (2nd ed.). Reading, MA: Addison-Wesley.

Schocken, D., & Roth, G. (1977). Reduced beta-adrenergic receptor concentrations in aging man. *Nature, 267*, 856-858.

Schow, R. L., Christensen, J. M., Hutchinson, J. M., & Nerbonne, M. A. (1978). *Communiation disorders of the aged*. Baltimore: University Park Press.

Schuknecht, H. F. (1964). Further observations on the pathology of presbycusis. *Archives of Otolaryngology, 80*, 369-382.

Schulz, U., & Hunziker, O. (1980). Comparative studies of neuronal perikaryon size and shape in the aging cerebral cortex. *Journal of Gerontology, 35*, 483-491.

Schwartz, D., Mayaux, M.-J., Spira, A., Moscato, M.-L., Jouannet, P., Czyglik, F., & David, G. (1983). Semen characteristics as a function of age in 833 fertile men. *Fertility and Sterility, 39*, 530-535.

Seals, D. R., Hagberg, J. M., Hurley, B. F., Ehsani, A. A., & Holloszy, J. G. (1984). Effects of endurance training on glucose tolerance and plasma lipid levels in older men and women. *Journal of the American Medical Association, 252*, 645-649.

Secord, P. F., & Jourard, S. M. (1953). The appraisal of body-cathexis. *Journal of Consulting Psychology, 17*, 343-347.

Sekuler, R., Hutman, L. P., & Owsley, C. (1980). Human aging and spatial vision. *Science*, *209*, 1255-1256.

Sekuler, R., & Owsley, C. (1982). The spatial vision of older humans. In R. Sekuler, D. Kline, & K. Dismukes (Eds.), *Aging and human visual function*. New York: Alan R. Liss.

Selmanowitz, V. J., Rizer, R. L., & Orentreich, N. (1977). Aging of the skin and its appendages. In C. E. Finch & L. Hayflick (Eds.), *Handbook of the biology of aging*. New York: Van Nostrand Reinhold.

Sever, P. S., Osikowska, B., Birch, M., & Tunbridge, R. D. G. (1977). Plasma-noradrenaline in essential hypertension. *Lancet*, *1*, 1078-1081.

Severson, J. A., Marcusson, J., Winblad, B., & Finch, C. E. (1982). Age-correlated loss of dopaminergic binding sites in human basal ganglia. *Journal of Neurochemistry*, *39*, 1623-1631.

Seymour, F. I., Duffy, C., & Koerner, A. (1935). A case of authenticated fertility in a man of 94. *Journal of the American Medical Association*, *105*, 1423-1424.

Shefer, V. F. (1973). Absolute number of neurons and thickness of the cerebral cortex during aging, senile and vascular dementia, and Pick's and Alzheimer's diseases. *Neuroscience and Behavioral Psychology*, *6*, 319-324.

Shephard, R. J. (1978). *Physical activity and aging*. Chicago: Yearbook Medical.

Shephard, R. J. (1981). Cardiovascular limitations in the aged. In E. L. Smith & R. C. Serfass (Eds.), *Exercise and aging: The scientific basis* (pp. 19-30). Hillside, NJ: Enslow.

Shephard, R. J., & Sidney, K. H. (1975). Effects of physical exercise on plasma growth hormone and cortisol levels in human subjects. *Exercise and Sports Sciences Reviews*, *3*, 1-30.

Shephard, R. J., & Sidney, K. H. (1978). Exercise and aging. *Exercise and Sport Sciences Reviews*, *6*, 1-57.

Sherman, B. M., & Korenman, S. G. (1975). Hormonal characteristics of the human menstrual cycle throughout reproductive life. *Journal of Clinical Investigations*, *55*, 699-706.

Sherman, B. M., West, J. H., & Korenman, S. G. (1976). The menopausal transition: Analysis of LH, FSH, estradiol, and progesterone concentrations during menstrual cycles of older women. *Journal of Clinical Endocrinology and Metabolism*, *42*, 629-636.

Sherwin, R. S., Insel, P. A., Tobin, J. D., Liljenquist, J. E., Andres, R., & Berman, M. (1972). Computer modelings: An aid to understanding insulin action. *Diabetes*, *21* (Suppl. 1), 347.

Shickman, G. M. (1981). Time-dependent functions in vision. In R. A. Moses (Ed.), *Adler's physiology of the eye* (pp. 663-713). St. Louis: C. V. Mosby.

Shock, N. W. (1961). Physiological aspects of aging in man. *Annual Review of Physiology*, *23*, 97-122.

Shock, N. W. (1977). Systems integration. In C. E. Finch & L. Hayflick (Eds.), *Handbook of the biology of aging*. New York: Van Nostrand Reinhold.

Shock, N. W. (1983). Aging of physiological systems. *Journal of Chronic Diseases*, *36*, 137-142.

Shock, N. W., Andres, R., Norris, A. H., & Tobin, J. D. (1978). Patterns of longitudinal changes in renal function. *Proceedings XII International Congress of Gerontology, Tokyo, August 20-25* (pp. 525-527). Amsterdam: Excerpta Medica.

Shoenfeld, Y., Udassin, R., Shapiro, Y., Ohri, A., & Sohar, E. (1978). Age and sex differences in response to short exposure to extreme dry heat. *Journal of Applied Physiology: Respiratory, Environmental, and Exercise Physiology, 44,* 1-4.

Sidney, K. H. (1981). Cardiovascular benefits of physical activity in the exercising aged. In E. L. Smith & R. C. Serfass (Eds.), *Exercise and aging: The scientific basis* (pp. 131-148). Hillside, NJ: Enslow.

Sidney, K. H., & Shephard, R. J. (1976). Attitudes toward health and physical activity in the elderly: Effects of a physical training program. *Medicine and Science in Sports and Exercise, 8,* 246-252.

Sidney, K. H., & Shephard, R. J. (1977a). Activity patterns of elderly men and women. *Journal of Gerontology, 32,* 25-32.

Sidney, K. H., & Shephard, R. J. (1977b). Growth hormone and cortisol—age differences, effects of exercise and training. *Canadian Journal of Applied Sports Sciences, 2,* 190-193.

Sidney, K. H., & Shephard, R. J. (1978). Frequency and intensity of exercise training for elderly subjects. *Medicine and Science in Sports and Exercise, 10,* 125-131.

Sidney, K. H., Shephard, R. J., & Harrison, J. E. (1977). Endurance training and body composition of the elderly. *American Journal of Clinical Nutrition, 30,* 326-333.

Siegel, W. C., Blomqvist, G., & Mitchell, J. H. (1970). Effects of a quantitated physical training program on middle-aged sedentary men. *Circulation, 41,* 19-29.

Silman, S. (1979). The effects of aging on the stapedius reflex threshold. *Journal of the Acoustical Society of America, 66,* 735-738.

Silver, H. M., & Landowne, M. (1953). The relation of age to certain electrocardiographic responses of normal adults to a standardized exercise. *Circulation, 8,* 510-520.

Silverman, C. A., Silman, S., & Miller, M. H. (1983). The acoustic reflex threshold in aging ears. *Journal of the Acoustical Society of America, 73,* 248-255.

Silverstone, F. A., Brandfonbrener, M., Shock, N. W., & Yiengst, M. J. (1957). Age differences in the intravenous glucose tolerance tests and the response to insulin. *Journal of Clinical Investigations, 36,* 504-514.

Sims, N. R., Bowen, D. M., & Davison, A. N. (1982). Acetylcholine synthesis and glucose metabolism in aging and dementia. In E. Giacobini, G. Filogano, & A. Vernadakis (Eds.), *Aging: Vol. 20. The aging brain: Cellular and molecular mechanisms of aging in the nervous system.* New York: Raven Press.

Singh, S., & Singh, K. S. (1976). *Phonetics: Principles and practices.* Baltimore, University Park Press.

Skinner, J. S. (1970). The cardiovascular system with aging and exercise. In D. Brunner & E. Jokl (Eds.), *Medicine and sport: Vol. 4. Physical activity and aging* (pp. 100-109). Baltimore: University Park Press.

Sklar, M., Kirsner, J. B., & Palmer, W. L. (1956). Symposium on medical problems of the aged: Gastrointestinal disease in the aged. *Medical Clinics of North America, 40,* 223-237.

Sleight, P. (1979). The effect of aging on the circulation. *Age and Aging, 8,* 98.

Smith, C. G. (1942). Age incidence of atrophy of olfactory nerves in man. *Journal of Comparative Neurology, 17,* 589-595.

Smith, E. L. (1971). *Bone changes with age and physical activity.* Unpublished dissertation, University of Wisconsin.

Smith, E. L. (1981). Physical activity: A preventive and maintenance modality for bone loss with age. In F. J. Nagle & H. J. Montoye (Eds.), *Exercise in health and disease* (pp. 196-202). Springfield, IL: C C Thomas.

Smith, E. L., & Reddan, W. (1977). Physical activity—a modality for bone accretion in the aged. *American Journal of Roetgenology, 126,* 1297.

Smith, M. B. (1974). Competence and adaptation. *American Journal of Occupational Therapy, 28,* 11-15.

Smith, R. A., & Prather, W. F. (1971). Phoneme discrimination in older persons under varying signal-to-noise conditions. *Journal of Speech and Hearing Research, 14,* 630-638.

Smith, S. E., & Davies, P. D. O. (1973). Quinine taste thresholds: A family study and a twin study. *Annals of Human Genetics, 37,* 227-232.

Smulyan, H., Csermely, T. J., Mookherjee, S., & Warner, R. A. (1983). Effect of age on arterial distensibility in asymptomatic humans. *Arteriosclerosis, 3,* 199-205.

Snyder, P. J., & Utiger, R. D. (1972a). Response to thyrotropin releasing hormone (TRH) in normal man. *Journal of Clinical Endocrinology, 34,* 380-385.

Snyder, P. J., & Utiger, R. D. (1972b). Thyrotrophin response to thyrotrophin releasing hormone in normal females over forty. *Journal of Clinical Encodrinology, 34,* 1096-1098.

Soerjodibroto, W. S., Heard, C. R. C., & Exton-Smith, A. N. (1979). Glucose tolerance and plasma insulin sensitivity in elderly patients. *Age and Ageing, 8,* 65-74.

Solnick, R. L., & Birren, J. E. (1977). Age and male erectile responsiveness. *Archives of Sexual Behavior, 6,* 1-9.

Sonstroem, R. J. (1984). Exercise and self-esteem. *Exercise and Sports Sciences Reviews, 12,* 123-155.

Sontag, S. (1972). The double standard of aging. *Saturday Review,* September 23, 1972, pp. 29-38.

Sorbini, C. A., Grassi, V., Solinas, E., & Muiesan, G. (1968). Arterial oxygen tension in relation to age in healthy subjects. *Respiration, 25,* 3-13.

Sowers, J. R., Rubenstein, L. Z., & Stern, N. (1983). Plasma norepinephrine responses to posture and isometric exercise increase with age in the absence of obesity. *Journal of Gerontology, 38,* 315-317.

Sparrow, D., Beausoleil, N. I., Garvey, A. J., Rosner, B., & Silbert, J. E. (1982). The influence of cigarette smoking and age on bone loss in men. *Archives of Environmental Health, 37,* 246-249.

Sparrow, D., Bosse, R., & Rowe, J. W. (1980). The influence of age, alcohol consumption and body build on gonadal function in men. *Journal of Clinical Endocrinology and Metabolism, 51,* 508-512.

Spector, A. (1982). Aging of the lens and cataract formation. In R. Sekuler, D. Kline, & K. Dismukes (Eds.), *Aging and human visual function.* New York: Alan R. Liss.

Spirduso, W. W. (1980). Physical fitness, aging, and psychomotor speed: A review. *Journal of Gerontology, 35,* 850-865.

Stamford, B. A. (1973). Effects of chronic institutionalization on the physical working capacity and trainability of geriatric men. *Journal of Gerontology, 28,* 441-446.

Stearns, E. L., MacDonald, J. A. Kauffman, B. J., Lucman, T. S., Winters, J. S., & Faiman, C. (1974). Declining testicular function with age: Hormonal and clinical correlates. *American Journal of Medicine, 57,* 761-766.

Stein, I. D., & Granick, G. (1980). Human vertebral bone: Relation of strength, porosity, and mineralization to fluoride content. *Calcified Tissue International, 32,* 189-194.

Sternbach, R. A. (1978). Psychological dimensions and perceptual analyses, including pathologies of pain. In E. C. Carterette & M. P. Friedman (Eds.), *Handbook of perception: Volume VIB: Feeling and hurting* (pp. 231-261). New York: Academic Press.

Sticht, T. G., & Gray, B. B. (1969). The intelligibility of time compressed words as a function of age and hearing loss. *Journal of Speech and Hearing Research, 12*, 443-448.

Stiles, M. H. (1967). Motivation for sports participation in the community. *Canadian Medical Association Journal, 96*, 889-892.

Strandell, T. (1963). Electrocardiographic findings at rest, during and after exercise in healthy old men compared with young men. *Acta Medica Scandinavica, 174*, 479-499.

Straus, B. (1979). Disorders of the digestive system. In I. Rossman (Ed.), *Clinical geriatrics*. Philadelphia: Lippincott.

Strauss, E. L. (1970). A study on olfactory acuity. *Annals of Otology, Rhinology, and Laryngology, 79*, 95-104.

Strauzenberg, S. E. (1978). On the specific effects of various physical activities on the cardiovascular and metabolic functions in older people. In F. Landry & W. A. R. Orban (Eds.), *Sports medicine* (Vol. 5, pp. 125-132). Miami: Symposia Specialists.

Strehler, B. L. (1977). *Time, cells, and aging*. New York: Academic Press.

Suiteri, P. K., & MacDonald, P. C. (1973). Role of extraglandular estrogen in human endocrinology. In R. O. Greep & E. B. Astwood (Eds.), *Handbook of physiology (Vol. 2, Part I)*. Baltimore: Williams & Wilkins.

Suominen, H., Heikkinen, E., Liesen, H., Michel, D., & Hollmann, W. (1977). Effects of eight weeks' endurance training on skeletal muscle metabolism in 56-70 year old sedentary men. *European Journal of Applied Physiology, 37*, 173-180.

Suominen, H., Heikkinen, E., & Parkatti, T. (1977). Effects of eight weeks' physical training on muscle and connective tissue of the M. vastus lateralis in 69 year-old men and women. *Journal of Gerontology, 32*, 33-37.

Suominen, H., Heikkinen, E., Parkatti, T., Forsberg, S., & Kiiskinen, A. (1980). Effect of lifelong physical training on functional aging in men. *Scandinavian Journal of the Society of Medicine, 14* (Suppl.), 225-240.

Surr, R. K. (1977). Effect of age on clinical hearing aid evaluation results. *Journal of the American Audiology Society, 3*, 1-5.

Surwillo, W. W. (1964). Age and the perception of short intervals of time. *Journal of Gerontology, 19*, 322-324.

Sutton, M. S. J., Reichek, N., Levett, J., Kastor, J. A., & Giuliani, E. (1980). Effects of age, body size, and blood pressure on the normal human left ventricle. *Circulation, 62* (Suppl. III), 305.

Swerdloff, R. S., & Heber, D. (1982). Effects of aging on male reproductive function. In S. G. Korenman (Ed.), *Endocrine aspects of aging*. New York: Elsevier Biomedical.

Sworn, M. J., & Fox, M. (1972). Donor kidney selection for transplantation. *British Journal of Urology, 44*, 377-383.

Szanto, S. (1975). Metabolic studies in physically outstanding elderly men. *Age and Aging, 4*, 37-42.

Takazakura, E., Sawabu, N., Handa, A., Takada, A., Shinoda, A., & Takeuchi, J. (1972). Intrarenal vascular changes with age and disease. *Kidney International, 2*, 224-230.

Takeda, R., Morimoto, S., Uchida, K., Miyamori, I., & Hashiba, I. (1980). Effect of age on plasma aldosterone response to exogenous angiotensin II in normotensive subjects. *Acta Endocrinologica, 94*, 552-558.

Talbert, G. B. (1977). Aging of the reproductive system. In C. E. Finch & L. Hayflick (Eds.), *Handbook of the biology of aging*. New York: Van Nostrand Reinhold.

Talbert, G. B. (1978). Effect of aging on the ovaries and female gametes on reproductive capacity. In E. L. Schneider (Ed.), *Aging: Vol. 4. The aging reproductive system*. New York: Raven Press.

Taylor, A. L., Finster, J. L., & Mintz, D. H. (1969). Metabolic clearance and production rates of human growth hormone. *Journal of Clinical Investigation, 48*, 2349-2358.

Teräslinna, P., Partanen, T., Oja, P., & Koskela, A. (1970). Some social characteristics and living habits associated with willingness to participate in a physical activity intervention study. *Journal of Sports Medicine and Physical Fitness, 10*, 138-144.

Thomas, D. J., Zilkha, E., Redmond, S., DuBoulay, G. H., Marshall, J., Russell, R. W. R., & Symon, L. (1979). An intravenous 133-Xenon clearance technique for measuring cerebral blood flow. *Journal of Neurological Science, 40*, 53-63.

Thomas, P. D., Hunt, W. C., Garry, P. J., Hood, R. B., Goodwin, J. M., & Goodwin, J. S. (1983). Hearing acuity in a healthy elderly population: Effects on emotional, cognitive, and social status. *Journal of Gerontology, 38*, 321-325.

Thompson, D. J., Sills, J. A., Recke, K. S., & Bui, D. M. (1980). Acoustic reflex growth in the aging adult. *Journal of Speech and Hearing Research, 23*, 405-418.

Thompson, E. N., & Williams, R. (1965). Effect of age on liver function with particular reference to BSP excretion. *Gut, 6*, 266-269.

Thompson, W. (1972). Correlates of the self-concept. *Dede Wallace Center Monograph*, No. 6, Nashville, Tennessee.

Thorne, N. (1981). The aging of the skin. *Practitioner, 225*, 793-800.

Thornbury, J. M., & Mistretta, C. M. (1981). Tactile sensitivity as a function of age. *Journal of Gerontology, 36*, 34-39.

Till, R. E., & Franklin, L. D. (1981). On the locus of age differences in visual information processing. *Journal of Gerontology, 36*, 200-210.

Tillman, K. (1965). Relationship between physical fitness and selected personality traits. *Research Quarterly, 36*, 483-486.

Timiras, P. S. (1972). *Developmental physiology and aging*. New York: Macmillan.

Tobis, J. S., Nayak, L., & Hoehler, F. (1981). Visual perception of verticality and horizontality among elderly fallers. *Archives of Physical Medicine and Rehabilitation, 62*, 619-622.

Toglia, J. U. (1975). Dizziness in the elderly. In W. Fields (Ed.), *Neurological and sensory disorders in the elderly*. New York: Grune & Stratton.

Tomlinson, B. E., Blessed, G., & Roth, M. (1968). Observations on the brains of nondemented old people. *Journal of Neurological Science, 7*, 331-356.

Tomlinson, B. E., & Henderson, G. E. (1976). Some quantitative cerebral findings in normal and demented old people. In R. D. Terry and S. Gershon (Eds.), *Aging: Vol. 3. Neurobiology of aging*. New York: Raven Press.

Tomonaga, M. (1977). Histochemical and ultrastructural changes in senile human skeletal muscle. *Journal of the American Geriatrics Society, 25*, 125-131.

Tonna, E. A. (1977). Aging of skeletal-dental systems and supporting tissues. In C. E. Finch & L. Hayflick (Eds.), *Handbook of the biology of aging*. New York: Van Nostrand Reinhold.

Treloar, A. E., Boynton, R. E., Benn, B. G., & Brown, B. W. (1967). Variation of the human menstrual cycle throughout reproductive life. *International Journal of Fertility, 12*, 77-126.

Tsitouras, P. D., Martin, C. E., & Harman, S. M. (1982). Relationship of serum testosterone to sexual activity in healthy elderly men. *Journal of Gerontology, 37*, 288-293.

Tuck, M. L., Williams, G. H., Cain, J. P., Sullivan, J. M., & Deuhy, R. G. (1973). Relation of age, diastolic pressure, and known duration of hypertension to presence of low renin essential hypertension. *American Journal of Cardiology, 32*, 637-642.

Tunbridge, W. M. G., Evered, D. C., Hall, R., Appleton, D., Brewis, M., Clark, F., Evans, J. G., Young, E., Bird, T., & Smith, P. A. (1977). The spectrum of thyroid disease in a community: The Whickham survey. *Clinical Endocrinology, 7*, 481-493.

Turkington, M. R., & Everitt, A. V. (1976). The neurohypophysis and aging with special reference to the antidiuretic hormone. In A. V. Everitt & J. A. Burgess (Eds.), *Hypothalamus, pituitary and aging.* Springfield, IL: C C Thomas.

Turner, J. M., Mead, J., & Wohl, M. E. (1968). Elasticity of human lungs in relation to age. *Journal of Applied Physiology, 25*, 664-671.

Turvey, M. T. (1973). On peripheral and central processes in vision: Inferences from an information-processing analysis of masking with patterned stimuli. *Psychological Review, 80*, 1-52.

Twomey, L., Taylor, J., & Furniss, B. (1983). Age changes in the bone density and structure of the lumbar vertebral column. *Journal of Anatomy, 136*, 15-25.

Tzankoff, S. P., & Norris, A. H. (1977). Effect of muscle mass decrease on age-related BMR changes. *Journal of Applied Physiology, 43*, 1001-1006.

Tzankoff, S. P., Robinson, S., Pyke, F. S., & Brown, D. A. (1972). Physiological adjustments of work in older men as affected by training. *Journal of Applied Physiology, 33*, 346-350.

Vekemans, M., & Robyn C. (1975). Influence of age on serum prolactin levels in women and men. *British Medical Journal, 4*, 738-739.

Venstrom, D., & Amoore, J. E. (1968). Olfactory threshold in relation to age, sex, or smoking. *Journal of Food Science, 33*, 264-265.

Verinis, J. S., & Roll, S. (1970). Primary and secondary male characteristics: The hairiness and large penis stereotypes. *Psychological Reports, 26*, 123-126.

Vermeuelen, A. (1976). Leydig-cell function in old age. In A. V. Everitt & J. A. Burgess (Eds.), *Hypothalamus, pituitary, and aging.* Springfield, IL: C C Thomas.

Verwoerdt, A., Pfeiffer, E., & Wang, H. S. (1969). Sexual behavior in sensescence. In E. Palmore (Ed.), *Normal aging* (pp. 282-299). Durham, NC: Duke University Press.

Verzar, F. (1966). Anterior pituitary function in age. In B. T. Donovan & G. W. Harris (Eds.), *The pituitary gland* (Vol. 2). Berkeley: University of California Press.

Vestal, R. E., Wood, A. J. J., & Shand, D. G. (1979). Reduced B-adrenoceptor sensitivity in the elderly. *Clinical Pharmacology and Therapeutics, 26*, 181-186.

Vijayashankar, N., & Brody, H. (1979). A quantitative study of the pigmented neurons in the nuclei locus coeruleus and subcoeruleus in men as related to aging. *Journal of Neuropathology and Experimental Neurology, 38*, 490-497.

Wagner, J. A., Robinson, W., & Marino, R. P. (1974). Age and temperature regulation of humans in neutral and cold environments. *Journal of Applied Physiology, 37*, 562-565.

Wagner, J. A., Robinson, S., Tzankoff, S. P., & Marino, R. P. (1972). Heat tolerance and acclimation to work in the heat in relation to age. *Journal of Applied Physiology, 33*, 616-622.

Wallin, B. G., Sundlof, G., Eriksson, B.-M., Dominiak, P., Grobecker, H., & Lindblad, L. E. (1981). Plasma noradrenaline correlates to sympathetic muscle nerve activity in normotensive men. *Acta Physiologica Scandinavica, 111*, 69-73.

Walsh, D. A. (1976). Age differences in central perceptual processing: A dichoptic backward masking investigation. *Journal of Gerontology, 31*, 178-185.

Walsh, D. A. (1982). The development of visual information processes in adulthood and old age. In R. Sekuler, D. Kline, & K. Dismukes (Eds.), *Aging and human visual function*. New York: Alan R. Liss.

Walsh, D. A., Williams, M. V., & Hertzog, C. K. (1979). Age-related differences in two stages of central percephial processes: The effects of short duration targets and criterion differences. *Journal of Gerontology, 34*, 234-241.

Wang, H. S., & Busse, E. W. (1975). Correlates of regional cerebral blood flow in elderly community residents. In M. Harper, B. Jennett, D. Miller, & J. O. Rowan (Eds.), *Blood flow and metabolism in the brain*. London: Churchill Livingstone.

Wankel, L. M. (1980). Involvement in vigorous physical activity: Considerations for enhancing self-motivation. In R. R. Danielson & K. F. Danielson (Eds.), *Fitness motivation* (pp. 18-32). Toronto: Orcel Publications.

Wantz, M. S., & Gay, J. E. (1981). *The aging processes: A health perspective*. Cambridge, MA: Winthrop.

Wapner, S., & Werner, H. (1965). An experimental approach to body perception from the organismic-developmental point of view. In S. Wapner & H. Werner (Eds.), *The body percept* (pp. 9-25). New York: Random House.

Warwick, R. (1976). *Eugene Wolff's anatomy of the eye and orbit*, 7th ed. Philadelphia: Saunders.

Watkin, D. M., & Shock, N. W. (1955). Agewise standard value for C_{In}, C_{PAH} and Tm_{PAH} in adult males. *Journal of Clinical Investigations, 34*, 969.

Weale, R. A. (1963). *The aging eye*. London: H. K. Lewis.

Weale, R. A. (1975). Senile changes in visual acuity. *Transactions of the Ophthalmological Societies of the United Kingdom, 95*, 36-38.

Weale, R. A., (1978). The eye and aging. *Interdisciplinary topics in gerontology, 13*, 1-13.

Weale, R. A. (1981). Human ocular aging and ambient temperature. *British Journal of Ophthalmology, 65*, 869-870.

Weale, R. A. (1982). Senile ocular changes, cell death, and vision. In R. Sekuler, D. Kline, & K. Dismukes (Eds.), *Aging and human visual function*. New York: Alan R. Liss.

Weaver, J. K., & Chalmers, J. (1966). Cancellous bone: Its strength and changes with aging and an evaluation of some methods for measuring its mineral content. I. Age changes in cancellous bone. *Journal of Bone and Joint Surgery (A), 48*, 289-299.

Webb, W. B. (1982). Sleep in older persons. Sleep structures of 50- to 60-year-old men and women. *Journal of Gerontology, 37*, 581-586.

Webster, S. G. P. (1978). The pancreas and small bowel. In J. Brocklehurst (Ed.), *Textbook of geriatric medicine and gerontology* (2nd ed.). New York: Churchill Livingstone.

Webster, S. G. P., & Leeming, J. T. (1975). Assessment of small bowel function in the elderly. *Gut, 16*, 109-113.

Weg, R. B. (1978). The physiology of sexuality in aging. In R. H. Davis (Ed.), *Sexuality and aging*. Los Angeles: The University of Southern California Press.

Weg, R. B. (1983). The physiological perspective. In R. B. Weg (Ed.), *Sexuality in the later years* (pp. 39-80). New York: Academic Press.

Weidmann, P., Beretta-Piccoli, C., Ziegler, W. H., Keusch, G., Gluck, Z., & Reubi, F. C. (1978). Age versus urinary sodium for judging renin, aldosterone and catecholamine levels: Studies in normal subjects and patients with essential hypertension. *Kidney International, 14*, 619-628.

Weidmann, P., DeMyttenaere-Bursztein, S., Maxwell, M. H., & DeLima, J. (1975). Effect of aging on plasma renin and aldosterone in normal man. *Kidney International, 8*, 325-333.

Weisbrodt, N. W. (1981). Motility of the large intestine. In L. R. Johnson (Ed.), *Gastro-intestinal physiology*. St. Louis: C. V. Mosby.

Weisfeldt, M. L. (1980). Left ventricular function. In M. L. Weisfeldt (Ed.), *Aging: Vol. 12. The aging heart: Its function and response to stress*. New York: Raven Press.

Werner, H. (1965). Introduction. In S. Wapner & H. Werner (Eds.), *The body percept* (pp. 4-8). New York: Random House.

Wessel, J. A., & Van Huss, W. D. (1969). The influence of physical activity and age on the exercise adaptation of women. *Journal of Sports Medicine and Physical Fitness, 9*, 173-180.

Wesson, L. G. (1969). *Physiology of the human kidney*. New York: Grune and Stratton.

West, J. B. (1977). *Ventilation blood flow and gas exchange* (3rd ed.). Oxford: Blackwell Scientific.

Whitbourne, S. K., & Weinstock, C. S. (1979). *Adult development: The differentiation of experience*. New York: Holt, Rinehart and Winston.

White, P., Hiley, C. R., Goodhardt, M. J., Carrasco, L. H., Keet, J. P., Williams, I. E. I., & Bowen, D. M. (1977). Neocortical cholinergic neurons in elderly people. *Lancet, 1*, 668.

White, R. (1963). Self-esteem, sense of competence and ego strength. *Psychological Issues, 11*, 125-150.

White, R. W. (1959). Motivation reconsidered: The concept of competence. *Psychological Review, 66*, 297-334.

Whitton, J. T. (1973). New values for epidermal thickness and their importance. *Health Physics, 24*, 1-8.

Wide, L., & Hobson, B. M. (1983). Qualitative difference in follicle-stimulating hormone activity in the pituitaries of young women compared to that of men and elderly women. *Journal of Clinical Endocrinology and Metabolism, 56*, 371-375.

Williams, R. L., Karacan, I., & Hursch, C. (1974). *Electroencephalography (EEG) of human sleep: Clinical applications*. New York: Wiley.

Willis, S. L., Blieszner, R., & Baltes, P. B. (1981). Intellectual training research in aging: Modification of performance on the fluid ability of figural relations. *Journal of Educational Psychology, 73*, 41-50.

Wilson, R. H. (1981). The effects of aging on the magnitude of the acoustic reflex. *Journal of Speech and Hearing Research, 24*, 406-414.

Witkin, H. A. (1965). Devlopment of the body concept and psychological differentiation. In S. Wapner & H. Werner (Eds.), *The body percept* (pp. 26-47). New York: Random House.

Wolf, E. (1960). Glare and age. *Archives of Ophthalmology, 64*, 502-514.

Wolf, E., & Gardiner, J. S. (1965). Studies on the scatter of light in the dioptric media of the eye as a basis of visual glare. *Archives of Ophthalmology, 74*, 338-345.

Wolfsen, A. R. (1982). Aging and the adrenals. In S. G. Korenman (Ed.), *Endocrine aspects of aging*. New York: Elsevier Biomedical.

Wright, T. W., Zauner, C. W., & Cade, R. (1982). Cardiac output in male middle aged runners. Journal of Sports Medicine and Physical Fitness, 22, 17-22.

Wylie, R. C. (1974). *The self-concept* (rev. ed., Vol. 1). Lincoln: University of Nebraska Press.

Yamagata, A. (1965). Histopathological studies of the colon in relation to age. *Japanese Journal of Gastroenterology, 62*, 229-235.

Yamaguchi, T., Hatazawa, J., Kubota, K., Abe, Y., Fujiwara, T., & Matsuzawa, T. (1983). Correlations between regional cerebral blood flow and age-related brain atrophy: A quantitatiave study with computed tomography and the Xenon-133 inhalation method. *Journal of the American Geriatrics Society, 31*, 412-416.

Yamaura, H., Ito, M., Kubota, K., & Matsuzawa, T. (1980). Brain atrophy during aging: A quantitative study with computed tomography. *Journal of Gerontology, 35*, 492-498.

Yarington, C. T. Jr. (1976). Presbycusis. In J. L. Northern (Ed.), *Hearing disorders.* Boston: Little, Brown.

Yin, F. C. P. (1980). The aging vasculature and its effects on the heart. In M. L. Weisfeldt (Ed.), *Aging: Vol. 12. The aging heart: Its function and response to stress.* New York: Raven Press.

Yin, F. C., Spurgean, H. A., Raizes, G. S., Greene, H. L., Weisfeldt, M. L., & Shock, N. W. (1976). Age associated decrease in chronotropic response to isoproternol. *Circulation, 54*(II), 167.

Yin, F. C. P., Raizes, G. S., Guarnieri, T., Spurgean, H. A., Lakatta, E. G., Fortuin, N. J., & Weisfeldt, M. L. (1978). Age-associated decrease in ventricular response to haemodynamic stress during beta-adrenergic blockade. *British Heart Journal, 40*, 1349-1355.

Young, R. W. (1976). Visual cells and the concept of renewal. *Investigative Ophthalmology, 15*, 700-725.

Young, J. B., Rowe, J. W., Pallotta, J. A., Sparrow, D., & Landsberg, L. (1980). Enhanced plasma norepinephrine response to upright posture and oral glucose administration in elderly human subjects. *Metabolism, 29*, 532-539.

Young, A., Stokes, M., & Crowe, M. (1982). The relationship between quadriceps size and strength in elderly women. *Clinical Science, 63*, 35-36.

Zellweger, H., & Simpson, J. (1973). Is routine prenatal karyotyping indicated in pregnancies of very young women? *Journal of Pediatrics, 82*, 675-677.

Ziegler, M. G., Lake, C. R., & Kopin, I. J. (1976). Plasma noradrenaline increases with age. *Nature, 261*, 333-335.

Zion, L. C. (1965). Body concept as it relates to self-concept. *Research Quarterly, 36*, 490-495.

Author Index

Subject Index